The Theory of Deliberative Wisdom

Basic Bioethics

Arthur Caplan, editor

A complete list of the books in the Basic Bioethics series appears at the back of this book.

The Theory of Deliberative Wisdom

Eric Racine

The MIT Press
Cambridge, Massachusetts
London, England

The MIT Press
Massachusetts Institute of Technology
77 Massachusetts Avenue, Cambridge, MA 02139
mitpress.mit.edu

The MIT Press would like to thank the anonymous peer reviewers who provided comments on drafts of this book. The generous work of academic experts is essential for establishing the authority and quality of our publications. We acknowledge with gratitude the contributions of these otherwise uncredited readers.

This book was set in Stone Serif and Stone Sans by Westchester Publishing Services. Printed and bound in the United States of America.

Library of Congress Cataloging-in-Publication Data

Names: Racine, Eric, 1976– author.
Title: The theory of deliberative wisdom / Eric Racine.
Description: Cambridge, Massachusetts : The MIT Press, [2025] | Series: Basic bioethics | Includes bibliographical references and index.
Identifiers: LCCN 2024023733 (print) | LCCN 2024023734 (ebook) | ISBN 9780262551618 (paperback) | ISBN 9780262381987 (epub) | ISBN 9780262381994 (pdf)
Subjects: LCSH: Wisdom—Philosophy. | Wisdom—Moral and ethical aspects.
Classification: LCC B1205.W5 R34 2025 (print) | LCC B1205.W5 (ebook) | DDC 223—dc23/eng/20241011
LC record available at https://lccn.loc.gov/2024023733
LC ebook record available at https://lccn.loc.gov/2024023734

10 9 8 7 6 5 4 3 2 1

EU product safety and compliance information contact is: mitp-eu-gpsr@mit.edu

To Jérôme, with whom I was able to develop the inspiration and vision for the ideas presented in this book and put them into practice from a young age to envision a better life and a better world for us and for others. Long ago you invited me to "bring a philosophy of light, the philosophy of inner being to those who seek it." You also showed me the ways of dialogue, both in form and in content . . . I will never forget you, dear friend.

Thank you my dear and numerous students and fellows, who in countless ways have made possible the development of the ideas put forth in this book. Your collaboration and dedication to articulate and explore some of these ideas means more than you can imagine. I have learned a tremendous amount with and from you. For each and every one of you, I foresee a brilliant future.

To Nathalie for her support and for her inspiration in the development of the ideas contained in this book and also for allowing me to implement them in the most difficult times to consider our well-being and that of our family. Thank you for being a companion with whom I continue to have so many enlightening and enriching experiences, to learn and grow as we lovingly reveal ourselves to each other. Thank you to our three children André-Anne, Gabrielle, and Élliot for being such stimulating conversation partners and for being inspiring and wonderful human beings as we all learn to be a family together through our experiences.

Contents

Preface

Emerging technologies in artificial intelligence and biomedical science create a range of possibilities for healthcare professionals and patients. How should we deal with them? Accept them all? Refuse them all? Critically analyze each of them for their potential positive use and contribution to human development? If so, how can we evaluate them to decide which technologies and which specific uses are valuable from the standpoint of human welfare and well-being? In another realm, social changes are bringing up questions about the meaning of human life and relationships. For example, many Western societies have historically maintained the model of nuclear families and monogamy in spite of significant challenges to this kind of arrangement, such as high divorce rates and considerable rates of infidelity. Could polyamory—as some have been advocating—and other similar kinds of family and relationship models encountered in many non-Western societies be ethically defensible or even preferable? Yet another example, capitalism seems to have brought our planet beyond its limit of sustainable ecological rejuvenation: How can this socioeconomic model be reoriented to promote human welfare in keeping with the planet's health? All these questions are partly moral questions insofar as they raise profound questions about how we view ourselves and engage with the world. They tend to vex some customary ways of thinking about health, relationships, society, work, and the economy, notably in light of their implications for human well-being and flourishing.

Moral problems such as these are sometimes difficult to define and discuss. They relate to what gives meaning and value to human lives. The ability of human beings to openly discuss problems of morality is a desirable trait in social evolution, both in communities and in individuals, because it allows us to find satisfactory resolutions to these problems, which in turn promotes human flourishing and growth. When a human being's ability to open-mindedly search for existential meaning in response to problematic situations is jeopardized, it constitutes a threat to their greater well-being. From an individual standpoint, a person unable to overcome situations that run counter to their own moral reasoning is likely to feel overwhelmed and even distressed by their inability to find meaningful resolution. If a culture consistently fails to create meaning in the face of adversity and to offer paths to the future, it is diminished in its ability to foster the flourishing of the individual and the community. This is precisely why ethics is an integral aspect of individual and collective self-renewal: it lends itself to the moral dynamism necessary to overcome the obstacles of collective life, and it offers fulfillment to the members of its communities, who face ever-changing conditions. In contrast, a static individual psychology or culture that cannot make sense of and surmount new moral problems is imperiled and risks decline, especially in a world faced with changes of great quantity and magnitude. Unfortunately, the views entertained about moral problems tend to be narrow and often reductionist (e.g., as mere conflicts of values or of ethical principles), restricting much of the potential for human growth within the understanding and resolution of these problems.

Writing from the standpoint of the early twentieth century, the American philosopher John Dewey lamented the lack of progress in "moral science," or what we now commonly call *ethics*. He thereby echoed a plight previously formulated by John Stuart Mill decades before. The importance of this observation cannot be overestimated. In contemporary societies, the resources dedicated to reflecting on morals and related matters are significant if we adopt an inclusive vision of what the questions of morality and their implications are. Consider all the problems related to valuations of life and human existence embedded in societal and legal problems (e.g., the intersection between ethics and fields such as health, law, economics, and politics) and the full range of contexts from which they emerge (e.g., healthcare, business, family, and education). Human beings engage in vast discussion and grant significant attention to moral problems (beyond the field of ethics *sensu stricto*), and the reasons for this are clear. There are tremendous costs and implications of human conflicts and unresolved moral tensions between individuals and between communities (e.g., legal disputes, community division, armed conflicts, inequalities, and exclusion). Thus, finding common visions of what is valuable is a motivation for moving forward in all aspects of our lives, but we often lack clarity in assessing whether our actions are aligned with the potential for reward and growth, especially for those engaged in a given action, change, or process. Developing advanced methods to analyze, reflect upon, and resolve moral tensions is therefore of fundamental importance.

In response to the existence of unavoidable moral problems, given constant social and technological changes, a wide array of scholarly work has broached relevant topics in ethics from within disciplines such as philosophy, psychology, sociology, neuroscience, and theology, among others. Reforms have been undertaken to introduce ethicists and ethics services into private corporations (e.g., ethics consultants) and public institutions, such as hospitals and parliaments. Personal growth and self-help literature focused on well-being and the search for meaning and related topics, as well as counseling services, overlap with ethics and show how much effort and energy are invested in this area. All of this is to say that human societies spend significant resources on discussing moral problems and attempting to implement reforms relevant to these discussions. However, we are slow to utilize all the intellectual and human resources we have with a clear sense of purpose and efficacy. Even if progress is achieved in the scholarship within academic ethics, there are clear gaps in our ability to put insights generated by ethics into practice. Perhaps we simply do not have the optimal tools or approaches to speak more openly about these problems and deal with them appropriately?

This book is an attempt to better define a form of ethics inspired by philosophical pragmatism and interdisciplinary studies in ethics. Its aim is to liberate human energies in the pursuit of open-ended ethical inquiries such that ethics—as a form of deliberative wisdom described in this book—is put at the forefront of our cultural and political goals, both in individual and public life. This orientation calls for openness, acknowledges that responsiveness is a feature of human flourishing, and this openness involves dialogical structures of knowing and interacting with others. Indeed, learning about the meaning of life and the value of human existence is not only an individual project but one in which we are engaged with others as co-learners and co-discoverers in the full breadth and depth of human experience, in its infinite manifestations and interpretations. My intention is that this book, based on a synthesis of scholarship from various disciplines, propels the development of more advanced theories and methodologies that help peacefully overcome disagreements about moral values, and support the development of creative, innovative ethical thinking that helps deepen the meaning of human existence. I recognize that a single book by a single author from a Western background will never alone fully achieve such a goal, but hopefully it can service and bring attention to this endeavor. This is my primary objective, and it joins many like-minded approaches in various fields and settings.

Acknowledgments

The Theory of Deliberative Wisdom is a project that grew over time and required several years of ongoing research, writing, discussion, revision, and editing. I am immensely grateful for the help and support that many people have offered along the way to making this project a reality.

At work, members of the Pragmatic Health Ethics Research Unit as a whole have offered collegial feedback on the project itself and several pieces of it. Given my iterative writing strategies, some prior versions of chapters were harder to follow and read, thus I am grateful for the patience, generosity, and open-mindedness of my students and research personnel. I want to also acknowledge a few members who reviewed several chapters and provided editorial assistance. I have in mind notably Wren Boehlen, Nicole Padley, Matthew Valiquette, Alice Côté, Audrey Francoeur, and Polina Gruschevska. Moreover, Grace Feeney read the entire manuscript to help me prepare the revised submission. Mathilda Ryon helped format the references and Anne-Sophie Dufour improved the figures. Because the ideas shared in this book evolved and grew as we pursued our work, I owe to this group a great deal.

With respect to the publication process, I am grateful for Philip Laughlin's commitment to this book as well as Arthur Caplan's encouragement. I extend my thanks to my colleague Walter Glannon for giving advice on the publication process. The initial review of an advanced draft of the manuscript in 2022 by anonymous reviewers commissioned by MIT helped me significantly edit the manuscript. The second review process of 2023 was thus much smoother. I want to acknowledge the constructive feedback and patience of all reviewers in this process. Few pieces of this manuscript have been discussed publicly, especially given the time of the writing of this manuscript, which made it impossible to meet in person and exchange, but

I want to thank the members of the Institute of Ethics, History and Theory of Medicine at Ludwig-Maximilian-University (LMU) Munich for having the opportunity to share some of the early ideas found in chapters 3–7 in 2019. I am also grateful for opportunities to present overviews of the current theory to philosophy students and bioethics students at the University of Montreal in courses from my colleagues Aude Bandini and Vardit Ravitsky. The writing of this work was also possible thanks to a career award from the Fonds de recherche du Québec–Santé.

At home and on a more personal note, I am grateful to Nathalie for her availability to listen to my wild ideas leading to this book and for her constructive feedback as we discussed much about everything. *Ibidem* to our children André-Anne, Gabrielle, and Élliot who all encouraged me in this project and challenged my ideas in our ongoing and lively family conversations. Our family life is a rich philosophical terrain. In my memories is Jérôme to whom I dedicate the spirit of this book. The quintessence of philosophical friendship is to discover the world and make sense of it together. True friendship is a genuine dialogue of existences in the spirit of deliberative wisdom. One is fortunate to have such a true friend in a lifetime; and fortunately, the spirit of friendship breeds new friendship, allowing life to continue despite the loss. I keep in mind my father who encouraged me to go further, to look to the future while not forgetting the past.

Introduction and Overview

Ethics, as explained and envisioned by this book, is a discipline concerned with understanding problematic situations involving human values and well-being, finding creative resolutions to these complex situations, and reflecting critically on how problematic situations are dealt with when envisioned in the context of the pursuit of a good life. As Socrates voiced at his trial, "The unexamined life is not worth living" (West and Plato 1979). Put differently, critical thought about the goals of human existence—as encouraged by ethics—is crucial to pursue an authentic and good life, that is, a flourishing life. However, despite its history, ethics often falls short of offering theoretical and material resources to carry out this colossal task. Theoretically, ethics is still rather fragmented. Many scholars of ethics (myself included) still work too much in isolation from, for example, sociological, economic, and psychological knowledge about human welfare and flourishing. For instance, we have only rather recently accepted the idea that the empirical study of human moralities can be relevant to the task of thinking about the good life and helping in the resolution of challenging human situations. In terms of material resources, ethics is undervalued and underrepresented (e.g., in social media), as evidenced by a lack of funding to the field of ethics itself, lack of human resources, and lack of genuine consideration in discussions of various aspects of our individual and collective lives. It is also the case that some shortcomings may originate outside of the field of ethics; a powerful ethics theory and methodology can be opposed because it calls into question established structures of power and social arrangements that are currently not conducive to human flourishing.

In this book, I take up the challenge of formulating an integrative theory and methodology to help understand and tackle moral problems, that is, problems that raise questions about the meaning of life and of human experiences in light of their contribution to well-being and flourishing. This framework is embedded in an account of moral agency that considers reflection on human values and the meaning of our attitudes and behaviors essential to the good life. In formulating this theory, I identify and elaborate on what I take to be key supporting concepts, including situational moral insight, ethical deliberation, and ethical enactment that all need to be better articulated theoretically and investigated in real-world settings. Perhaps more importantly, I draw connections between somewhat disparate ideas and concepts to propose a functional and dynamic model of ethics theory, a living ethics theory, that is deeply connected to lived experience and that offers lifelong learning and an evolving outlook on ethics theory and methodology. In doing so, I pull from my own personal life and from the context in which I live and work (Québec, Canada), both in its unique local aspects that sometimes orient the choice of cases and examples and in its undeniable international aspects that demonstrate we are all interconnected. Some of the key ideas supporting this synthesis are not entirely new and build extensively from established scholarship. What I hope to provide is greater coherence between different traditions, disciplines, and methods, such that we advance in the pursuit of interdisciplinary ethics research and practice, and that some ongoing perplexities (e.g., the role of empirical research in ethics) find firmer footing. In many ways, the current book moves away from some more traditional debates (e.g., the is-ought problem) or some more traditional formulation of debates (e.g., the is-ought as a dichotomy or logical fallacy), notably because of its inspiration from pragmatism's instrumentalist epistemology, which offers ways of reconsidering dualisms and dichotomies.

While preparing this book, I reviewed a wide array of literature outside of my primary field (bioethics) to enrich my understanding of the different components of ethics. There are no doubt places where experts in certain fields will find fault with my understanding, and I welcome this criticism in order to envision the theory presented here as a first version of a model that needs further refinement and inquiry. Every statement in this book is meant to be tested and evaluated, (in)validated, and revised in light of these developments. Nothing written in this book should be taken for granted or thought of in isolation from one's experience of the world and the social realities we confront.

The premise of this book is inspired by the idea advanced by John Dewey, that ethics is an interdisciplinary metadiscipline, one guided by a practical goal of using knowledge from many different fields to guide human action toward flourishing. As a pioneer of pragmatist ethics, he proposed that ethics is not a province of its own. In ethics, knowledge is used for the purpose of flourishing, a purpose that must constantly evolve with human beings and societies, and thus, must constantly be called into question. Ethics and its practitioners can (and should) mobilize the resources of other domains of knowledge to enable humans to pursue eudaimonic goals, that is, goals that are connected to the pursuit of the good life, the flourishing life. When it does so, ethics instrumentalizes knowledge about human behavior and psychology, giving this knowledge a practical purpose in achieving these goals.

As a scholarly field informed by knowledge of human psychology and behavior, ethics, as envisioned in this book, helps make these goals explicit for the purposes of testing.[1] It uses methods such as deliberation to allow its users to tap into the rich experiences of human beings, mobilize them to enrich our understanding of what the good and flourishing life is, and foster creative future-oriented thinking toward this end. A single author's knowledge only extends so far in developing such a tentative theory and methodology. Some may find my treatment of certain theories and concepts cursory, but my primary intention is not to engage with academic debates but rather to offer a constructive orientation for the future. Given this, I hope that my treatment of other theories is nevertheless charitable and defensible.

The Theory of Deliberative Wisdom is structured as thirteen chapters that provide background motivations for the theory,[2] offer an overview of its components, introduce its key concepts, and discuss some of their limitations. Each chapter is meant to stand alone but to also constitute the building block of a theory. This theory provides an account of situated moral agency, proposes human flourishing as its focus and the knowledge thereof as deliberative ethical wisdom, and then describes in more detail aspects of situational insight necessary to introduce an account of scenario-based deliberation and ethical imagination and the related concept of ethical enactments and their evaluation. Before presenting an overview of the book, I must first explain a rather unconventional distinction of crucial importance between morality and ethics. I hope readers may forgive this digression in the introduction, but this distinction is at work in some of the chapter titles and chapter summaries that would otherwise appear inconsistent.

Morality is the domain of lived experience that structures how our thoughts, feelings, and behaviors are embedded in valuations and values regarding our own welfare, that of others, and the pursuit thereof. Human moralities—plural because they are numerous—rely on more implicit ways of thinking about and using human agents' valuation processes. They are usually organized in coherent systems of ideas about what is right and wrong and are deeply connected to our identities as persons, because what we value and pursue based on our interests defines us individually and collectively, insofar as they characterize the meaning of our existences. Moralities are perpetuated in a wide array of social practices across generations, while also being adapted by every person who needs to make choices about living their life, hence the role of ethics to support such choices. As a social species, human beings rely on traditional moralities handed down through generations to guide themselves, raise future generations, and set functional expectations for each other's behavior in all aspects of life, from work to sexuality. Morality is a constituent of human existence, and few human beings truly lack basic moral sense, though they may not enact a given morality consistently and may have impairments affecting basic components of morality (e.g., specific deficits of empathy in some instances of psychopathy).

Ethics, on the other hand, is a discipline that tries to glean a scientific understanding of human morality (e.g., its nature, its mechanisms) and specific moralities, especially in the context of difficult moral situations (e.g., when multiple moralities are at play or when any particular morality does not have a clear answer) to guide the use of our basic moral sense. Different cultures and civilizations have surely evolved different forms of ethics (and have inescapably upheld moralities), but I situate the origins of ethics in ancient Greek philosophy. This tradition developed in communication with other traditions of the ancient world, a point worth keeping in mind in the spirit of intercultural dialogue favoring exchange and learning. Some ethical inquiries—which I call the scientific study of human moralities—are more empirical and are used to understand (e.g., identify, characterize, explain) moral experiences. Others (which I call ethics in a narrower sense) are more normative and try to reconstruct action orientations in light of human flourishing. Ethics investigates openly whether certain social and moral practices (e.g., capitalist economy based on profit-making, dietary regimes with important environmental and animal welfare implications) are actually practices conducive to human flourishing. The fallibility of all human moralities is explicitly recognized in ethics, and like in other sciences, fallible knowledge is ascertained though inquiry. Accordingly, a scientific, open-minded, and experience-based approach is encouraged by ethics, promoting critical thought about our valuations and values, including those that are implicitly at the core of human moralities and those generated by a deliberative exercise. While morality is inescapable, ethics is optional in the sense that commonly accepted moral valuations and values embedded in our lives and that of our societies can remain largely and even completely unquestioned (e.g., dogma) without any intervention from ethics.

The distinction between morality and ethics is used throughout this book as systematically as possible to differentiate phenomena. For example, I distinguish moral awareness from ethical awareness and moral theory (theory explaining moral behavior and psychology) from ethics theory, which attempts to refurnish moral orientations with solutions based on research, deliberation, reflection, and so on. Although this distinction does not imply a staunch duality—and the barriers between morality and ethics are admittedly porous—it does allow for investigation and discussion of these phenomena within two different frames of reference. This distinction corresponds roughly to the contemporary distinction between system 1 cognition (automatic, intuitive, and unconscious thinking) and system 2 cognition (slow, controlled, and analytical thinking) introduced by the psychologist and mathematician Amos Tversky and psychologist and economist Daniel Kahneman. This duality between morality (as a system 1) and ethics (as a system 2) is central to classic pragmatism, as seen in Dewey's emphasis on habits and the crucial role of learning. It is at work in basically all aspects of the proposed theory and is embodied in its purpose to accompany people from where and who they are as persons, to who they want to be and become as flourishing persons.

Chapter 1: Situated Moral Agency and the Ethical Life

To introduce the user and agent perspective central to this theory, the first chapter provides an account of moral agency and the role of ethics therein. Following Dewey's insights and recent literature in cognitive science on embodied, embedded, enactive, and extended (4E) cognition, I argue in favor of a transactional ontology that envisions experiences as transactions between organisms (human beings) and their environments. Experiences are, by their very nature, situated—though the contextual and dynamic aspects of experience may sometimes elude agents who entertain a view of their agency as isolated from context. Human existence consists of situated experiences, the memory of them, and the projection of them into the future. The active role we have in our existence as agents may be dampened by alternate visions of persons as mere passive individuals who fail to appreciate the fundamental role of agency in enacting their lives and thinking about their moral aspects. Indeed, contextual factors can become normalized in everyday life such that they seem static and immutable. Freeing ourselves from such static and fatalistic visions is important to liberate the creativity needed to open deliberation to possible futures, as required by ethical deliberations. By taking heed of the abilities they have to act and envision situations differently, moral agents gain more freedom to act and the ability to change their situations. I conclude this chapter by highlighting that general methodological ethics orientations follow from this account of moral agency, favoring approaches that are (1) situated and situating, and that foster (2) narrativity and meaningfulness, (3) reflexivity and fallibilism as well as (4) holistic accounts of outcomes, (5) nonreductionist inquiry, and (6) participation and deliberation. These are six orientations I strive to follow and which I come back to when I discuss the procedural aspects of the evaluation of enactments (chapter 13).

Chapter 2: Deliberative Wisdom and Human Flourishing

This chapter proposes to reconcile the rather intellectual nature of ethics with the fulfillment of more existential goals such as personal growth and development. I allude to the ancient idea that the good life is a flourishing life and explain how the flourishing life can be more perspicaciously and concretely understood with advances in contemporary psychology. I pull insights from Aristotle's view of eudaimonia (the good life) as a flourishing life, from Dewey's account of ethics as a form of critical reflection on our habits in light of their ability to foster growth, and from recent scholarship on human flourishing, especially Carol Ryff's multidimensional model of eudaimonia that integrates Dewey's concept of growth and other important dimensions of flourishing. In this chapter, I explain that ethical wisdom represents the knowledge and ability to discuss, identify, enact, and evaluate courses of action in light of their ability to foster growth and flourishing of oneself and others. Deliberative wisdom is defined more precisely as a form of nonauthoritarian and participatory ethical wisdom, a path to ethical wisdom that recognizes the need for open discussion about what is deemed the most valuable courses of action and goals to pursue. Paths to human flourishing vary, and so there is a need to learn from each other about what is most valuable in life. When the handling of moral problems is framed accordingly and ethics considered fallible, testable, and doubtable, the successful and unsuccessful attempts at responding to situations contribute to insights on what constitutes human flourishing.

Chapter 3: Situated Moral Experience

In this book, situations are presented as the basic units of understanding of human behavior and human psychology. They play a key role because of how they capture moments of existence as context-bound transactions between the agent and their environment. This chapter pulls from literature in social psychology and the social sciences to present the concept of situations and their most important features, including that (1) situations comprise an agent, an environment, and their transactions; (2) situations are multidimensional (e.g., human, social, physical dimensions) units of human experience; (3) situations can change along with our interpretation and knowledge of them; and (4) mechanisms of situational recall (i.e., episodic memory) factor into moral experience and ethical reasoning. Methodological implications following from these features of situations are described in the context of ethics research and ethics practice, highlighting the importance of situated understandings of human moral behavior.

Chapter 4: Morally Problematic Situations: Primary Objects of Moral Experience and Ethics

This chapter transitions from a situational account of moral agency to an argument about the situated nature of moral matters as such. It is the first of four chapters dedicated to important characteristics of morally problematic situations. Standard ethical analysis and theory have often emphasized that dilemmas, choices, and decisions are at the core of ethics. However, these need to be situated such that they capture moments of existence as transactions between the agent and their environment. Adopting morally problematic situations as units of analysis for moral experience and ethical analysis helps mobilize the resources of the long-standing social and psychological investigation of human situations. This chapter proposes a definition of morally problematic situations, explains how morally problematic situations are part of a broader typology of moral matters, and explains how certain methodological orientations correspond to the need for an understanding of situations as morally problematic situations.

Chapter 5: Morally Problematic Situations: Their Open-Ended Nature

One of the struggles of empirical investigations of human moralities has always been the definition of the "moral domain," that is, what counts as being within what we call "moral" and "morality," and what is outside this purview. Numerous debates have surfaced about the nature of this distinction and its justification. Because of its commitment to experiential and thus necessarily plural understandings of human morality, the theory of deliberative wisdom invites us to step away from theoretical and methodological efforts to define precisely the domain of human morality, and therefore, the moral dimension of morally problematic situations. Although this may sound circular or relativistic, it is actually a recognition that various attempts to define the domain of morality rely on conventions and habits. Moral practices are not necessarily ethical and, unfortunately, much discussion in this area has conflated morality and ethics with venous attempts to narrow the domain of morality in a way that conforms to a given understanding of ethics at a given time. In contrast, the theory of deliberative wisdom invites us to recognize the inherent pluralism of human moralities and distinguish this from the more unifying quest of critically examining these moralities in light of human flourishing, which is the task of ethics proper. Accordingly, when investigating human moralities, an open stance must be taken to reflect how different moral agents and cultures envision morality and moral practices. Great care should always be taken to bring to light and mitigate biases, which are unavoidable, given the situatedness of all actors, projected in studies of human moralities.

Chapter 6: Morally Problematic Situations: Their Experiential Nature

Chapter 6 further explains that it is the experiential aspects of morally problematic situations that make them such. Again, in contrast to attempts to ground moralities in certain specific and well-defined values and rational principles, and which sometimes overrationalize the practices embedded in implicit social attitudes, the proposed theory invites a recognition of the experiential nature of human morality. This means that understandings of human morality unavoidably differ between human groups with similar or shared values—although the actual homogeneity of human groups should not be overestimated. Even between individuals, including those who are part of groups sharing traditions such as religious traditions, political traditions, and linguistic traditions, there is diversity in moral beliefs, attitudes, and practices due to differences in individual experience. The experiential nature of moral existence and moral problems encountered in this respect is the starting point of subsequent ethical deliberation and ethical enactment. It is also that to which ethics must return to enrich human experience, meaningfully tackle moral problems, and accompany those involved in the path of pursuing ethical enactments. Empirical research on moralities is described as necessary to attend to the features of situations and to the broad spectrum of what are considered moral situations. Valuations are inherent to moral experience and to deliberation about moral experiences, but the willingness to understand the experiential aspect of moral situations (e.g., distress, discomfort) does not imply the justification of that experience (e.g., as justified and legitimate distress and discomfort). In contrast to scientific studies of morality, ethics seeks to enrich moral experience and support growth in the capacity to deal with morality and moral problems such that what is initially experienced as distress and discomfort may need to be reframed or even criticized. I acknowledge that "empirical ethics" is often a mixture of scientific studies of morality and of traditional (normative) ethics (for further clarification, see chapter 5 and table 5.2).

Chapter 7: Morally Problematic Situations: Actionable and Participatory Units of Situational Insight

Chapter 7 ventures out in claiming that a situational account of moral problems best suits the purpose of an action-oriented pragmatist epistemology. Given the nature of human thinking, notably mechanisms of episodic memory, our current and past experiences are structured as situated moments of existence. In contrast to the terminology and related epistemologies of "events" or "facts," which tend to neutralize situations and explain away moral agency, the vocabulary of situations corresponds to the view that human beings are agents in their own lives and can act upon and within situations. Before becoming facts or events, situations are experiences. This is a slight but significant conceptual nuance that helps convey how coming to understand and resolve moral problems can be an enriching and empowering human experience, because it deepens the understanding of moral situations, and thus, of human moral existence, since situations are constitutive of our existence (as chapters 1 and 3 explore). Understanding human realities as situations, and not only as mere facts we observe or events that happen to us or that we attend, also helps mobilize human and material resources within the purview of human action and supports the use of participatory methods. Although this suggestion could give the impression that human beings are condemned to act situationally, this does not prevent the consideration of macro-level social phenomena and the role of human action therein. The situational gaze has the merit of connecting human thinking with human action via agents who actively construct and act upon their (moral) worlds. This chapter explains the idea that situations are actionable units of human experience and that an agential account of moral situations inspires participatory resolution strategies. Seeing morally problematic situations in this light helps in understanding that learning about situations empowers moral agents. Moral situations also serve as participatory and experiential benchmarks to which we can measure successes and failures.

Chapter 8: Moral and Ethical Awareness

Chapter 8 introduces and distinguishes moral and ethical awareness. Moral awareness is the quotidian, habitual, and commonly relied on process of identifying or recognizing potentially problematic situations. This is a fallible process deeply embedded in how we see the world and our fellow human beings. There is no doubt great inter- and intragroup variability in what we are sensitive to and what we feel raises important issues for our welfare, our flourishing, and that of those we care for. The premise underlying the theory of deliberative wisdom is that moral awareness can be better understood, cultivated, and nurtured. It can also be blunted, dismissed, and demolished. When the latter occurs, there are significant implications for human welfare, such as when soldiers are encouraged to commit war crimes and healthcare professionals are forced to work in distressful conditions. The result is psychological and moral distress, as well as moral injury, clearly showing the importance of human morality.

In contrast, I refer to ethical awareness as a deliberately cultivated ability to identify moral problems based on critical reflection on the strengths and limitations of our moral awareness in properly identifying situations that have an impact on human flourishing. Ethical awareness represents a more advanced form of moral awareness that questions the justification and appropriateness of unproblematized and implicitly held forms of moral awareness. Ethical progress can be accomplished when ways of being aware lead to greater human flourishing, though we must be very careful in how we envision human flourishing to ensure that it actually captures how those concerned view their own well-being. Indeed, the perverted attitude of imposing on others what is in their best interest is central to, for example, inherently unethical colonialism and sexism. In this chapter, I offer the suggestion that we can cultivate and enhance moral awareness based on a better understanding of the barriers and biases that dampen our moral awareness and make us oblivious to the interests and welfare of others, that is, ethical awareness, consistent with the above distinction between morality and ethics.[3]

Chapter 9: Barriers to Moral Awareness and Gateways to Ethical Awareness

To nurture and cultivate moral awareness, we need to understand the processes by which we might actively attempt to ignore or deny the existence of moral problems. This denial can be rooted in self-interest or protection of one's ego, enabling agents to avoid questions about their own habits, unpleasant emotions and feelings related to these situations, and questions about the appropriateness of attitudes and behaviors in those situations. I argue that acknowledging the existence of barriers (including affective, cognitive, interpersonal, and social barriers) to recognizing moral problems opens up strategies to a cultivated ethical awareness. The guiding light is to ask whether and how disagreements about the existence and nature of moral problems are more deeply rooted in the lack of recognition of each other as persons and of our interests to pursue satisfying lives. Some barriers to questioning our moral awareness in light of human flourishing may be reliant on unreflective behavior, while others are the result of purposeful and self-interested neglect.

Chapter 10: Scenario-Based Ethical Deliberation and Ethical Imagination

This chapter offers an account of scenario-based ethical deliberation inspired by Dewey and my mentor, Hubert Doucet, emeritus professor at the Université de Montréal and pioneer of bioethics in Canada. In contrast to some current theories on moral deliberation (e.g., discourse ethics, deliberative democracy) that have been criticized for their idealistic and rationalistic overtones (e.g., the idealization of rational consensus on norms), deliberative wisdom calls for real-world and open-ended dialogical processes. The theory of deliberative wisdom moves away from what some call a "judgment model," or top-down approach, to espouse instead participatory and deliberative processes. The concept of scenarios is clarified and its connection to a situational account of moral agency unpacked. Action-scenarios designate possible courses of action generated via deliberation in response to morally problematic situations. They represent the options available in response to a given situation. (When action-scenarios are selected as preferred responses to a situation, they take the name of response-scenarios.) I then contrast scenario-based deliberation with other kinds of approaches, such as standard principle-based ethical analysis. For instance, the analysis of ethical principles—a mainstay in academic ethics—is treated as one of the many important dimensions of ethical deliberation that need be considered within the broader context of responding to the situation. Ethical imagination is explained, and its fundamental role stressed in the forward-looking nature of ethical thinking. Imagination is a necessary resource to generate potential responses to morally problematic situations through deliberation (i.e., action-scenarios) and to envision their implications.

Chapter 11: Key Moments in Ethical Deliberation

An ethical deliberation typically comprises a number of moments, which are not necessarily sequential steps. In this chapter, I explain three major moments, all of which require that the existence of a moral problem be at least partially recognized based on mechanisms of situational insight (chapters 3–7) and moral awareness (chapters 8–9). These moments are (1) broadening and deepening the understanding of the situation via stakeholder consultation and face-to face discussion, (2) envisioning of action-scenarios (based on the processes of imagination and dramatic rehearsal), and (3) reaching a judgment or decision based on the comparative evaluation and analysis of scenarios and their ethical merits. Regarding the final step, two complementary criteria representing two attitudes toward human flourishing (ethical acceptability and ethical praiseworthiness) are presented and discussed to generate a more concrete strategy to evaluate action-scenarios. For each of these three steps, more detailed practical information is provided on how they can be achieved and supported.

Chapter 12: Ethical Enactments

Influential accounts of ethics practice frequently employ models of application and implementation that imply a number of problematic assumptions. They presume that moral truth is known to some (e.g., experts in ethics) and needs to be applied or implemented rather mechanically by users of ethics knowledge. The development of codes of ethics, ethics guidelines, best practices, and ethics education are frequently envisioned in such mechanical terms. To properly reflect on the fallibility of ethics knowledge and its situational and deliberative nature, however, greater room must be provided for learning-as-we-go and learning-as-we-experience. Otherwise, users of ethics knowledge risk missing out on collateral cues and learnings needed to avoid the potential authoritative and disempowering use of ethics knowledge, if used without discernment. This chapter explains the concept of enactment (and of ethical enactment in particular) to reflect the embodied and contextualized nature of ethics knowledge. Enactment suggests that the logic guiding action is embedded within an understanding of the situation to act out of a projected story or scenario, consistent with the scenario-based model of ethics deliberation laid out in chapter 10.

Chapter 13: Evaluating Ethical Enactments

In this final chapter, I explain how the evaluation of ethics-oriented enactments is still in its infancy, notably because of disciplinary limitations and biases favoring theoretical and abstract forms of ethics. Accordingly, I first discuss the importance of evaluation in ethics and especially in the context of a living ethics theory such as this. Then, I present substantive and procedural guideposts that can constitute the backbone of an evaluative process in ethics. The substantive guideposts are developed and operationalized based on the concepts of ethical acceptability and ethical praiseworthiness introduced in chapter 2. The procedural guideposts are meant to counterbalance the substantive guideposts and are inspired by the methodological orientations described in chapter 1 (e.g., holistic accounts of outcomes, nonreductionist inquiry, and participation and deliberation).

1 Situated Moral Agency and the Ethical Life

An ethics theory is not only a theory about desirable principles and values such as justice or autonomy. It is more fundamentally—if it is to be applicable in and connected to real-life situations—also a theory about human behavior and about the role of human morality in communities and individual lives. In this sense, ethics should not be severed from psychology, the social sciences, or other fields of the humanities and to some extent the biological sciences. Ethics, as I propose to envision it following Dewey, is a theoretical and practical science that should build from empirical studies of human morality stemming from these fields. This chapter on situated moral agency provides a cursory account of moral agency that fits into this broader view of an interdisciplinary ethics. Relying on a transactional ontology and an instrumentalist epistemology, I explain in this chapter how human existence is made up of experiences. These experiences are situated as they consist of transactions between agent(s) and environment(s). Therein valuations are constitutive of the moral life. Experiences make up our existences, and when we make sense of experiences within our existence, they typically take a narrative form. Pulling from long-standing eudaimonic orientations in ancient ethics theory, I explain how contemporary psychology supports the importance of a morally meaningful life and how a flourishing life is a desirable existential narrative, owing in part to its benefits to psychological and physical health. Ethics, as I define it in this context, attempts to guide our thinking and behavior toward greater human flourishing. It is both a scientific and a practical discipline that uses knowledge from various fields to empower individuals and communities as full agents in their lives. Ethics is often solicited in the context of difficult situations, but this is not its exclusive focus since our whole life narratives are subject to scrutiny and questioning in light of human flourishing. Based on this overview of situated moral agency, a general methodological orientation is sketched for the theory of deliberative wisdom as a form of ethics that honors the meaning-making contribution of human morality. It favors approaches that are (1) situated and situating and that foster (2) narrativity and meaningfulness, (3) reflexivity and fallibilism, (4) holistic accounts of outcomes, (5) nonreductionist inquiry, and (6) participation and deliberation.

> We must not begin by talking of pure ideas,—vagabond thoughts that tramp the public roads without any human habitation,—but must begin with men [human beings] and their conversation
>
> —Charles Sanders Peirce

An ethics theory is partly a theory of human behavior, human psychology, and human sociality, but it is also different from theories in the behavioral, psychological, and social sciences. Ethics is not satisfied by generating explanations and descriptions of human moral behavior; it also explicitly and deliberately engages with human morality from the point of view of the agent. It thus considers human existence in light of what is thought to be desirable to nourish and grow within our existence as human beings. In other words, and as I will explain in this chapter, ethics designates the mobilization of knowledge toward our flourishing. Ethics is often part of the aspirations of other disciplines (e.g., psychology) and can be found within them; some approaches in psychology are much more explanation and description oriented (e.g., neuropsychology), while others resemble ethics and its goal of supporting existential learning (e.g., humanistic psychology).

Given that ethics is concerned with how we can use knowledge to better human lives, ethics theory[1] is confronted with questions about the nature of human happiness, flourishing, or other proposed cardinal orientation. It needs to explain why this orientation is meritorious, attractive, beautiful, satisfying, and so on. It must explain our motivations for moving in such a direction, what we gain or lose in this process, what the best ways of bringing about this orientation are. It must help make sense of how we can enact visions of what matters and what is valuable. Those engaging with such questions (e.g., engaging in reflection about these questions generally, wrestling with these questions while encountering specific moral problems, keeping these questions in mind in daily life or when engaging in ethics activities) need to grapple implicitly or explicitly with questions related to the basic nature of human existence and what fulfills human beings. This can be seen as a lofty and ambitious enterprise, but it may be inescapable and interwoven in our daily reflections on what it means to be a good colleague, a good citizen, a good partner, a good parent, a good grandchild, a good friend, and so on. Silence also speaks in the sense that ethics theories (e.g., more procedural ethics theories) that do not explicitly state their stance on these matters tend to assume a number of implicit or a priori beliefs and/or make implicit problematic claims about human nature and the desirability of certain goals. For instance, some normative ethical theories propose that the quintessence of ethicality is respecting one's duty toward other human beings (Kant 1785/1998) or maximizing utility and welfare for the greatest number of individuals (Mill 1863/2001). These are important principles, but we also need to understand how these ideals fit within our understanding of human psychology and behavior such that we can progress in enacting such cardinal orientations.

In thinking about and assessing what is ethical, ready-made guidelines are attractive due to their apparent applicability to countless situations. This has led to a common view that ethics theories need to provide ethical justifications considered "foundations" (Sherwin 1999) offering unshakable, universal, acontextual, absolute justifications.[2] However, using ethical principles as indubitable foundations has often been found to be a less than useful strategy (see discussions in Hoffmaster [2018] and Hoffmaster and Hooker [2009]) and even dangerous, as extreme examples of deontology (e.g., Halberstam 1988) and utilitarianism (e.g., Singer 2011) show. More modest *contextualized* ethical justifications and appeals to ethical principles are generated through, for instance, individual self-reflection or group discussion, with an appreciation of what is at stake in a situation, how the deliberation occurs in that context, including the limitations that this induces. Appreciating that our justifications for our actions are relative to their implications means that what we value is contextualized and makes sense at a given time, in a given location, and in a given social context. As John Arras wrote from a pragmatist standpoint, "The study of morality . . . ceases to lust after timeless foundational principles in order to ask what actions and forms of social organization will best foster the flourishing of our biological and social natures" (Arras 2002). In line with this quote, the starting point for a pragmatist theory of ethics like the theory of deliberative wisdom is experience, and ethics should aim to support a flourishing life for individuals in their communities, which, as I will explain, is integral to ethical existence.

To expose some of the more philosophical assumptions about human nature and the nature of moral experience at the core of the theory of deliberative wisdom, this chapter provides an overview of a transactional account of situated moral agency and the ethical life. This is an account that claims that experiences are fundamental transactional moments of existence where organisms (human beings) and environments co-constitute each other.

Transactional Ontologies, Situations, and Situated Moral Agency

Following insights of pragmatism (Brinkmann 2011; Rauthmann and Sherman 2019) as well as recent work in ecological psychology (Gallagher 2014; Gibson 1977) and 4E cognition, that is, embodied, embedded, enacted, extended cognition theory (Gallagher 2014; Gibson 1977), I wish to situate deliberative wisdom as a process of adaptation of the organism to its environment. Because ethics has tended to focus on the exercise of ethical judgment and ethical reasoning, and less so on how the exercise of this reasoning and judgment is carried within a person, and that a person is in a physical and social environment, this more comprehensive and holistic point of view invites the reconsideration of common ontological assumptions about agents and agency as it relates to ethics.[3]

Influential strains of ethics have tended to depict agents atomistically as individuals, but pragmatist theory presents the view that agents are transacting individuals. Here, transaction[4] represents a view by which agency is envisioned as a continuous process of acting through (i.e., to trans-act) an environment. Dewey explains that transactions differ from interactions. Interactions rely on the idea that two independent entities can influence each other. For example, he cites Newtonian physics as being premised on the understanding of the interactions between different independent bodies. In contrast, Albert Einstein's physics proposes that time and space are not independent properties of nature, but rather that they point to relationships between things from the point of view of the observer (Brinkmann 2011). Because organisms are constituted by their abilities to act on their environments, and these environments are that to which the organism and its abilities respond, there is a co-constitution of organisms and environments. In this sense, a fuller account of human agency must always adopt what I will call (and elaborate further on, notably in chapters 3–7) a situational gaze by which agency and possibilities of action are understood in relationship to an environment.

In a general sense, and again following Dewey, the concept of situation refers, in this book, to the fact that human existence is always occurring in context. More precisely, a situation designates the transaction of a human being with its environment. This transaction gives way to lived experience such that a situation is always composed of an agent (a human being), an environment, and the lived experience of the human being in that given space and time. In other words, specific moments or episodes of human existence can be identified as situations. Incidentally, etymologically, "situation" comes from the Latin (via Old French) *situare* (to locate, to place), which comes from *situs* (a place, a position). In the early days of social psychology in the 1920s, a situation represented a unit that emphasized the idea that human behavior was a function of environmental and personal factors. Thus, the study of human behavior needed to take into account the fact that agents *qua* agents interpreted their existence and life events. These ideas grew in the work of Kurt Lewin (1946), as well as in the philosophical work of Dewey, and the social science of George Herbert Mead (1934/2015) and Herbert Blumer (1969) (e.g., in the form of symbolic interactionism).[5]

The concept of situation helps bring insights from transactional ontology to the study of human behavior. Because situations capture the transactional nature of human existence, it is a move away from the "psychology of a separate and independent consciousness" (Dewey 1922, p. 85) because "to get a rational basis for moral discussion we must begin with recognizing that functions and habits are ways of using and incorporating the environment in which the latter has its say as surely as the former" (Dewey 1922, p. 15). Applied to the study of human morality, this means that what Dewey describes as habits (our personal and behavioral traits, our *habitus*, our dispositions to act in certain ways, a notion also central to Pierre Bourdieu's [1977] sociology of practice) need to be thought about in relationship to the environment or, following such a transactional ontology, as relationships to environments. Just like the habit of breathing involves both air and lungs, an example often used by Dewey, the meaning of human words, for example, requires biological and physical conditions (e.g., to produce speech in various forms), as well as human relationships in which meaning is shared and validated. Both breathing and speech reflect transactions with different aspects of the environment. Accordingly, habits, including our moral habits, need to be understood *environmentally*, or as I will write more often, *situationally*, to stress the psychosocial and lived aspects of human situations.

Instrumentalist Epistemologies

When transactional ontological insights are also applied to the meaning of our ideas, a form of instrumentalist epistemology is yielded. Therein, ideas express the meaning of transactions and are also part of the way we transact with the world such that their relevance can be assessed based on their ability to express this transaction. In this way, ideas are tools or instruments. This is the instrumentalism advocated by Dewey, notably consistent with the view that our ideas are themselves products of transactions and that their meaning should be understood in terms of how they express fitness or lack thereof between organism and environment. In other words, our ideas are not purely representations or denotations of the world; they are how we make sense of our transactions with the world. Accordingly, our ideas are moving affairs, generated in context by agents who are embodied, whose thoughts and minds are shaped by environments; and environments also become part of what we call "mind," following the contemporary extended mind thesis (Brinkmann 2011; Foucart 2012/2013).

The scientific method provides the ability to fix the meaning of our ideas in terms of what Dewey called their "warranted assertability," that is, the conditions that make them true. In other words, "clear" ideas, as advanced by Charles Sanders Peirce, are those that can be defined in terms of their implications, their functional role, their outcomes (Peirce 1878). That our ideas become a function of their outcomes is that to which the famous the pragmatic maxim refers (Peirce 1878).[6] This epistemological and methodological (methodological in the sense of guiding us in how we can make ideas clearer) orientation is at the root of pragmatism as a philosophical tradition (Misak 2013) and, when taken to heart, carries radical transformations in the field of ethics where the meaning of ideas have only been partially fixed in light of their implications. This in fact leads to the proposal of what Peirce described as an experimental philosophy and a laboratory philosophy. In the domain of human morality, this calls for an experimental form of ethics and ethics laboratories where ideas are tested for their outcomes. As readers will see, the ethical knowledge gleaned from experimenting is what I describe under the concept of deliberative wisdom.

Dewey explains how the evolution of knowledge has trended toward this continued instrumentalization of ideas. For example, philosophers and scientists (e.g., Heraclitus, Parmenides, Aristotle, Galileo, Linné, Darwin, Einstein) have long been puzzled and defied by the challenge of understanding physical, biological, and social realities, which are always in flux (e.g., the moving celestial bodies, the growth and death of biological organisms). In response, knowledge was for a long time considered to rely on the concept of unchanging essences based on Plato's influential theory of forms. The example of mathematical knowledge served therein as a powerful model to which all bodies of philosophical knowledge (including ethics) would conform or find inspiration in, to find, for example, general foundational ethical rules (Rota 1991). At the same time, we have often come to realize that what we thought to be stable, fixed categories can be better explained in terms of relationships and functional properties. For example, while we think that physical objects are stable, concrete, and basically constant throughout time, we have learned that the entire physical universe is expanding, including everything in it (Einstein 1979/1997). Although we long thought that biological species represented fixed, enduring entities, we have learned that they evolve by function of ecological pressures (Darwin 1859). A similar analysis can be applied to social behavior and language more specifically, where the meanings of our ideas were thought of as having fixed and clear definitions; but now, we have accounted for the generation of rather stable but evolving meanings with conceptual prototype theory (Rosch et al. 1976) and semantic metrics (Rumelhart, McClelland, and PDP Research Group 1986; McClelland and Rogers 2003). Current machine learning relies on this ability for meanings to be formed and explained in terms of exposure to features and patterns of the world through experience-informed classifications. Such observations about the progression of science[7] to explain realities based on their relationships and their outcomes was integrated by pragmatists like Peirce into logic (e.g., the pragmatic maxim), William James into psychology (e.g., consciousness as a flux), Dewey into philosophy (notably politics, education, and ethics with his "instrumentalism"), and Mead into social science (e.g., the interactionist development of self).

The ability of ideas to reflect transactions is a warrant of their truth and meaning (Dewey 1941). Following this pragmatist insight, the role of scientific inquiry is to understand parts of the world based on their transactions and generate refined ideas that build on the features of realities to explain and (re)furnish our views of the world with additional layers of understanding as well as uses of our ideas built on these more advanced meanings (Peirce 1877). In other words, science makes of our ideas increasingly powerful tools for more precise and focused transactions for which the outcomes become more predictable. Dewey often used water as an example to illustrate this point. From an experiential standpoint, water carries meanings of being notably potable, transparent, and liquid, and for some may even be synonymous with life. But when understood as H_2O, a chemical molecule, additional uses of the idea of water are allowed: water can now be understood and used as a solvent, can facilitate the production of its solid form, and can be differentiated from other clear liquids based on chemical tests (Dewey 1925, 1948). Wilfrid Sellars described these kinds of relationships between our more ordinary views of realities and science as a process by which the "scientific image of world" enriches our "manifest image of the world" or experiential view thereof (Sellars 1963). Our everyday experience of the world is already meaningful; it is rich with insight and understanding as suggested by Peirce's statement in epigraph. But science is a process by which we take ideas about certain phenomena and put them to the test, to refine the meaning of our ideas. In several domains, the evolution of science has challenged our common sense essentialist understandings and exposed how finer grain knowledge of our realities needs to take into account dynamic interactions and relationships with and within the world, thus challenging our armchair psychological essentialism (for further explanation on psychological essentialism, see Cimpian and Salomon [2014]).

Transactions, Biological Life, and Human Existence

Biological life is a continuous process of transaction between an organism and its environment, and health can be described as a status of equilibrium or preserved homeostasis of the organism in its environment (Billman 2020). Human existence, the life of human beings that have the ability to make sense of their transactions with the world through language and meaning making, is different from the mere biological aspects of the life of the human organism as a physiological entity. Obviously, the biological aspects of life are necessary to support human existence overall but there are rare situations where these can come apart. Such is the case of the condition colloquially called brain death[8] (and more precisely death determined by neurological criterion) first described by the French physicians Pierre Mollaret and Maurice Goulon. In this form of *coma dépassé*,[9] human beings lose consciousness completely and are unable to transact with their environment—apart from some forms of physiological transactions ensuring basic homeostasis—as meaning making beings, such that human existence ceases (Mollaret and Goulon 1959). Some aspects of biological life, at least in some rudimentary and supported form (e.g., with the help of mechanical ventilation and artificial hydration and nutrition) can be maintained, but the transactions of a human being become transactions of a human body only.

Human existence is a life as it is experienced beyond the mere biological understanding of life as a biological transactional process. Human existence can be narrated and communicated to be made sense of (e.g., in the form of narratives about the meaning of life) (Wong 1998). In other words, human existence is a meaningful and meaning-making activity, while biological life reflects the more basic functioning of the organism that supports this activity. Making sense of life does not require explicit or constant contemplation of the meaning of life—it is embedded in everyday thinking and habits—as unreflected transactions with our environments—and is something that we all do naturally, without necessarily consciously, continuously, reflecting on our experience of the world. Furthermore, the meaning of one's life is not individualistic, because there is a considerable range of social transactions that shape human beings as a social species. Thus, the meaning of life and of our ideas about life take into account their implications on others, because our identities and meaning-making activities are intimately connected to the transactions that involve and bear not only on ourselves but on others, too (Baumeister 1991). They also rely on a medium, language, which is a profoundly social phenomenon. In other words, human existence is profoundly relational.[10]

Human Existence as a Meaning-Oriented and Meaning-Making Activity

These clarifications about the nature of biological life and of human existence bring me to stress how human existence is structured as a quest for meaning (Zika and Chamberlain1992). The ability for human beings to produce meaning through transactions, to make sense of life events, and to relate to others and find directions are crucial aspects of human flourishing as a cardinal idea about life and its purpose (Ryff 2014a; Ryff and Singer 2008). There is now mounting evidence that individuals who have a sense of purpose and report living a fulfilling life are also those who are in better physical and mental health (Friedman and Ryff 2012; Ryff and Singer 1998; Urry et al. 2004). Importantly, the literature on resilience shows that the ability to make sense of difficult life circumstances is protective, while the inability to make sense of events has the opposite effect giving credence to Nietzsche that "He who has a why to live for can bear almost any how" (Nietzsche, *Twilight of the Idols*, 1889/1997 as cited in Frankl [1985]). Biological life needs to be sustained with its own basic physiological requirements, but we have come also to understand that physical health also interacts with mental health such that wellness needs to be thought about holistically, including the social and political aspects of health (Ruger 2005). When an individual lacks the resources and confidence to see past a given obstacle and transact satisfactorily, anxiety and other feelings of inadequacy and despair arise (Wegener et al. 2007). One of the hallmarks of depression is the inability to see past one's current situation and to envision how the current situation makes sense within a life narrative that will continue past a given life episode (McFarland and Vasterling 2017; Murphy et al. 2019; Schacter, Addis, and Buckner 2007). Beck's depression cognitive triad[11] also explains how negative views about oneself (e.g., "I am not a good or worthy person") support negative views about the world (e.g., "No one likes me or appreciates me"), which support negative views about the future (e.g., things will never change and keep becoming worse) (Beck et al. 1979).

Existence can become unbearable when difficult situations are seen as insurmountable. Suicide is often colloquially described as being a permanent "solution" to a temporary problem, but this is not necessarily how a suicidal person sees it. Persons who come to contemplate suicide have difficulty projecting themselves beyond a certain period of their life; they cannot see through their situation because of intense pain, psychological suffering, marginalization, or discrimination, among other factors (Hirsch et al. 2006). Because of their distress, they cannot generate a positive enough interpretation of their situation, such that their orientation to the future is undermined (Hirsch et al. 2007). A positive orientation toward the future is related to the ability to make sense of life and project oneself in the future; it reflects the ability to integrate the past and current self with the future self (Erickson 1968; Johnson, Blum, and Cheng 2014; Seginer 2009).[12]

By describing human flourishing in relationship to a positive future orientation of self-development and meaning making, I may seem to induce that the causes of withering are simply related to human thinking and human moods per se such that "fixing" these thoughts and moods would "solve" the problem. It is true that, for example, cognitive behavioral therapy for mood disorders can be an effective strategy for some patients (Fennell 2012; Gelenberg 2010). But such orientations can lead to psychological reductionism of human existence and life circumstances, which is far from my intention. The causes leading to the inability of positive existential meaning making are many and can be physiological, social, and political. The difficult circumstances that some people experience (e.g., unemployment, poverty, divorce, illness) may cause insurmountable challenge and pain, but not all facing such circumstances—even with similar life trajectories—adopt the same orientation (Witten, Savahi, and Adams 2019). Nevertheless, the ability to overcome challenges is not only a matter of willpower or positive orientation. It is shaped and deeply embedded in the life circumstances and access to resources and support system that allow people to overcome difficulties (Johnson, Blum, and Cheng 2014). Some people can afford to take the time to find meaning in difficult situations, have access to support that helps them see the bigger picture, but some do not. For example, a wealthy person becoming unemployed could transform this situation into an opening to new possibilities, a new beginning to reassess their life trajectory. However, a person who desperately relied on that job might not have the same capacity to face the situation in a constructive light because they are struggling for their very survival. This reflects the depth of the mutual constitution between individuals and their communities because a person who is disenfranchised and disempowered in a given community (e.g., a state where support for unemployed people is minimal or inexistent) is further challenged by unfortunate life circumstances (Paquin and Lévesque 2014; Raynault, Côté, and Chartrand 2013). Likewise, human relationships are important because the sense of belonging and participation is empowered by the ability to make meaning out of our situations and the desire to live a meaningful life (Clark, Biggeri, and Frediani 2019; De Jaegher and Di Paolo 2007). This is in fact a distinct argument in support of democracy as an inclusive, participation-fostering political community (Ruger 2005).

Thus, meaning making is an integral part of human existence. Life without meaning is difficult. Losing one's meaningfulness (e.g., through relationships, occupation, community) is a very difficult process that impacts mental and physical health. When positive meaning is found, human beings shine and reflect on others their well-being and the meaningfulness they find in their existence. Being able to find meaning, to conduct one's life with a given orientation, is a crucial positive aspect of human existence, and its overall contribution to psychological and physical well-being is increasingly evidenced. Pioneers of humanistic psychology (positive psychology) heralded this importance of well-being to global health and flourishing (Sloan et al. 2017). In the next chapter, I explain that this state and process of thriving has now been studied by psychologists (as "flow"[13] [Csikszentmihalyi 1975, 1990], "flourishing," etc.) and constitutes that to which ethics needs to contribute in the form of deliberative wisdom. As I will explain, morality is the set of rather commonly agreed on socially acceptable and desirable behaviors, beliefs, and attitudes shared in a group, while ethics attempts to critically examine these behaviors, beliefs, and attitudes in light of their ability to contribute to a meaningful, flourishing life for individuals and communities.

Biological Human Life without Meaning and Human Existence

Human existence is profoundly relational because so many human transactions involve environments made up of human beings. Moreover, the language we use to express and make meaning out of our lived experiences is an intersubjective technology that only exists when there is a community of speakers to listen, understand, and share these meanings. Biological human life can take place without human existence (e.g., brain death), but certain aspects of human existence—the meaning created—can continue to live without supporting biological life.[14] Metaphorically, human existence continues to exist after passing away, but it happens in the eyes of others, that is, in their memories of the transactions with a deceased person.

Many agree that a merely biological life (confined to physiological transactions) with no relational aspects of existence—despite the value we attribute to human life—is not worth living because it is meaningless and purposeless. It is not a first-person existence anymore; it is simply the survival of a body without awareness or meaningful and meaning-making transactions. There has been a growing medical, legal, and ethical consensus that such a life can be forfeited (American Academy of Neurology 1989; Quality Standards Subcommittee of the American Academy of Neurology 1995) if this was the preference of the patient or the proxy decision-makers (Racine et al. 2010). For example, patients who are in a vegetative state (now called unresponsive wakefulness syndrome [Laureys et al. 2010]) have by definition lost the ability to interact and communicate with others, as traditionally assumed (Multi-Society Task Force on PVS 1994a, 1994b). There comes a time when the person who is wakeful (e.g., preserved sleep–wake cycles due to preserved function of the brainstem) is conclusively and irreversibly unaware due to damage to the cortex and/or the thalamus, which serves as a relay, integrator, and processor of information from the brainstem to the cortex (Bernat 2006a). In such cases, the value of living a biological life without human existence, that is, life without having the *experience* of human life, becomes a crucial ethical and existential question.

Some religious doctrines have tightly clung to the concept of the sanctity of human life, arguing in favor of preserving life in cases where human existence has ceased and only biological life remains (John Paul II 2004), although even in a tradition like Catholicism, this is debated extensively (Tollefsen 2008). Such a stance based on religious or other doctrines can represent how customary morality, which by default calls for a highly valuable *prima facie* principle of respect for human life and taken to be a fixed and static domain (e.g., by duty to preserve tradition and replicate doctrinaire ways of envisioning human welfare), is unable to adapt to new situations and reimagine human welfare. It is incapable of envisioning a flourishing life based on actual outcomes. However, this kind of stance can lead to significant harm since, for example, pain processing is still not well understood in patients suffering from unresponsive wakefulness syndrome. Maintaining life at any expense could lead to dire situations where patients suffer because effective pain treatment is difficult without communication (Demertzi et al. 2012). As I will explain, ethics is opposed to such stagnant moral imaginaries. Ethics refuses that that the power to imagine one's existence and what a flourishing life is should be taken as a given or appropriated by political, religious, or social authorities. Engagement in thinking about one's existence is essential to the idea of existing as a meaning-making human being. Taking away this ability or formatting it in preset ways is the opposite of freeing existential possibilities, because a flourishing life is a life of self-actualization, not of the actualization of others per se. On this account, there is something deeply contrary to the nature of human existence to tell others what is best for them or decide for them. However, it is true that we learn from others, and, as we exist in relation to others, we explore and enact our existences with others.

Despite a certain medical, scientific, legal, and ethical consensus about refusing or withdrawing life-sustaining treatments for patients in such neurological states, there is still some hesitancy about the previously described situation. Human existence is *kept alive* metaphorically through the memories of having lived and transacted with someone. Memories can be triggered by the mere sight or remembrance of that person even though no more meaningful transaction occurs for the person in such a state.[15] The question thus becomes whether a human being in such a situation should be maintained biologically in life without having any meaningful existence except being there to maintain the memories and existence of others. Like many, I believe that this choice is personal but that meaningful transactional existence is really what counts, not mere biological life. Importantly, because of new findings about preserved awareness in a subset of these patients (perhaps 20 percent according to current research) (Monti et al. 2010; Owen et al. 2006), it may be prudent to preserve attribution of human existence to patients in those states (Blain-Moraes, Racine, and Mashour 2018) because of the uncertainty surrounding the diagnosis, as well as the entanglement of human existence and biological life.[16] Nevertheless, it remains the case that many (e.g., in survey studies on the vegetative state [Gray, Knickman, and Wegner 2011]) find the limited prospects of meaningful transactions too bleak to continue living.

Morality as an Essential Component of Existential Meaning

Morality designates the set of attitudes and behaviors we hold and exercise about what we think is right and wrong, better and worse. Such valuations about right and wrong, better and worse, are inherently part of our experience of the world (Prairat 2014). (Following Dewey, they take the form of prizings [when they are more subjective in nature] in contrast to appraisals and values [when more intersubjective in nature][17] [Prairat 2014; Ralston 2010]). This is why morality is always present in what we think and do, even if this remains implicit and assumed, that is, what is sometimes called implicit moralities (Strohminger et al. 2014). Morality is an essential component of existential meaning because it captures how we feel and believe we should behave and think about each other; it points to what we value and cherish, what we despise and abhor. When valuations become more established, as enduring values, they help us integrate events and experiences in the process of shaping who we are (Hitlin 2003). Interestingly, in a series of psychological experiments, Nina Strohminger and Shaun Nichols have shown that the values people hold are considered—in comparison to other domains of personality such as memories, desires or preferences, and perceptions—to be one of, if not the most important factor defining a person (Strohminger and Nichols 2014).

There are likely clear evolutionary reasons why social beings like humans have developed moralities, notably altruistic and cooperative behaviors, as well as fitting cognitive and affective equipment (e.g., empathy, role taking, theory of mind) that make these dispositions and behaviors possible (Hoffman 1981). Morality is not the opposite of self-interest or egoism because, to act morally, one must find the motivation to do so, and that motivation involves some form of enlightened self-interest wherein the recognition of others plays a central role in rewarding the agent, and in enabling them to act in accordance with beliefs and values (Fang et al. 2017; Wang et al. 2018). Our sense of acting morally, notably because of the social mechanisms of reputation and of attribution of blame and praise, is reinforced such that we integrate norms (through the process of socialization [Durkheim 1893, 1925]) and project our own behaviors and psychological well-being (e.g., our self-esteem) in our ability to be reputed as moral persons (Ricoeur 1992). To be considered a liar, a thief, untrustworthy, or unfair is no way to gain admiration, trust, help, or cooperation from others. Force becomes the only way of persuading others to obey and comply when moral esteem disappears. Accordingly, morality plays a pivotal role in our appreciation of ourselves and of others, which is essential to establish and support cooperation and altruism.

Ethics as Distinct from Morality

Few authors make a strong distinction between ethics and morality, as I pointed out in the introduction of this book, but it is of crucial importance for a theory like the theory of deliberative wisdom. From an etymological standpoint, both terms (one derived from ancient Greek, the other from Latin) describe a similar phenomenon. Their etymological roots (*ethos* and *mores*) designate the body of beliefs, behaviors, and attitudes shared by a community of people about what represents the character or nature of a good or bad person. However, I use—like a few others (e.g., Hartman et al. 2020)—the concept of morality to designate, at a more primary level, social habits and norms that guide human behaviors on a regular, everyday basis. My use of the term underscores the habitual, uncritical, unreflected-upon acceptance of these social habits and norms. Morality encompasses the domain of these habits. In contrast, ethics describes the critical assessment of our human moralities in light of their conduciveness to human flourishing.

While morality designates something like a system 1 in modern-day cognitive science parlance (Kahneman 2011), ethics represents a system 2, a deliberate, effortful process of making sense of our moralities and the conflicting moral intuitions we have. It requires effort and time, which we do not always have. This is the value of morality as a system 1: it allows us—as a set of heuristics and rules of thumb—to make reasonably sound decisions that fit with our cultural environments and meet the expectations regarding our different social roles. In contrast, what we consider to be valuable from an ethical standpoint is grounded in what we observe and experience, judged in light of what we project to be desirable but as being open to questioning.

Morality—at least in the basic sense in which I am using this term as a more instinctive and unreflected set of attitudes and behaviors—can be dangerous and unethical. What is and has been sometimes considered moral at a given time and place can be tremendously detrimental to the well-being of others. For example, human beings' sense of morality, of wanting to do the right thing and thinking that they are right, can lead to a sense of duty to harm others, for example, those who are seen as enemies and obstacles to a certain view of what is good, right, and preferable. There was a chilling sense of higher duty in the Nazi regime's attempt to "purify" the Aryan race. Leaders of German society (e.g., physicians) felt strongly compelled by the movement they joined in disproportionate numbers (Shevell 1999). The persecution of non-Aryans was embedded in a very problematic and perverse morality bound to confused ideas about biological purity and a sense of cultural superiority (Hitler 1925). The national-socialist morality was indefensible from an ethical standpoint.

Human moralities, especially those heavily driven by politics and religion, have been used to manipulate human beings and pit them against each other such that they view members of, for example, other ethnic, linguistic, racial, or national groups as less valuable than them. Furthermore, moralities—as forms of shared and socially enforced attitudes and behaviors—can impose views and circumvent the ability of people and communities to think for themselves and develop their own capacities, which is a key aspect of human flourishing. As Mihaly Csikszentmihalyi writes, "Repression is not the way to virtue. When people restrain themselves out of fear, their lives are by necessity diminished. Only through freely chosen discipline can life be enjoyed and still kept within the bounds of reason" (Csikszentmihalyi 1990). I thus depart from common morality theorists (e.g., Clouser 1995) who attribute a strong or quasi-foundational role to common morality, although I do agree that ethics is rooted in morality and our moral experiences of the world. It starts from moral experience but does not simply comfort it.

As discussed in more detail later in this book (chapter 5), morality is best envisioned as a domain with fuzzy and evolving boundaries. What was considered to be moral, centuries and even just a few decades years ago (e.g., holding racist beliefs, holding homophobic attitudes), is now immoral, and vice versa. These changes may be perplexing, because they seem to indicate that we do not know what is moral and fall prey to relativism. In response, conservatives in groups and societies tend to long for established moral truths and the principles of yesterday, sometimes even believing morality to be fixed or predetermined or God given. Though not all change is for the better, our understanding of what is more conducive to human welfare and flourishing is nonetheless constantly evolving, notably because of the plurality of human experiences. Furthermore, our material conditions make it such that some possibilities and contexts (e.g., advent of new reproductive technologies creating the possibility of novel family arrangements) have changed human lives, creating a wide range of new situations in work, family, and other social settings. Ethics furthers this plasticity and adaptability of human moralities with the project of shaping morality to align it with the pursuit of a life explicitly oriented toward human flourishing.

Ethics is also different from the scientific study of human moralities, although it pulls from this body of scientific knowledge. I designate as the scientific study of morality, efforts developed in different academic disciplines (e.g., sociology, anthropology, psychology, neuroscience), to understand and explain human moral behaviors. These efforts to understand human moralities or components of them (e.g., how people cooperate, empathize), their mechanisms, especially in real-life settings, are important and provide a more comprehensive outlook on all aspects of human moralities. Ethics differs from these efforts insofar as it attempts to use experiential and scientific knowledge about human moralities to engage in a process of critically considering whether one's existence or one's dealing with a given situation is conducive of human flourishing. Ethics is explicitly practical in intent and method as it attempts to guide human behavior, especially when conflicting moral intuitions surface in a given situation. This is not a goal necessarily shared by more descriptive and explanatory studies of human moralities. For example, moral psychology, the sociology of morality, or the anthropology of morality do not claim to guide human behaviors for the better. They attempt to explain phenomena such as cooperation, altruism, diversity of moralities, and changing moralities, but they do not examine critically to see which one should be adopted. Taking distance from these behaviors is important as some of them (e.g., empathy) can appear to be inherently good but can actually undermine other valuable attitudes and behaviors (e.g., being empathic and feeling closer to someone can undermine the ability to be just) (Decety and Cowell 2015). Hence, when I describe ethics in shorthand as a science, it is in a charitable sense much like medicine is called a science but is in fact a combination of theoretical and practical knowledge.

The somewhat different and diverging intents of those involved in the study of morality (e.g., social psychologists, anthropologists) and those who think about difficult moral situations and help others deal with them (ethicists) have come to create a chiasm between them (Flanagan 2009). The hybrid field of empirical ethics represents an attempt to bridge scientific research on human moralities with the work of ethicists. However, there is still considerable diversity in foci, objectives, and methods in empirical ethics (Davies, Ives, and Dunn 2015), with some work in the field pulling from more dialogical methods (Ives 2014). I hope that this book contributes to the use of insights from scientific studies of human (moral) behavior and psychology to support the development of a more robust and informed ethics.[18]

By formulating this wish, I do not allude to any form of crude scientism or positivism that would entail that ethics has to be reduced to conform to some understanding of morality. There is a significant amount of influential scholarship (e.g., Herbert Spencer's social Darwinism [Spencer 1892], Edward O. Wilson's sociobiology [Wilson 1975]) that has opted for overly simple models of how we could build a more ethical society if we only promoted such and such tendency (e.g., empathy, altruism, compassion, justice) of human psychology or behavior. Ethics is not only a reinforcement of certain cornerstones of human moralities. It is a creative meaning-making process; it cannot be reduced to the crude application of moral thinking. Ethics requires the ability to imagine resolutions to moral problems while maintaining coherence with one's values in dealing with difficult moral situations. It is an exercise in projecting oneself and reinventing new equilibria and new ways of thinking and behaving. It is a mode of habit reconsideration and reformation. As a creative meaning-making activity, ethics is explainable in biological or in psychological terms because it taps into psychological processes involved in, for example, supporting self-esteem and the search for a flourishing life as rewarding experiences. But better understanding such mechanisms does not take away ethics' specificity as an exercise in using such knowledge practically.

Ethics as Oriented toward Flourishing

Human moralities have existed for as long as human beings have existed because social animals cannot exist without regulating and adapting their behaviors in function of each other. Nonhuman moralities are also found in more rudimentary form in a range of other intelligent social species (Waal 1996). Human moralities capture a wide range of coherent insights about, and can be explicitly framed around, the pursuit of a collective or individual form of well-being (e.g., in various moral precepts and behaviors). In contrast, ethics has a clear history. This field was born (at least in the Western tradition) when a number of philosophers, particularly in Athens, started to discuss and critically examine moral beliefs about the good and fulfilled life. Is this a life dedicated to the pursuit of wealth and fame? A life of prestige and high reputation? A life of pleasure? Most of these thinkers framed their quest as the search for a life of flourishing, an orientation they strived to define in various ways, as shown by the various schools of thinking that promulgated different accounts of human flourishing. The ancient Greek term for this fulfilled life is eudaimonia, which is now more often translated as human flourishing although it had been frequently translated inaccurately as happiness, which does not do justice to the idea (for an explanation, see Ryff [1989] and Ryff and Singer [2008]). Debates on the actual equivalence of the terms continue, but it suffices for our purposes to have a translation that ensures that ancient wisdom on eudaimonia can be pulled from and brought to light in a way that preserves its deeper meanings as a lifelong project.

Eudaimonia designates a life accompanied by a good (*eu-* in Greek) spirit (*daemon* in Greek), because a flourishing life was judged to be guided by such a good spirit. For example, Socrates claimed to be accompanied by a "divine something," an inner voice, who warned him about mistakes but never told him what to do (West and Plato 1979). Irrespective of the more mystical components and terminology associated with this account of the good life, the ancient Greek philosophers, and subsequently Roman philosophers, wanted to put the idea of the good life to the test (Parry and Thorsrud 2021). The eudaimonic life was neither a life of pure pleasure and distraction, nor merely fame, but a life of self-cultivation, of growth, of improvement, all of which could be most appropriately translated as a life of flourishing (Parry and Thorsrud 2021). Interestingly, this distinction between the hedonistic and the eudaimonic life is now supported by contemporary psychology (Baumeister et al. 2013). This quest for the good and flourishing life under the auspices of what we call ethics, was not simply theoretical in nature; it was a practical quest. Discussion about human flourishing was an embodied practice; questions were directed at the first person and asked what the good life for a person and their fellow citizens in mind was. The practice was meant to guide one's thinking and behavior in concrete situations and act as a compass for one's life. In other words, ethics was and is a science and a practice of human flourishing, an idea that pervades the theory of deliberative wisdom; this is the object of the next chapter.

Ethical Agency as Engagement with Flourishing

Within the scientific study of human morality, moral agents designate individuals as they embody specific moralities, but ethics is concerned with the ethical agent. The ethical agent is a moral agent who is in the process of questioning and reflecting on human moralities, including their own. Hence, ethical agency here designates an active and engaged way of thinking about moral life that brings to light the valuations inherent to behaviors (including entrenched habits), as well as the search for self-fulfillment and the realizing of individuals' own potential and that of others, which has a clear bearing on individual well-being. Ethical agency is not only about being accountable for one's behaviors or for being able to provide rational justification for behavior. The life of the ethical agent is a creative one; a moral agent moves from being in the passenger's seat to the driver's seat of their moral life. This kind of life cannot be that of a moral agent and especially not of the moral patient[19] (Pluhar 1988). It is a life in which values are envisioned and enacted through actions that touch upon one's self and identity as a person with values and interests (Hitlin 2003), and wherein the loss of one's sense of moral identity and moral integrity paves the way for unethical and immoral behavior (Knoll et al. 2016; Tenbrunsel and Messick 2004). Clearly, the capacity to act as an ethical agent is not magical. It is limited by context, human and material resources, and the disposition of every individual, especially as a system 2 process. It may also be domain-specific such that is easier for some people to talk openly about certain moral aspects of their lives and more difficult in other aspects.

Ethics Is about Solving Situated Moral Problems and Also Maintaining a Positive Self and Narrative

Situations—especially morally problematic situations, because they confront us with our limits as transacting beings—have a potential transformative, life-changing character for ethical agents. These difficult situations concern existence and meaning in life; they entice an openness to learning and engaging in activities that broaden one's idea of meaning in life (Bakracheva 2019). The focus of ethics (Zizzo, Bell, and Racine 2016), notably pragmatist ethics (e.g., Miller, Fins, and Bacchetta 1996; Racine 2016), on difficult moral situations (e.g., difficult decisions in end-of-life for patients in an unresponsive wakefulness syndrome, complex issues related to corporate responsibility and conflicts of interest) should not be interpreted as implying that ethics is only about solving such specific situations. Ethical agency is a lifelong project, that is, a sense of self, authenticity and integrity, and search for existential meaning (Frankl 1985). Although the theory of deliberative wisdom emphasizes situations and the situated nature of moral agents and the importance of situations as units of experience that represent transactions between the agent and their environment (see chapters 3–7), there is still an agent that exists between these situations, an agent whose existence we cannot lost sight of. For instance, what situations are considered to be morally problematic situations and which situations we remember (because of the mechanisms of memory and their biases; see chapter 3) shape who we are and scaffold our self-identify within and across experiences (Schacter, Chiao, and Mitchell 2003).

Agents have identities across situations and whose sense of continuity, whose sense of self, built as a self-narrative, endures over situations (Ricoeur 1992). When memory collapses or is challenged, the sense of self can be lost momentarily of permanently. This explains the considerable fear toward Alzheimer's disease, which is not only a loss of memory but a loss of sense of self, for oneself and for others (Simm et al. 2017). This loss of memory, self, and ability to project oneself generates immense challenges (Naue 2008) and partly explains attempts to go as far as to commit suicide before the advent of the disease (Davis 2014). Importantly, efforts to sustain this sense of self (e.g., through prompting autobiographical memory) could help alleviate the pain caused by this disease (El Haj, Gallouj, and Antoine 2019) and help agents make sense of their existence and maintain a narrative (Baloyannis 2019).

Narratives provide a powerful way of making sense of human existence[20] as developed in different fields of inquiry such as psychology (Sarbin 1986), social science (Polkinghorne 1988), medicine (Charon 2001), and ethics and philosophy (Schechtman 1996). To narrate one's existence represents the ability to speak from the first-person perspective about the meaning of one's life, how one fits into the world, and about how one's existence is bound to the existence of others. It represents a profound way of making sense of one's existence because it entails the ability to reflect on oneself (reflexivity) and one's values. In this light, the ethical agent, that is, the moral agent engaged in reflecting on their behaviors and revising them critically to pursue a flourishing and meaningful life, is a narrative ethical agent because narrative is the form by which the moral life is examined in the context of ethics. Ethical narrativity represents the ability to make sense of existence (including the role of morality therein) in light of an orientation toward growth and meaning making. It is an active and continuous pursuit because the story of such an existence is one that cannot be assumed to be complete and closed to further learning in the face of difficulties. Furthermore, narratives are situated because they are inextricably tied to the events encountered in and created by one's existence.

Although they may appear so, reflexivity and narrativity are not intrinsically ethically defensible in nature, since they could be used for nefarious purposes (e.g., using convincing national narratives to persuade a group to undertake criminal behaviors against members of other ethnic groups). Some narratives can be profoundly unethical and lead to the disempowerment of minorities, for example, by "justifying" racism. To contribute to ethical action, these narrative-making abilities and dispositions need to be exercised with the intent to improve one's contribution to the welfare of others and to foster more meaningful experiences for oneself and for others; they need to be channeled to cultivate the narrative of flourishing lives (for extensive elaboration on this, see Ricoeur [1992], Schechtman [2014], and Wong [2014]).

Ethics and Narrative Editing

Knowledge about human morality sheds light on a specific aspect of the transactional nature of human existence, such that existence is potentially enriched, and narratives opened to future learning and to reorientations based on this knowledge. Some unchecked problematic assumptions about ourselves, our self-understanding, and our relationships with others hold considerable importance and lead to premature conclusive views. Consider, for example, common meta-narratives about how achievement is the affair of the individual, of their independent work and effort, irrespective of context and social factors, observed by Owen Flanagan (2009). In many respects, I share Flanagan's analysis that certain assumptions (e.g., about desert and merit, about voluntariness of human behavior), as they are typically understood in the US and elsewhere, are problematically atomistic. In this social context, ethics needs to pull from knowledge about human behavior to critically examine American settled moral beliefs and assumptions so that we can align them more productively with the goal of human flourishing and free them from unproductive and partial ideas.

Most often, revisions to our ideas and habits do not entail their outright rejection but involve better understanding their place within a broader framework of appreciation, that is, understanding their strengths and limitations in light of their implications. For example, regarding a common American metanarrative, a pragmatist attitude does not call for an end to certain principles and virtues such as individuals rights and respect for autonomy. However, it may involve a change in how we understand them in a broader ecology of ideas to include other considerations (Martela 2015). For ethics to become a "moral science," as advanced by Dewey, it cannot ignore or neglect insights about human social and moral behavior because "What cannot be understood cannot be managed intelligently" (Dewey 1922, p. 3). Accordingly, contrary to the sometimes uncritical glorification of narratives as authorities on who and what we are (e.g., our self-narratives embedded in broader social and political narratives), pragmatism invites us to critically revise our understandings of ourselves in terms of what we contribute to human flourishing and the fallibility of our habits in light of knowledge. This orientation is consistent with theories showing that an open worldview, wherein values are questioned, and open-mindedness fostered, are constitutive of a healthy, adaptive, and productive self (Ryff 2014a).

Self-narratives thus need to be seen as fallible and responsive if persons remain open to questioning their impact on others. In stating this, I recognize that there is a very fine line between promoting self-reflection about self-interpretations and self-narratives versus discrediting the narrative of someone else or of oneself. The process of invalidating narratives—narrative erasure—when someone claims to know better than another about the meaning of their life and life experiences—is an existentially disruptive process seen in racism, discrimination, and sexism; for example, when it is claimed that these phenomena do not exist (Hall and Fields 2015). It is sometimes said that the scars that cannot be seen are the hardest to heal; shattering, dismissing, or invalidating someone's narrative has profound psychological implications. Overstressing how certain stories are incorrect, false, or misleading can be profoundly destructive to the sense of self. Wars are often foremost wars occurring between diverging narratives, that is, divergent accounts of history, such that wars are justified on divergent interpretations of who is and was right, about who has legitimacy to claim rights and lands, about who is entitled and who is not, and even on attempts to invalidate and erase the narratives of opposing nations (e.g., the 2022 Russian invasion of Ukraine). Those who have the power to shape social and political narratives thus have tremendous power, but there is a political responsibility in striving to promote welfare within the exercise of this power such that different political communities can flourish and interact peacefully. The importance of learning from another's narrative rather than immediately invalidating it is therefore paramount, especially because similar thoughts and words can be used either to undermine a person and their strengths to overcome difficulties or rather support a person in seeing some difficulties more clearly and accompanying them in developing a more accurate and lucid narrative about who they are. In sum, flourishing human beings are persons whose existence is purposeful and meaningful, as is revealed in some narrative forms. But meaning and purpose can become dogma and doctrinaire if unchecked by a sense of their impact on others. This process thus involves editing our own self-perception and self-narratives as we transact with environments and produce meaning resulting from experiences, including how others view us and react to the stories we articulate about ourselves.

Methodological Implications of a Situated Account of Moral Agency for Ethics Theory

In this chapter, I described a situational account of moral and ethical life to introduce as a starting point before engaging in further theoretical development about ethical existence (e.g., human flourishing) and the situated nature of difficult moral problems. In forthcoming chapters, I make more concrete and specific methodological recommendations on how we should understand and tackle morally problematic situations; how we should generate empowering thoughts and behaviors that are conducive to growth and human flourishing, deliberation, and imagination; and how we should evaluate outcomes of ethics-guided action using participatory methods. Based on the current overview of situated moral agency, a general methodological orientation can be sketched for the desirable features of ethics theory and methods in keeping with this account of ethical agency. This is an orientation pursued by the theory of deliberative wisdom as a form of ethics that honors the meaning-making contribution of human morality.

- **Situated and situating**: Given the emphasis on the situated nature of experience and existence, methodologies for understanding, resolving, and evaluating resolutions of moral problems need to capture the situated and embodied nature of moral experience. This implies a view of the moral agent as being in situation. This is a cornerstone of chapters on situations and morally problematic situations (chapters 3–7) but also infuses how mechanisms of awareness and processes of deliberation, imagination, and enactment are later described in chapters 8–11.
- **Narrativity and meaningfulness**: To make sense of our existence, the narrative, story-like form is an established strategy and orientation supported by millennium traditions of oral, and more recently, written history. Narratives allow individuals and communities to recount who they are, where they come from, who they want to be, and where they want to go. Storytelling is a fundamental way of making sense of the fact that human beings have memories of what they have experienced and use their ability to project their existence (and the existence of those around them, i.e., their communities) using scenarios (scenarization). As I describe later (chapter 10), scenarios are devices that allow the projection of situations in the future to explore their implications. As an orientation, narration

and narratives call for methods that allow people to share their "making sense" of situations and their experience (chapter 6) and their ability to project into the future (see chapter 3 on the role of memory in future-oriented thinking).

- **Reflexivity and fallibilism:** Since ethics theory is oriented to human flourishing, it is open to revisiting how situational sense-making can be profoundly changed by confrontation with new situations. Self-reflection (reflexivity) represents the ability to re-situate oneself and reorient one's narrative in light of its outcomes and implications, including how dealing with situations in certain ways may enrich or impoverish our perspectives of ourselves and others. Like other orientations (e.g., participation, narration), the reflective orientation is not intrinsically ethically praiseworthy because it can be used in detrimental ways that disempower and diminish others. Approaches that make room and time for reflection, self-reflection, questioning and self-questioning, and open discussion (Walker 1993) can help support reflexivity (see example of self-reflection tool developed in neonatal care in Racine et al. [2017a]). In the deliberative account of human flourishing provided in chapter 2, I stress how human flourishing must be understood open-endedly to allow for continued growth and learning about ourselves and about what matters in life. It is also in this light that evaluating the outcomes of enactments is described as an ongoing learning process (chapter 13).
- **Holistic accounts of outcomes:** A situated account of moral agency as described in this chapter calls for concepts and methods that are connected to how moral agents live morally problematic situations and how they make sense of them from an existential standpoint. This is partly the subject of chapters 4 and 6 and is reflected in how deliberation and imagination (chapter 10) as well as enactments (chapter 11) are intended to return to human experience as it is understood holistically.
- **Nonreductionist inquiry:** A corollary of a commitment to a holistic account of human existence and experience is to ensure that ethics inquiry does not commit to various forms of disciplinary reductionisms (Callahan 1973). This is why I often state that my use of the term science does not imply scientism but a commitment to the value of knowledge about human experience. In my treatment of situational insight and morally problematic situations (chapter 5) and mechanisms of moral and ethical awareness

(chapters 8 and 9), I go at length to avoid various forms of reductionisms that could entail the restriction of moral considerations to a single set of aspects (e.g., cognitive, affective).

- **Participation and deliberation:** A transactional account of ethical agency implies that persons are agents. Agents create their existence through their behaviors and their ways of thinking. This does not mean that they exist in siloes; quite the contrary. But agency helps understanding how human beings are action oriented from a physical (e.g., the structure of our bodies), biological (e.g., the functioning of our organs), and social standpoint (e.g., the structure of our families, work environments). Tapping into and facilitating human beings' capacities as agents is thus of fundamental importance. This is now well recognized in various literatures about well-being and mental health where participatory and deliberative methods[21] have gained greater recognition (Abma, Voskes, and Widdershoven 2017; Clark, Biggeri, and Frediani 2019; Harris 2018; Montreuil and Carnevale 2018). The ability to act upon one's situation, to project one's existence into the future, to control one's environment and existence, to be open to new situations, and to be empowered by acting upon situations are all signs of positive psychological and social health and functioning (see chapters 3 and 11). They stand in contrast to passive experiences and feelings of marginalization, disempowerment, disenfranchisement, stigma, alienation, exclusion, powerlessness, underdevelopment, closure to change, and so on (see also chapter 7).

Conclusion

In this chapter, I reviewed a number of insights in psychology, social science, and the humanities that support an account of situated moral agency in which morality and moral meaning making are central to the understanding of both moral agency and ethical life. Existence is not merely biological life. It is human life proper, and, within human existence, valuations are inherent to human behavior. These valuations can be implicit, such as in our habits, or they can be explicit and reasoned as values (e.g., autonomy) and principles, relying on the respective of certain values (e.g., respect for autonomy). Ethics represents the attempt to critically revise valuations and values as embodied in our thoughts and behaviors, in light of their contribution to human flourishing based on their outcomes, in the broad sense of this term, including outcomes for the agent's progression toward an ethical life and narrative. A situated account of agency leads to a number of methodological orientations for ethics—approaches that are situated and situating and that foster narrativity and meaningfulness, reflexivity and fallibilism, holistic accounts of outcomes, nonreductionist inquiry, and participation and deliberation. In the next chapter, I advance an account of deliberative wisdom, that is, an understanding of ethical wisdom as the theoretical and practical knowledge of human flourishing that recognizes the transactional nature of our own human experiences and that of our social and interpersonal lives. The transactional nature of our own ethical ideas and their purpose as tools of human flourishing implies that their development and exploration is strongly supported by a deliberative or dialogical approach.

2 Deliberative Wisdom and Human Flourishing

Moral problems are experienced as difficult situations, but at the same time, because they carry us to our limits, they bring to the forefront opportunities for self-reflection and self-knowledge of one's values and the impact of one's actions on others. In this respect, ethics is a reflective and creative activity akin to scientific and artistic inquiry that cultivates a positive orientation to the moral aspects of our lives and toward the experience of morally problematic situations. In ethics, persons engage in creative thinking to resolve fundamental human obstacles, thereby gaining a sense of growth and self-fulfillment, because of the abilities they exercise and develop, as well was what they learn throughout this process. Aristotle described the exercise of this practical judgment as *phronesis* and embedded this effort into the quest for human flourishing, that is, eudaimonia. These concepts originate in ancient times, and they capture a fundamental truth about the profound existential roots of ethics and the orientations of an ethical person. Nowadays, our lived worlds have expanded in societal and technological complexity and present human beings with changes that are greater and faster than ever. In response, practical judgment on matters of human morality may need to frequently sidestep traditions. Customary habits often need to be refined or simply changed. Fortunately, in the face of changes and the wide array of situations that existence bestows, a person can envision the future through the lens of the progressive acquisition of *wisdom*, namely skilled practical judgment aligned with the ability to pursue a flourishing life, and encourage such an orientation in others. In keeping with insights from the tradition, in this chapter, I position human flourishing as the focal point of human development and of ethics, which is concerned with living the good and meaningful life. This helps explain why ethics can help one acquire and cultivate ethical wisdom, defined as the ability to gain insight into human flourishing through re-situating and realigning oneself (and/or others) toward the pursuit of different dimensions of human flourishing (e.g., positive relationships, meaningful life, autonomy) as described in more contemporary literature on the topic, which I also review. In the context of this book, ethical wisdom is envisioned as inherently deliberative or dialogical such that it is through openness and discussion that this wisdom is grown. Likewise, the view of the goal of ethics put forward in *The Theory of Deliberative Wisdom* is holistic and integrative in that not only are the acts of individuals important, but the personalities and characters of moral agents are also paramount. Being good and doing good capture different facets inherent to a comprehensive understanding of ethics, because the resolution of morally problematic situations involves actual individuals on whom problems have a potentially profound existential impact. What the active understanding and resolution of moral problems does to a person is often missing from ethics theories, but it is crucial to reflect on the human motivations that can encourage engagement with ethics. The centrality of this focus and orientation toward deliberative wisdom as a path to flourishing explains why this chapter figures at the beginning of this book.

> When the wise man points to the moon, the fool looks at the finger.
> —Chinese proverb

> I regard it as the chief duty of the State to protect the individual and give him the opportunity to develop into a creative personality.[1]
> —Einstein, *The World as I See It*, 1932

In my first monograph, *Pragmatic Neuroethics*, I described how mainstream ethics tends to focus on the criterion of ethical acceptability, that is, whether a proposed action or technology is acceptable to others. I encouraged thinking about how ethics should also integrate a more substantive criterion of ethical praiseworthiness, that is, whether a proposed action or technology is made morally laudable by contributing to human welfare and flourishing (Racine 2010). The quotation by Einstein in the opening of this chapter alludes to these two ideas in the context of the fundamental missions of the state—and of human communities more generally—to protect the individual and safeguard their interests (ethical acceptability) and to promote their full well-being as a creative being who self-actualizes (ethical praiseworthiness). Both are necessary and complementary goals in the pursuit of a flourishing life for oneself and for others because the flourishing of individuals is connected to that of human communities and the cultivation of positive relationships with others.

Several authors have described how mainstream ethics has frequently prioritized thin, procedural, individual foci (the ethically acceptable) and left aside thick, substantive individual and communal aspects of ethics and human flourishing (the ethically praiseworthy) (Berlin 2002; MacIntyre 1984; Taylor 1992; Walzer 1994/2019; Williams 1985). For example, Isaiah Berlin (2002) speaks of negative and positive liberty[2]; Bernard B. Williams

(1985) and Michael Walzer (1994/2019) contrast thin (evaluative) concepts and thick (descriptive and evaluative) concepts following the work of Clifford Geertz (1973); and Charles Taylor (1992) analyzes the tension between an individual focus on equal recognition of persons and the deeper sources of our identities. These distinctions capture different concerns, but they allude to two contrasting orientations: one centers on the protection from harms to the individual in a narrow sense and the more individualist aspects of well-being in which actions and situations are either acceptable or not, while the other points to the realization of individuals' views of what is valuable and contributes to the well-being of others such that actions and situations can be praiseworthy. The theory of deliberative wisdom builds from efforts to open ethics up to deeper considerations related to human flourishing captured by the concept of ethical praiseworthiness and key dimensions of human life. It is an effort to capture ideas of positive liberty and the thick and deeper sources of ourselves as they relate to our interpersonal relationships and the communal aspects of our lives. However, it does not discount an orientation toward the ethically acceptable, because both orientations are complementary lenses that are part of a richer and broader account of ethics.

The concern with integrating the two orientations of ethics (ethical acceptability and ethical praiseworthiness) as part of an account of ethics oriented toward human flourishing is omnipresent throughout this book; there is a sense in which ethical praiseworthiness expresses more cogently the essence of a flourishing life as part of both the goal and the journey of human life. However, ethical acceptability is also very important because it is a way of envisioning this journey as being respectful of others and acceptable to others, hopefully in ways that support mutual learning. The conjugation of these two orientations is meant to be carried in productive dialogues where unproductive dichotomies between the thick and the thin, the individual and the communal, are avoided by cultivating an integrative focus on the analyses of courses of action (see chapter 10). This focus operationalizes this reconciliation and makes ethics more usable to instill and support eudaimonic orientations in facing questions that challenge the meaning of our existences. I come back to these ideas in the chapter about the nature of ethical deliberation and of ethical enactment (see chapter 12).

Overall and in the context of trying to broaden the scope of ethics, human flourishing—a translation of the Greek concept of eudaimonia—constitutes a positive, integrative, and defensible orientation for ethics theory despite,

I will argue, some warranted philosophical skepticism and some unwarranted and unfair pessimism.[3] Human flourishing includes autonomy and self-determination, which tend to be well and perhaps even overly emphasized in the Western tradition,[4] as well as the contribution of personal growth, finding purpose in life, and fostering positive relationships with others. These dimensions of human flourishing have been investigated extensively in contemporary psychology literature (Ryff and Singer 2008). Importantly, contemporary literature uses a plethora of terms to describe human flourishing: the pioneers of humanistic and positive psychology Carl Rogers and Abraham Maslow, for example, speak of an actualization tendency and self-actualization; Alfred Adler refers to "the useful side of life" to designate a positive and constructive orientation in human life (Dwornik 2003), while Victor Frankl describes this as "self-transcendence" (Frankl 1966).

In searching for resolutions to moral problems, human beings are in a sense forced—if they want to agree and live together—to contemplate what they consider to be greater, more defensible goods worth pursuing both individually and collectively. Importantly, this overture creates an opportunity for deeper thinking of what is valuable and what makes human flourishing possible. It is an occasion to ask what are the kinds of experiences and situations that allow humans to be guided by intrinsically motivating goals (Deci and Ryan 1985), in the pursuit of flourishing. Thus viewed, ethics is not chiefly an exercise in applying an established ethical truth (in the form of principles or codes), as it has been criticized for sometimes becoming (see critiques of this influential view in Altman [1983], Berlin [2002], Hoffmaster [2018], Miller, Fins, and Bacchetta [1996]). Rather, ethics is an inquiry, an opportunity to discover the more warranted, more justified ways of thinking and behaving, better able to contribute to the development of human potential. In other words, morally problematic situations—sometimes viewed primarily as decisions, choices, or dilemmas to solve (e.g., Beauchamp and Childress [2001], Shaw [2005], and see critical analysis in Racine [2016])—are also opportunities to reconsider, rethink, and possibly change our habits in ways that align them more powerfully with our own growth and that of others. Human flourishing is thus constituted and shaped by our continued experiences of orienting our views and behaviors toward open-ended and testable goals informed by the difficult situations to which we are confronted. It is therefore not a fixed end. Also, it is not a narrow or individual quest because it is supported by what we learn from others. Indeed, within an open-ended and empirically grounded ethics inspired by pragmatism, ethical wisdom is inherently deliberative (or dialogical) and participatory, in keeping with the transactional nature of human experience and existence (Brinkmann 2011) (see also previous chapter), including the experience of morally problematic situations and the moral life (Gallagher 2014).[5] It is in this sense that I find favorable the use of wisdom to describe a kind of knowledge which is not only to be found in academic or scientific expertise in a narrow sense, but also in the knowledge gleaned from life experiences. In other words, deliberative wisdom is intended to democratize and open up the idea of knowledge about human flourishing.

Such a proposal about the deliberative nature of knowledge on human flourishing follows important threads of ancient philosophy's accounts of eudaimonia (Hadot 2002; Parry and Thorsrud 2021). Yet the proposal remains relevant given the need to understand in much more detail the nature of human flourishing and the habits and practices conducive to flourishing (Ryff 2014a; Seligman and Csikszentmihalyi 2000). Although central to ancient ethics, the focus on human flourishing and wisdom almost disappeared from mainstream ethics (e.g., see an historical analysis in Anscombe [1958]) but it resurfaced in humanistic psychology at the beginning of the twentieth century, then receded with the advent of positive psychology (Seligman and Csikszentmihalyi 2000). Modern research in psychology defines, operationalizes, and measures phenomena such as "human flourishing," "well-being," "intrinsic motivation," and "wisdom." This research can always be contested insofar as different definitions of these phenomena exist and are bound to bring challenges to empirical work. However, being able to measure, observe, and assess these phenomena can help further our understanding of them and of their real-world relevance.

It is noteworthy that the scholarly tradition that was initiated in psychology in the 1920s about optimal human performance and existence (e.g., Adler, Maslow, Rogers, Erik Erikson), to which Dewey could also be ascribed (Dwornik 2003), was reinvigorated in the past few decades. After a post-WWII decline in popularity, it has become a thriving field of inquiry, under the auspices of "positive psychology" (Seligman and Csikszentmihalyi 2000). This field is described as "positive" not because it offers a series of recipes for making people happy, but because it focuses on the positive functioning of individuals instead of their deficits, their failures, their psychological and mental health problems. Like other fields (e.g., medicine and even ethics), psychology was judged dominated by thinking about negative aspects of human existence. In response, positive psychology intends to renew the tradition of thinking about personal development and human flourishing. In this respect, it reconnects with ancient ethics theory (Ryff 1989) and pioneers such as Adler. Various key concepts are used to investigate and discuss positive functioning, such as that of "flow" (Csikszentmihalyi 1990), happiness, and flourishing (Ryff 1989), with some important differences and some overlaps too between those concepts. Because of this book's focus on ethics, the concept of human flourishing offers a promising bridge between this scholarship in psychology and ethics as seen in the work of Alasdair MacIntyre, Amartya Sen, and Martha Nussbaum, which has rejuvenated eudaimonism (MacIntyre 1984; Nussbaum 1997; Sen 1999). Indeed, the focus on positive functioning provides a realistic psychological motivation preserving a concept of self, based on concerns for personal integrity and authenticity (Knoll et al. 2016). The revival of the orientation toward positive human functioning remains incomplete. Importantly, our contemporary societies are still in need of major changes to better support and develop the potentialities of all human beings (Sen 1989), and ethics could position itself more explicitly in relationship to human flourishing.

This chapter expounds the view that deliberative wisdom is the theoretical and practical knowledge about human flourishing. First, I provide an account of ethical wisdom as being inherently deliberative, that is, deliberative wisdom as based on the transactional nature of human experience. In doing so, I explain why ethics has moved away from human flourishing as its focal point, partly because of the adoption of a positivist stance on what constitutes knowledge (e.g., as detached and objective knowledge) and factors that reinforce the demand for authoritative ethics knowledge. This has led ethics away from experience and existence and its roots therein. I argue for the value of a renewal of this focus. I pull from Aristotle and Dewey to lay this foundation from an alternate view. Second, I show how current scholarship on human flourishing, notably Ryff's integrative six-component model of human flourishing,[6] helps situate deliberative wisdom within a rich body of knowledge in ways that allow for the further specification of insights about ethical praiseworthiness and ethical acceptability when approaching the tasks of ethical deliberation and of ethical enactment (see chapters 10 and 12). My treatment of the topics covered in this chapter is not exhaustive, and my goal throughout this book is not to provide a novel account of wisdom, human flourishing, or ethical deliberation per se. Rather, it is to make interdisciplinary connections between important ideas and scholarship that bring theory to practice in a way that helps bridge thought and experience in human ethics to offer a workable connection between ethics and the pursuit of a form of open-ended wisdom about human flourishing, that is, a form of open-ended eudaimonism.

Human Flourishing as the Focal Point of Ethics

The Place of Human Flourishing in Ethics

The topics of wisdom and human flourishing might appear a bit disconnected from contemporary ethics for those who are familiar with the field.[7] Why should ethics theory be concerned with human flourishing? Is ethics not fundamentally about resolving difficult moral decisions and choices, about defining appropriate and justified lines of conduct based on acknowledged and defensible principles? The concerns of modern-day workers, practitioners, and stakeholders may seem to be only loosely connected with the general topic of human flourishing. How could any view or definition of human flourishing help guide human conduct? Are we not replacing a complex problem with an even more complex one by following such a path? There are so many ways to think about the good and better life—how could we possibly envision this as an orientation or terrain of discussion that would offer any progress? Skeptics certainly have a strong footing here. At the same time, engagement with questions of human morality unavoidably bring up questions about the value of certain ideas and experiences. This or that way of doing things best embodies a value like autonomy or respects the principle of autonomy. But why does autonomy matter? What does it add to human experiences? Is it always valuable? Is it valuable for everyone?[8] Thus given these difficulties, can we really do away with human flourishing and views about the good life when engaging with moral questions through the lens of ethics?

The contemporary reluctance to envision human flourishing and related wisdom as focal points of ethics theory is worth contemplating because these goals bring up issues about personal identity, that is, the kinds of persons we really are and the kinds of lives we really want to live. A focus on flourishing is difficult without thinking about our own lives and personal life philosophies, which may cause discomfort. It is easier and more comfortable to focus on more technical issues, such as discussing the ethical nature of our intents or the ethical nature of our actions and their consequences in an allegedly more objective and detached way. Radical questions about the good life, the flourishing life may also seem to lead to answers that are much less generalizable and bring out one's singularity such that relativism is in play. If this kind of life is good for me (e.g., valuing family relationships, valuing professional commitments and good work), what guarantees that the same life is good for others? Are the learnings I take away also applicable beyond just myself?

But with the advantage of thinking about ethics as a third person endeavor comes the peril of its inapplicability to the first person such that we cannot make sense of how it would help us as persons or would also help other persons make sense of and surmount their own difficult situations. Some have already remarked that several influential ethics theories have rarely been applied in real-life scenarios. Are some of the sophisticated ethics theories based on ethical deliberation (Habermas 1991, 1999) able to inform and guide practices (Willette 1998)? Is John Rawls's seminal theory of justice (Rawls 1971) able to guide the workings of real-world institutions in assessing their merits? What about specific situations such as when a justified emphasis on respect for autonomy (e.g., neonatal care) ends up causing significant anxiety to mothers (Orfali and Gordon 2004)? What does this mean for our ethical theorizing about autonomy? What does it mean if important and seemingly well-justified theories are not connected or not easily connectable to real-world experiences and not evaluated in light of experience such that we cannot really say whether they enrich our lives concretely?[9] What would it mean for ethics theories to be able to account for and incorporate such experiences?

In response to such concerns, the psychologist and philosopher Frank Martela describes a paradigmatic change for ethics and philosophy, where ethics theory is envisioned as a tool and ethicists (or philosophers) as tool makers (Martela 2017). Therein, human flourishing is that for which ethics theories are instruments, where wisdom is the ability to "instrumentalize" knowledge toward this goal. Here, human flourishing must stand for a more open-ended idea than the more teleological accounts of at least some of the ancient thinkers. It must grant room to the meaning-making abilities of human beings and their wide-ranging life experiences and narratives. Thus, despite its apparent contemporary elusiveness, the idea of human flourishing can be reimagined as central to ethics as a quest to develop knowledge about flourishing. I come back to this in the next section of this chapter, but accordingly, in contrast with common views that wisdom is a peripheral concern, wisdom should instead be envisioned as the focal point of ethics given that it represents knowledge of human flourishing.

Wisdom

Definitions and views of wisdom have rarely been clear, sometimes purposefully. Plato's description of Socrates' search for the wisest man in Athens (in the *Apology*) stresses that the wise (namely Socrates) are those who know the magnitude of what they do not know. More amusingly, Diogenes is depicted as walking around carrying a lantern on a sunny day looking to find self-sufficient, free, and happy individuals without ever finding one. Reflecting the cynicism that he became famous for, Diogenes ironized, "It takes a wise man to discover a wise man." Likewise, in Dewey's account of growth, which I will come back to in greater detail, humility is a crucial component of wisdom, and serves a constructive and important regulatory role. He writes:

> Humility is not a caddish self-depreciation. It is the sense of our slight inability even with our best intelligence and effort to command events; a sense of our dependence upon forces that go their way without our wish and plan. Its purport is not to relax effort but to make us prize every opportunity of present growth. (Dewey 1922, p. 289)

Defining wisdom by the negative and considering wisdom a form of humility has a very long tradition (Ryan 2014). Humility is still widely considered a central component of wisdom in many theories (Bangen, Meeks, and Jeste 2013), and this is reflected in the open-endedness and fallibility of deliberative wisdom. However, wisdom is not comprised solely of humility but also includes positive knowledge about human life, although the recognition of the fallibility of that knowledge is intrinsic to wisdom, more so even for a form of deliberative wisdom.

As the historical discussions point to, wisdom is hard to define because it is an encompassing term that connects with a wide range of practical and existential realities. It describes a general attitude or disposition of persons who exhibit appropriate judgments and behaviors in ways that reflect a profound and holistic understanding of contexts in which these judgments and behaviors are made, such that these persons are able to adjust their thinking to yield optimal outcomes. This is why Aristotle's term for practical wisdom as *phronesis* is often translated as "prudence," which captures circumspection and carefulness. A wise person is also able to take distance from habitual ways of thinking and behaving by being critical—and even by being marginal in their views—to enact and exhibit more reasonableness than usual. Accordingly, a wise person has a certain kind of knowledge, adaptable knowledge that is mobilized toward finding ways of successfully handling situations even if they depart from conventions and typical reasoning. Finally, wisdom describes an ability or competence that includes but is not exhausted by ethics and its components (e.g., ethical awareness, ethical deliberation, ethical enactment). Wisdom can be exercised and cultivated in different domains (e.g., business and work, family life and intimate relationships, public service and politics) and any person is bound to be wiser in some domains and less wise in others. Pulling from contemporary psychology research, Carolyn Aldwin offers the following instructive definition:

> Wisdom is a practice that reflects the developmental process by which individuals increase in self-knowledge, self-integration, nonattachment, self-transcendence, and compassion, as well as a deeper understanding of life. This practice involves better self-regulation and ethical choices, resulting in greater good for oneself and others. (Aldwin [2009, p. 3] as cited in Trowbridge and Ferrari [2011, p. 90])

Modern research on wisdom in general—beyond the context of ethics—has yielded a number of useful findings. First, the construct of wisdom grasps an ability, distinct from intelligence and personality, measurable with tools with acceptable psychometric features (Ardelt 2003; Baltes and Staudinger 2000; Baltes et al. 1995; Bangen, Meeks, and Jeste 2013; Pasupathi, Staudinger, and Baltes 2001; Smith and Baltes 1990; Staudinger, Lopez, and Baltes 1997; Webster 2003, 2007, 2010). This finding points to Aristotle's distinction between *phronesis* (practical wisdom) and *sophia*, which designates a form of more theoretical knowledge. It also supports an account of wisdom which looms over traditional academic knowledge. Second, higher measures of wisdom are associated with greater scores for self-efficacy, generativity, ego-integrity, autonomy, developing purpose, managing emotions, and other positive psychosocial values, alongside lower scores for psychopathology measures, as well as less substance use and violent behavior in adolescents (Bangen, Meeks, and Jeste 2013; Brown and Greene 2006; Greene and Brown 2009; Perry et al. 2002; Webster 2003, 2007, 2010). Wisdom is associated with good mental health, prosocial attitudes and behaviors, and strong reasoning skills, whether based on self-assessment or assessment by others (Bangen, Meeks, and Jeste

2013). These findings also support the idea that practically wise persons live in serenity and know how to live well, an idea dear to ancient thinkers (Parry and Thorsrud 2021). Thus, current findings suggest that wisdom is a complex construct that defies one-dimensionality but that is nevertheless a human reality that can be studied, measured, compared, assessed, and cultivated as a concrete trait. Presently, much of this research remains somewhat silent on the procedural and substantive role of wisdom in ethics, perhaps given that most of this research is conducted under the umbrella of psychology.

Deliberative Wisdom and the Pursuit of Human Flourishing

Ethical wisdom is theoretical and practical knowledge about how to live the good life; it is wisdom in the context of moral affairs. It designates the ability to understand moral situations and find responses that foster human flourishing in oneself and others (Tiberius 2008). In other words, an ethically wise human being is one who achieves and strives to continuously achieve a certain level of moral awareness coupled with an ability to deal with ethical issues in morally problematic situations, including within their relationships (e.g., within work life, family life, and community life) on an ongoing basis such that human flourishing is fostered. Ethical wisdom is displayed by not relying uncritically on conventions and received ideas about the good life and the life worth living, but rather defining these through the cultivation of reflections on what on what truly matters for us and for others (Tiberius 2008). In the context of ethics; in contrast to wisdom exercised in the context of, for example, business or politics; the pretention of being reasonable and balanced may need to be calibrated to the complexity of moral matters and our uncertainties about the outcomes of our actions. This kind of humility is particularly relevant following an instrumentalist epistemology (see chapter 1) where truth will be found in the implication of our ideas, implications which are by to be learned from experience (Martela 2015; Racine et al. 2019). At the same time, an instrumentalist outlook invites us to step back from some of our everyday practices or habits and ask more concrete questions about principles and values at stake in quotidian life and habits (Tiberius 2008).

Ethical wisdom is cultivated when moral problems are recognized as important opportunities to exercise this wisdom. This is a difficult step to take because we can become blinded by the broader experiential context of interpersonal situations when our own outlook is focused on technical or practical aspects of situations. In such instances, we approach moral aspects of situations with a technical mind frame; we lose sight of the human dimensions of problems. This has been well described by those like Zygmunt Bauman and Leondis Donskis who have studied patterns of moral blindness and ethical fading (Bauman and Donskis 2013; Tenbrunsel and Messick 2004). The first phenomenon designates the process of becoming oblivious to the moral aspects of situations and of our lives; the second of deceiving ourselves in thinking our intents are morally good even when they are not. At other times, those who become aware of the morally problematic nature of a situation, such as political whistleblowers, like Edmund Dene Morel (decrying slavery in Congo in the late nineteenth century), Peter Buxtun who exposed the Tuskegee Syphilis Experiments, or Frances Haugen (denouncing abusive practices of Facebook and its deliberate fueling of division and hatred), call upon us to recognize a morally problematic situation and act upon it. However, these people can find themselves marginalized, despised, and criticized; we often fail to value their moral courage and recognize their extraordinary contribution to showing ethical gaps and the need to cultivate ethical wisdom (I come back to this point in chapter 8).

Deliberative wisdom is a more specific understanding of ethical wisdom according to which deliberation is intrinsic to knowledge about the flourishing life. This is because the understanding of such a life, in keeping with a transactional ontology (chapter 1), is never complete and requires openness to experience and dialogue, the potential of being proven wrong, and the need to learn from others. It is in this sense that Todd Lekan states that ethics theory is a deliberative practice because it is fundamentally through discussion and dialogue that we progress in ethics (Lekan 2006). In other words, given that the flourishing life is the ultimate goal of ethics, and that knowledge of human flourishing can only be partial and provisional, ethical wisdom is best envisioned as deliberative, not as a static or determined once and for all. Interestingly, in a presentation about Quebec and Labrador First Nations' views on well-being and flourishing, representatives of the association of these nations, Marjolaine Sioui and Nancy Gros-Louis McHugh (Sioui and McHugh 2021), explained how they envisioned well-being as better-being (*mieux-être*).[10] This subtle but very significant nuance reflects that well-being is an aspirational process toward betterment, a point also made by many contemporary theorists of human flourishing. This point is also central for a deliberative account of ethical wisdom, because otherwise we could fall prey to static accounts of what human flourishing means for a specific group of people at a specific time. We would too easily foreclose questions about whether a given account of flourishing is inclusive and relevant when we do not cash it out (Cotkin 1985) to see if it matters for others and in our current situations.

In sum, the concept of deliberative wisdom is astonishingly simple yet radical. It indicates that the search for reasonable attitudes and behaviors conducive to human flourishing are advantageously carried out when we recognize that human lives are fundamentally constituted by situated and transactional moral experiences. Recognizing the transactional nature of human existence (see chapter 1) calls for a corresponding dialogical or deliberative nature of ethical wisdom. The ability to enact creative problem solving is facilitated by in-depth understandings of the transactional nature of moral situations and moral experience both generally and in specific cases. The theory of deliberative wisdom provides a specific articulation of what ethics is and where such a process of learning is central to human flourishing. However, it remains itself only a tool that does not in itself guarantee any particular outcome of success, because all tools—in ethics and beyond—have their limitations. This also means that a deliberatively ethically wise person is not an aristocratic, self-imbued, stiff, rigid, or authoritative person who wields ethical knowledge as a weapon, but rather an open-minded, creative, and flexible person who is able to see opportunities for positive experiences where others might not. This is partly the case because knowledge of human flourishing can never be a kind of knowledge that forecloses eventual challenges and learning from others.

Flourishing and Human Experience

Based on theoretical and practical controversies and debates about the good life, human flourishing may seem to be an especially broad concept that may not be amenable to clarity and specific direction, and practical utility. At the same time—and I come back to this point in the second section of this chapter—despite these precisions, human flourishing remains firmly grounded in experience. Flourishing springs from and needs to speak to human experience. It is an orientation toward life as it is lived.

Following Peirce and Dewey, I adopt the view that concepts are tools or instruments that allow us to undertake theoretical work that starts by identifying intended conceptual meanings based on purposes or functions, then see if those concepts are instruments in service of intended purposes, all while reflecting critically on goals and purposes pursued (Racine et al. 2019). Take for example the desires we have (e.g., for love, friendship, professional success); we do not cease to pursue these goals because we lack semantic clarity regarding their nature or meaning. They are not meaningless; though they may be unverifiable by virtue of particular criteria of meaning, we can still think about them in intelligible ways and employ these words to describe some of life's most valued experiences and relationships. The words and concepts used to describe these values take root in experience; they are tools to describe, express, and share the meanings of habits and practices. At the same time, they are also projections of the kinds of experiences we seek and aspire to have. As we navigate between experience and aspirations, our understandings of our experiences and existences are enriched, that is, additional layers of meaning are gleaned from the back and forth between who we are and what constitutes us and who we desire to be. Thus, just like other ideas, the idea of human flourishing is best understood in terms of its implications for the kind of life we seek to live (Peirce 1878).[11]

Experience is soil in which human beings flourish or wither. The diversity in the experiences that give us the sense of flourishing may inhibit attempts to agree on one single account of well-being. But that is not the sole purpose of such a concept; it is rather to serve as a springboard for asking whether our life experiences are actually worthwhile and valuable in the search for better-being. The open-ended nature of the concept is thus to be accepted as a constitutive feature of the meaning of life itself. Envisioning human flourishing as being the object of dialogue is more like a mutual learning process that enriches those willing to question their habits than a mere transfer of knowledge or the imposition of universal dogma.[12] This kind of open-ended, dialogical understanding of human flourishing builds on the writings of Aristotle and Dewey, notably. It is thus worth identifying some of the major linkages, even if briefly.

Aristotle and Dewey: Predecessors to Deliberative Wisdom

Although I cannot elaborate extensively on the historical predecessors to deliberative wisdom, the contributions of Aristotle and Dewey (and their interconnections) need to be underscored to situate this concept.

From Aristotle and other ancient thinkers, we can take the idea that philosophy is a living and lived practice where ethics has a primordial place, as described by the historian of ancient philosophy Pierre Hadot. In contrast to the previously described reluctance of modern philosophy to engage with questions related to life and existence, ancient philosophical teachings were discussed in small groups (schools) and applied to the participants' own lives. The ancient thinkers did not fear engaging with what we would now consider personal life choices (e.g., social occupation, sexual behavior, respect for public authority and political orientation, dietary choices). These kinds of questions were deemed essential to thinking critically about one's own life and flourishing (Hadot 2002) because the wellness of the mind and person (the *psyche*) was the focal point of ethics theory (Parry and Thorsrud 2021). This focal point was oriented toward the actual experience of human life in its concrete and situated forms. Hence, ancient ethics entertained a deep and genuine connection with personal acts and behaviors. Socrates' call for existential self-examination cited in the introduction of this book is embodied and illustrated in its extreme form by his willingness to drink poison instead of fleeing from punishment. Other examples to cite would be Diogenes who flouted conventions of all sorts, even going as far as eating and masturbating in public (both offensive practices in Athens) to show the absurdity of certain social conventions. He lived in a large ceramic jar because he promoted a life of frugality and autarky consistent with his understanding of nature. His ideas about detachment were further expanded on by the Stoics. Epicurus and his followers (Epicureans) were interested in the good life and promoted a life of pleasure, with the understanding that some pleasures are much more worthy, given the important distinctions between groundless, natural, and necessary desires (Parry and Thorsrud 2021).

Ancient philosophers—despite their different views of what the good life is—were particularly interested in the relationships between different forms of human excellence (*aretês*, the virtues[13]) and human flourishing. They tended to agree that there is an essential, perhaps unequivocal, relationship between eudaimonia and the cultivation of human excellence; hence, why this kind of philosophy is largely described as eudaimonism (Parry and Thorsrud 2021). Aristotle proposed perhaps the now best known eudaimonic philosophy in which flourishing is bound to ethical excellence. Aristotle describes clearly the endpoint of ethical wisdom and the ethical life as eudaimonia. Translations of eudaimonia as happiness fail to denote the fact that Aristotle viewed the good life as a life of self-realization, in line with the teachings of Socrates. Aristotle distinguishes the more prudent attitude or disposition in deciding and acting properly across different circumstances (*phronesis*) from the higher level of knowledge (*sophia*) and insight into scientific principles explaining the world, including those that sometimes serve no immediate practical use or self-interest.[14] *Phronesis* is the disposition, accompanied with just and sound reasoning, oriented toward action regarding what is good and bad for human beings. It designates the experience-acquired ability—in contrast to *sophia*—to discern what is most conducive to well-being and flourishing. *Phronesis* is about finding the best thing to do in certain circumstances through sound and informed deliberation. Aristotle cites the Athenian politician and reformer Pericles as an example of someone who can find and pursue what is advantageous to himself and to others (Aristotle 1999). In contrast to this figure, he cites the pre-Socratic philosopher Anaxagoras and the mathematician and scientifically minded pre-Socratic philosopher Thales as having *sophia*—as they were both knowledgeable and possessed transcendental wisdom—but lacking *phronesis*, because they did not take into account their own interests as they were greatly motivated to pursue scientific knowledge. *Phronesis* is thus synonymous with ethical wisdom as defined previously in this chapter. Additionally, Aristotle's account of *phronesis* entails the ability to deliberate soundly.

Interestingly, although many would not immediately attribute an interest in the goal of eudaimonia to pragmatism, Dewey's ethics helped revitalize the ancient Greek focus on the more profound and existential aspects of ethics as the search for the good life.[15] It is also clear that Dewey was heavily influenced by Aristotle, with whom he holds many intellectual affinities (Chambliss 1993; Fahey 2010).[16] Like other thinkers enshrined in the tradition of enlightenment (such as David Hume, Jean-Jacques Rousseau, and Immanuel Kant), Dewey agrees that the pursuit of one's potential and individual freedom are paramount (Gouinlock 2002). This is in keeping with an ideal of an autonomous life and grasped by Kant under the maxim of *Sapere Aude* (Kant 1991), a phrase taken from the Roman poet Horace. However, Dewey was also influenced by Hegelianism (notably via the founders of pragmatism, Peirce and William James) (Misak 2013) and the writings of Karl Marx (Campbell 1988). He adopted a developmental—and even dialectical—perspective, while acknowledging the obstacles to self-realization posed by material conditions, social class, poverty, lack of education, and compromised health (Gouinlock 2002, p. 90).[17] Accordingly, to describe the ideal of an autonomous life of self-fulfillment wherein moral challenges and conflicts exist but can be tackled,

Dewey uses the more organic concept of "growth" to formulate the process of self-actualization and flourishing (Pekarsky 1990), much in the spirit of humanistic psychology of the time (Dwornik 2003). The process of growth, to cite James Gouinlock's synthesis of Dewey, "is one in which the organism enhances its ability to participate in its environment" (Gouinlock 1975, p. 238) and is therefore, like Aristotle's account of eudaimonia and *phronesis*, action-oriented. Growth is an ideal of human agency in that it "characterizes the life of human beings insofar as they are engaged in making sense of their situations, deliberating, acting, and then using the results of their actions to confirm, expand, or refine the understandings, dispositions, and skills that guide their intellectual and practical lives" (Pekarsky 1990, p. 291). Thus, growth, following Dewey, is neither simply an achievement nor a mere process. There is an end to ethics, and it resides in the harmonious agreements and adaptations, which give way to a serenity and a sense of efficacy, that accompany the liberation of growing powers in humans.

Like Aristotle's emphasis on sound reasoning and the ability to deliberate, the ideal of ethical growth in Dewey's thinking[18] is intimately connected with the ability to engage in ethics deliberation and pursue open-minded inquiries about the good and right thing to do (Gouinlock 1978). Because Dewey describes ethics as existing within a plurality of goods, he recognizes the limited ability we have in identifying a priori the actual nature of "good" and "better" courses of action, hence the importance of being open to the outcomes of actions and evaluating what has been achieved when action is guided by an ethics-focused process. This commits Dewey—like the ancient philosophers—to a form of living ethics theory[19] in which continual learning is intrinsic to the process of pursuing a better and more meaningful life. In human beings, and with respect to human morality, the capacity for growth involves flexibility and adaptability, where behavioral and cognitive rigidity arise as potential obstacles for growth. Flexibility nevertheless entails the possibility of a regression in adaptive abilities of an agent in face of new situations. Growth is more than simply gaining an intellectual understanding of a situation or more control over one's environment. Dewey elucidates this difference by explaining that growth is accompanied by an increasing appreciation of its fuller scope and meaningfulness, which he describes as a greater and more profound experience of a given situation (Pekarsky 1990, p. 288). In this way, growth signals a more encompassing frame of mind in which the full breadth of experience becomes apparent and wherein one's understanding is expanded in scope and not simply augmented in power. It is a sense of understanding the connections between different events in ways that may have eluded us before. Growth in meaning is not gained necessarily only by the accumulation of knowledge but it is primarily evidenced by what it offers in human experience. This experiential connection of ethical wisdom to human experience is a cornerstone of deliberative wisdom, much like this idea was previously developed by thinkers such as Aristotle and Dewey.

Deliberative Wisdom and Modern Theories of Human Flourishing

In this section, building on existing scholarship, I briefly review key insights from the traditions of humanistic and positive psychology on human flourishing and how they connect deeply and directly with ethics, even though these fields are often construed as being distinct and separate. A more practical and existential movement, humanist psychology, is highly concerned with human development and human flourishing. Its development took place in the mid-twentieth century with the writings and clinical work of figures like Maslow, Rogers, Adler, and Erikson, among others. After an initially strong development, this movement lost some of its momentum with the rise of more "scientific psychology" influenced by neuroscience. However, this constructive orientation and interest in human flourishing has been reignited more recently by a new movement in positive psychology (Seligman and Csikszentmihalyi 2000). I cannot review the scope of writings on human flourishing produced by psychology, but I want to propose and illustrate that there is a bridge—a potentially very important and rewarding one—between ethics and psychology. If ethics is envisioned as a science that elaborates promising futures that bring about human flourishing, then psychology tells us many useful things about human flourishing, including its depth, diversity, and concreteness. Ideally, the insights gleaned from this exploration invite productive synergies between these disciplines and the tools they offer in the hope of understanding more clearly and concretely what human flourishing is.

To support the theory of deliberative wisdom, Carol D. Ryff's model of human flourishing is particularly compelling because it draws on extensive literature reviews in humanistic psychology and philosophy, and is thus already an integrative and interdisciplinary model. Ryff incorporates wide-ranging insights and writings within a multidimensional (six-factor) model that includes (1) self-acceptance, (2) positive relations with others, (3) purpose in life, (4) personal growth, (5) environmental mastery, and (6) autonomy (Ryff 1989; Ryff and Singer 2008). Table 2.1 provides an overview of the model and illustrative connections that can be made between dimensions of human flourishing and existing ethics scholarship.[20] It can scaffold an integrative, inclusive, rich, and open-ended account of ethics.

According to Ryff's theory, as well as that of others, human flourishing designates a state of being that is often spelled out in reference to a number of attributes such as autonomy, self-confidence, and self-assurance based on one's abilities to self-actualize. Even though human flourishing can be measured at a given time (e.g., Ryff and Singer 2008) as a kind of metameasure capturing the meaningfulness and the orientation of one's existence, it is not simply a state akin to satisfaction or happiness measured at a given time (Baumeister et al. 2013). Rather, measures of human flourishing designate an evaluation of one's own trajectory and a sense of fulfillment in connection to one's intrinsic motivations, sense of self, and commitment to others as seen in Aristotle and Dewey's writings, which stress that eudaimonia and growth are lifelong processes.

Despite the integrative and cardinal value of the concept of human flourishing, it has important limitations, especially when a certain account of flourishing is envisioned as authoritative or universal. Any account of human flourishing and its dimensions is embedded in the ways that we think of flourishing, reflecting all kinds of positional biases based on gender, culture, cognitive profile, socioeconomics, and more (Rodogno, Krause-Jensen, and Ashcroft 2016). Furthermore, not everyone can equally access the conditions that support human flourishing, thus bringing charges of elitism against flourishing-oriented views of ethics (Willen 2022). This is a complex issue. More resources do not equate to more flourishing and human development, as Amartya Sen has repeatedly argued—individuals vary in their needs for resources to develop themselves and develop their capabilities (Sen 1989). Importantly, various dimensions of flourishing do not presume certain objective standards of what accomplishments are but refer to the meaningfulness of what one does and believes.

Despite the many barriers preventing individuals and groups from fully attaining the goal of human flourishing, the concept serves a double purpose.

Table 2.1

Dimensions of human flourishing and connections to ethics theory

Dimensions of human flourishing*	Illustrative connection to ethics theory
Self-acceptance: "The need to have positive self-regard" (Ryff and Singer 2008, p. 20). "A kind of self-evaluation that is long-term and involves awareness, and acceptance of, both personal strengths and weaknesses" (Ryff and Singer 2008, pp. 20–21).	This dimension reflects the need to work within one's own situation(s) and with one's own strengths. It is a recognition that ethics has a deep personal role to play in the search for happiness and meaning in life. Paul Ricoeur's ethics, which tackles the notion of self-esteem, describes the importance of an ethics that makes sense to real people (Ricoeur 1992). Self-acceptance and acceptance by others have also been central to disability ethics, which has expanded the understandings of flourishing lives and questioned assumptions about normalcy and about what a flourishing life means.
Positive relationships: "There is a near universal endorsement of the relational realm as a key feature of how to live" (Ryff and Singer 2008, p. 21).	This dimension has been well examined by feminist theorists who have argued for relational views on autonomy and the self, arguing that relationships have value that autonomy can sometimes undermine and challenge (Mackenzie and Stoljar 2000). Positive relationships are also fundamental within community-oriented or community-sensitive social thinkers like Charles Taylor (Taylor 1989), Michael Sandel (Sandel 1998), Alasdair MacIntyre (MacIntyre 1984) and Michael Walzer (Walzer 1994/2019).
Personal growth: "Involving a continual process of developing one's potential" (Ryff and Singer 2008, p. 21).	Although Dewey's ethics potentially hint at all six dimensions of Ryff's model, growth is likely that which most quintessentially reflects what is ethical (Pekarsky 1990). Growth is also the aspect of Ryff's model closest to Aristotle's understanding of eudaimonia. Personal growth in the form of actualization tendency and self-actualization is central to the theories of Abraham Maslow (Maslow 1970), Carl Rogers (Rogers 1961), and other early pioneers of humanistic and positive psychology.
Purpose in life: "Creating meaning and direction in life" (p. 22).	This dimension is central in hermeneutic theories and narrative approaches to ethics (Gadamer 1989), as well as pragmatist approaches that stress the importance of experience (Dewey 1925, 1934/1958, 1938/1986). Narrative and hermeneutical theories emphasize the importance of meaning and of meaning-making activities in human life. They point to the need to understand moral experience as being partly about the meaning of life and provide a framework to envision existence as a narrative in building and giving meaning and direction to moral agent(s).

Environmental mastery: "Being able to manipulate and control complex environments" (Ryff and Singer 2008, p. 22). "Although this area of well-being appears to have parallels with other psychological constructs, such as sense of control and self-efficacy, the emphasis on finding or creating a surrounding context that suits one's personal needs and capacities is unique to environmental mastery" (Ryff and Singer 2008, p. 23).	Theorists of development and alienation like Georg Wilhelm Friedrich Hegel, Karl Marx, and Jürgen Habermas have stressed how being prevented from acting upon one's life situation is a fundamental harm to human beings. Dewey stressed how human beings can live "consummatory experiences" when they are able to respond to new challenges and exercise their skills in ways that transform situations accompanied by rewarding feelings of liveliness and fullness in existence (Dennis and Powers 1974). Literature on learned hopelessness has also evidenced how important it is to have a sense of self-efficacy, of feeling like one is an agent in their own life with a level of control over one's environment (Peterson, Seligman, and Maier 1995; Seligman 1975; Wegener et al. 2007). The capabilities approach also stresses the importance of being able to develop one's potential concretely (Nussbaum 1997; Sen 1989).
Autonomy: "One does not look to others for approval but evaluates oneself by personal standards" (Ryff and Singer 2008, p. 23).	This dimension is well recognized in principles of respect for autonomy and individual rights philosophies. Kant developed this orientation and made the case for why a human life should be a life guided by reasons that we choose (and that should also serve as guide for the life of others). Autonomy is a condition for the ability to orient oneself in life (Racine et al. 2021) and could be a condition for the development of other meaningful aspects of human flourishing as developed more recently in self-determination theory (Deci and Ryan 1985; Ryan and Deci 2000).

*According to Ryff's theory. Direct quotes in the first column are used to illustrate Ryff's six dimensions.

First, it prevents falling prey to what Sen has described as "the circumstantial contingency of desires," that is, to be satisfied simply by what we know best or what we have calibrated our expectations to (Sen 1985, p. 197). Other authors refer to this phenomenon as adaptive preferences "that adjust to the low level of functioning one can actually achieve" (Nussbaum 1997, p. 283). In other words, the very concept of human flourishing is a placeholder or a prompt inviting us to reflect on what is the meaning of life and what counts as valuable projects, relationships, and engagements. This aspirational aspect of human flourishing prevents us from accepting only what we know best and what we know we can actually achieve only with existing means and resources as described by Sen and Nussbaum. Without an aspirational vision for what we seek, we risk settling for less, not formulating goals for ourselves and our societies.

Second, having even an imperfect and open-ended account of what is important in life allows us to criticize the obstacles to realizing what is important. Respect for autonomy, acceptance of and valuing pluralism are nice thoughts and principles, but why do they really matter? An account of how and why these aspirations connect to psychological and social needs of human beings helps explain why they matter and why ethics is a process by which we seek to enrich our understanding of these valued things. A purpose of human flourishing is thus to help critically assess certain habits and practices in light of their detrimental or positive impact on human beings.

A concept of human flourishing is an orientation for human development informed but uninhibited by the circumstantial contingency of desires. However, this concept needs to be humble and remain open to what we do not know and to what we will learn, especially from dialogue with others. Envisioning human flourishing in this open-ended and nondefinitive way corresponds to Dewey's proposal that ethics is much more about learning than teaching, and that human flourishing is not merely a state but a journey. Acknowledging the richness and plural forms of human flourishing does not repudiate the concept, but rather enriches our understanding of what gives meaning to human life within an ever-growing and expanding scope of humanity. In other words, we need to approach ethical wisdom about flourishing as deliberative wisdom. In this light, Ryff's model is a strong starting point and is the result of committed and diligent work, despite always being subject to revision and improvement.

Dimensions of Human Flourishing

Self-acceptance Self-acceptance designates the ability to live with an acceptance of one's strengths and weaknesses, such that the projected self is not disconnected from one's real, lived self. Self-acceptance paves the way for a productive engagement with the world, including how one builds positive relationships and a strong and authentic personality. Ryff and Burton Singer describe that this acceptance reflects "the need to have positive self-regard" (Ryff and Singer 2008, p. 20) and designates "a kind of self-evaluation that is long-term and involves awareness, and acceptance of both personal strengths and weaknesses" (Ryff and Singer 2008, pp. 20–21). Self-acceptance helps reinforce the idea that human flourishing—although deeply related to concern for others—is ultimately rooted in the self. When a person's preferences, values, and orientations differ from socially enforced norms, self-acceptance can be difficult due to stigma and marginalization generated by processes of othering that undermine the ability of a person to accept who they are. Thus, self-acceptance is realized on a social level, as well as an individual level. Others have a role in the self-acceptance of others; it can be easier for one to accept oneself when they are also accepted by others.

This particular dimension of human flourishing reflects the need to work within one's own situation and with one's own strengths and weaknesses, because one's actualization as a person begins precisely with who one is and who one wants to become. Self-acceptance is a reflection that ethics has a deeply personal role to play in the search for happiness and meaning in life. Ricoeur's ethics, which incorporates the notion of self-esteem, aptly describes the importance of an ethics that makes sense and brings meaning to individuals' lives and to those of others (Ricoeur 1992). Upholding well-founded esteem for oneself is premised on the genuine attribution of such worth and therefore invites an orientation toward a life that warrants praise, lending to the awareness and acceptance of oneself as a person having both strengths and weaknesses. When a person is marginalized by their community (e.g., due to culture of origin, language, sexual orientation, disability), self-acceptance can be a difficult process. Whether this process is made difficult by external factors (such as discrimination) or internal factors (such as perceived and internalized weaknesses), the growth, perspective, and meaning that arise from the journey toward self-acceptance can be significant and potentially rewarding. Individuals who have cultivated, through hardship and resilience, a positive self-regard can become persons who have a great deal of human understanding (e.g., empathy, compassion) to offer others. This said, creating obstacles to self-acceptance is not the recipe for generating resilience or wellness. It is instead cultivated by recognizing and assuming one's individuality and sense of belonging by fostering a connection with oneself, which is paramount in thinking about what human flourishing is (e.g., what is it if not about oneself?) and its importance (e.g., why is it important if not related to one's own development?). As the other dimensions of flourishing show, human flourishing is in fact mostly enacted and directed toward others. However, in order to be able to give to others, a person must be able to realize who they are and what they can offer such that their self-acceptance is connected to the ability to help others.

Positive relationships Human morality and ethics (as the reflective form of morality) are almost always about positive relationships with others and their welfare. As Winston said, "We make a living by what we get; we make a life by what we give."[21] Ryff and Singer write, "There is a near universal endorsement of the relational realm as a key feature of how to live" (Ryff and Singer 2008, p. 21). Being able to contribute to the welfare of others is clearly characteristic of the flourishing individual (Kim et al. 2020; Moradi, Van Quaquebeke, and Hunter 2018). Most everyone will easily recognize the profound importance and impact of having social support (e.g., family, friends) to physical and psychological well-being. The value placed on human relationships explains partly why helping and education professions are so highly valued and bring deep, intrinsic motivation for practitioners in these domains (Pellegrino 2001); prosocial behavior (including work oriented directly toward the welfare of others) brings meaning to life (Dutton, Roberts, and Bednar 2010; Klein 2017; Nelson et al. 2016). In return, these types of professions (e.g., nursing, medicine) are highly esteemed internationally (Little 2018) by the broader public through recognition of the commitment and the value of the services they offer (although this esteem is also tainted by social status associated with these professions, which are factors unrelated to altruistic motivations). Not everyone has access to or participates in the same types or number of positive relationships as some of the previously listed professions, but the multiple roles played in life (e.g., as a parent, a caregiver, a friend, a volunteer, a life companion) illustrate the salience and significance of positive relationships. Likewise, but perhaps even more so than other dimensions of human flourishing because they directly concern others, positive relationships are not only a means but an end as captured in Kant's beautiful second formulation of the categorical imperative: "So act that you use humanity, in your own person as well as in the person of any other, always at the same time as an end, never merely as a means" (Kant 1785/1998, p. 38).

Positive relationships are fundamental within community-oriented and community-sensitive social philosophies encountered in the writings of philosophers such as Taylor (1989), Sandel (1998), MacIntyre (1984) and Walzer (1994/2019), but most clearly in those of feminist scholars who have pioneered care ethics (ethics of care), such as Carol Gilligan (2001) and Nel Noddings (1982). The latter have argued for relational views of autonomy and the self. These feminist theorists have also indicated how relationships have value in and of themselves, and that an overly narrow focus on individual autonomy can undermine and challenge positive relationships (Mackenzie and Stoljar 2000). Generally speaking, feminist movements—although partly motivated by concern for the autonomy and well-being of women (but also issues of social justice)—are a challenge to the emphasis on autonomy and independence that has led to a loss of focus on cultivating positive and meaningful relationships, the value in helping others, and so on. They are a direct challenge to a certain individualistic account of autonomy (Mackenzie and Stoljar 2000). Some early ideas on positive relationships (e.g., empathy, sympathy) are also at the core of utilitarianism and consequentialism (e.g., Hume, Adam Smith, Jeremy Bentham, Mill),[22] although the contribution to the welfare of others can be lost when benefits are calculated less as humanistic values and more in terms of economic wealth and prosperity. In sum, positive relationships have been acknowledged in ethics theory as core to human morality and are increasingly evidenced as a constituent aspect of human flourishing, because helping others is deeply rewarding for the helping person (Kim et al. 2020).

Personal growth Growth designates "a continual process of developing one's potential" (Ryff and Singer 2008, p. 21) and calls for a long-term outlook on the process by which one accomplishes the things that bring increased meaning to their life and to the lives of others. Ryff and Singer state, "Of all the aspects of well-being, it is personal growth that comes closest in meaning to Aristotle's eudaimonia (*sic*), as it is explicitly concerned with the self-realization of the individual" (Ryff and Singer 2008, p. 21). Growth is also a concept utilized by Dewey to describe a flourishing life, and it encompasses most, if not all, of Ryff and Singer's dimensions of flourishing, because Dewey's account of growth encompasses the ability to develop autonomy, to gain control over one's environment, and to actualize one's potential (Pekarsky 1990). The value of growth is related to the fact that self-actualization brings a form of satisfaction that is much beyond instant gratification; it involves achieving the height of our potential. (Here, Aristotle would have alluded to virtues or concrete forms of excellence.) Growth encompasses a sense of progression and fulfillment of being able to cope with different situations, as well as a form of action that involves broadening one's perspective(s) and embracing the experiences that life offers.

Although Dewey's ethics potentially hints at all dimensions of Ryff's model, growth is likely that which most quintessentially reflects what is ethical from a pragmatist perspective, as examined previously in this chapter (Pekarsky 1990). Ryff also believes this to be the most defining trait of flourishing individuals. This line of thinking has been pursued by Maslow, Rogers, and other early leaders in positive psychology who have written about self-actualization and fully functioning persons [Maslow 1970; Rogers 1961]). Mainstream ethics has—with some exceptions—tended to stay away from ideas of growth and its relation to the good life, but the resonance of this idea has been maintained in popular culture; the meaningfulness of the idea has never vanished (e.g., in the immense personal growth literature) (Effing 2009). Progressive philosophies that have emphasized development and enfranchisement from alienation (e.g., Hegel, Marx, Dewey, Jean Piaget, Habermas, Sen, Nussbaum) all stress, in some way, the importance of growth in contrast to neglecting the development of one's being and perspectives. When dealing with difficult moral problems, "We can and often do learn and grow when our cares are put to the test. . . . morally troubling experiences may well help us to develop a deep awareness and affirmation of the things we care about and, indeed, of who we are" (Tigard 2019, p. 7). This is an idea cherished by pragmatists and central to an inquiry-based view of ethics (Martela 2015, 2017).

Purpose in life Having a sense of direction and orientation is another way to articulate what is typically described as purpose in one's life. Not having purpose in life often refers to periods of existence wherein one lacks motivation and the ability to appreciate the value of oneself and others.

Such situations refer to a lack of enjoyment in everyday activities alongside the ability to contribute to the welfare of others and learn from them (Zika and Chamberlain 1992). Feeling purposeless is a very difficult experience because it undermines the recognition that we are agents who desire self-actualization and want to contribute to the welfare of others. Purpose is the motivation that provides the energy to mobilize who we are toward flourishing. Purpose in life is related to processes of intrinsic motivation; it is the metanarrative that ordains other, more specific motivations, while always being nurtured by the pursuit of specific projects and continuous existential and experiential learning (Kolb 1984). Purpose in life is not a static idea or orientation; it is nourished by the contribution one enacts to maintain and develop this purpose on a regular basis, much like eudaimonia more broadly. Experiences that reveal deeper purpose in life are meaning-making moments insofar as meaningfulness suffuses not only a single situation but multiple moments of one's existence. In these moments, there is a sense of regained vigor and purpose, and the impression of gaining a new outlook toward perhaps the same situations. Periods of uncertainty in self-purpose provided by life-changing situations or reflections (e.g., professional orientation, sexual orientation, philosophical or religious orientations, health and illness) are usually unpleasant or disturbing, even if they can constitute new trajectories or renewal in one's life (Hammell 2004). Because the meaning of our lives is so tied up with what we live with others and how we experience the world with them, changes in significant relationships with others (e.g., family relationships, romantic relationships, friendships) are often moments where purpose in life is challenged and, when there is a rupture, give a sense of losing purpose or a trajectory in life.

Purpose in life is a central concern in hermeneutic theories and narrative approaches to ethics (Gadamer 1989, 1991; Ricoeur 1992), as well as existentialism (de Beauvoir 1947/2011; Sartre 1946/1996). These approaches stress, albeit in different ways, the importance of developing a genuine sense of why certain actions, beliefs, and relationships matter (or do not) based not on social conventions, but on whether purpose and coherence are found within one's existence and life experiences (Hartman et al. 2019; Kestenbaum 1992; Montreuil and Carnevale 2018). Existentialism is perhaps the movement that most clearly stressed the active role that individuals need to have in creating their own purpose such that, in their own words, "existence precedes essence," rather than the opposite (Sartre 1946/1996). In *Pour une morale de l'ambiguité*,[23] Simone de Beauvoir explains how our freedom (our ontological freedom) is a fundamentally human condition that generates significant ambiguity in how we ought to act (de Beauvoir 1947/2011). But this freedom is not a condemnation because ethics offers a constructive orientation to guide us. Ethics calls for the perpetuation of freedom and opportunities for others; in doing so, human beings transcend themselves and generate meaning, a point also made by other notable authors (Frankl 1966). In this, de Beauvoir shows how human beings, women and men, differ in their ability to live the life they want and define themselves and act in agreement with who they want to be because of social, political, and economic constraints created by patriarchal societies. Early humanistic psychology (e.g., Adler, Maslow, as cited previously) also pulls from similar existentialist ideas and emphasizes the active role of agents in defining the purpose and the orientation of their lives. Although this is less often recognized, James and Dewey heralded similar ideas about how an agent's action represents the projection of their being and most concretely expresses and contributes to the meaning of their life (Fesmire 2003). Dewey writes:

> The choice at stake in a moral deliberation or valuation is worth of this and that kind of character and disposition. . . . In committing oneself to a particular course, a person gives a lasting set to his own being. Consequently, it is proper to say that in picking this object rather that, one is in reality choosing what kind of person or self one is going to be (Dewey 1985, pp. 274, 287). In short, the thing actually at stake in any serious deliberation is not difference of quantity, but what kind of person one is to become, what sort of self is in the making, what kind of a world is making (Dewey 1983, p. 150). (Dewey [1983, 1985] as cited in Jacobs [2004, p. 220])

Environmental mastery Freedom without the ability to act is an incomplete and interrupted process (Racine 2017) such that the ability to enact choices is of paramount importance to the realization of human agency (Racine and Dubljević 2017). Accordingly, environmental mastery designates "being able to manipulate and control complex environments" and represents the concrete ability to act upon one's situation (Ryff and Singer 2008, p. 22). Ryff and Singer explain that "Although this area of well-being appears to have parallels with other psychological constructs, such as sense of control and self-efficacy, the emphasis on finding or creating a surrounding context that suits one's personal needs and capacities is unique to environmental mastery" (Ryff and Singer 2008, p. 22). This concept brings a comprehensive framework to understanding the connections between the individual and their surroundings, as well as the actions of the agent on their physical and social environment in line with a transactional account of human agency (Brinkmann 2011). It also acknowledges the role of the environment in structuring the kinds of possibilities afforded to the agent (Gibson 1977).

The value of being able to act concretely on one's living conditions (e.g., economic, social, political) has been a focus of theorists of development and alienation like Hegel (1993), Marx (Marx, Engels, and Stedman Jones 2002), Dewey (1922), and Habermas (1997). They stress how being prevented from acting upon one's living conditions is a fundamental harm to human beings (Ryan 1973). In contrast, consummatory experiences are accompanied by rewarding feelings of liveliness and fullness in existence, when individuals are able to respond to new challenges and exercise their skills in ways that transform situations (Dennis and Powers 1974). The ability to overcome challenges and resolve them brings contentment, satisfaction, and a sense of empowerment. To achieve environmental mastery, one has to be free and able to act in the first place. However, environmental mastery is limited by a range of barriers and inequalities related to social and physical conditions (e.g., lack of adaptations to physical or mental disabilities). Such experiences of not being free to act toward changing one's situation is an experience of agential limitation and vulnerability. This points directly to the social dimensions of alienation (and of liberation) so adamantly discussed by Karl Marx, Rosa Luxemburg, and Friedrich Engels (Poulin 1997). Contemporary literature on learned hopelessness has also evidenced the importance of having a sense of self-efficacy (Ryan 1975; Wegener et al. 2007). The capabilities approach also stresses the importance of having an environment suitable to exercise one's agency since it is not focused on "what people feel about what they do, but about what they are actually able to do" (Nussbaum 1997, p. 285).

Although the term "environmental mastery" may sound overly technical, it designates a familiar component of how well-being is seen across many settings and cultures (e.g., Quebec and Labrador First Nations report that the "capacity to act" is fundamental to their view of well-being [Sioui and McHugh 2021]). It designates a precious dimension of human flourishing and one that, when taken away, leads to situations of injustices that are self-evidently harmful. Environmental mastery was and is at the core of feminist movements and the desire to have access to inclusive social environments. It is central to activists on disability rights' call for greater justice and access to opportunities and to physical and social environments that are adapted to different bodies. Importantly, situations where human beings are prevented from exercising their agency and accessing important material and social goods can lead to terrible human conflicts and tensions. Wars are declared when opportunities (e.g., political, religious, economic freedoms) and resources (e.g., natural and financial resources) are taken away or access to them is denied.

Autonomy Human flourishing and autonomy are deeply enmeshed to the point where autonomy is sometimes viewed as being synonymous with the good life, especially in the Western world. However, flourishing describes a broader state and process of self-actualization that brings holistic satisfaction, whereas autonomy is partly a condition of agency in the process of human flourishing (Racine et al. 2021). Autonomy is a goal (e.g., to become autonomous) and a capacity, because it provides the opportunity for generating meaningful life experiences. It has a domino effect: once you have the capacity to be autonomous, you attain autonomy (as a goal), which gives you the capacity to be more autonomous. In ethics, autonomy typically alludes to the more abstract ability to establish rules, principles, values, or orientations for oneself, namely, to be self-governing rather than follow the dictates of others or of society (Dahl and Shapiro 2015; Dworkin 1988), in opposition to heteronomy. However, in daily experience, which I have researched across a variety of clinical and research settings, autonomy often designates a capacity that is less theoretical and more connected to decision-making; it is the ability to be provided with information to make choices and the conditions that allow someone to make decisions freely and voluntarily (Bogossian, Majnemer, and Racine 2020; Racine et al. 2013; Zizzo et al. 2016). Thus, valuing autonomy does not necessarily entail spelling out which values, principles, or rules of conduct are to be followed in given situations following highly sophisticated ethics theory, because deliberation is only one of several subcomponents of autonomy (Racine et al. 2021). Rather, autonomy designates the process of being able to exercise one's multifaceted capacity to make decisions and choices, which are always embedded in contexts (Racine et al. 2021). Autonomy is thus instantiated in real experiences, in making real choices that involve real goods. Hence, autonomy and environmental mastery are associated with each other when autonomy is envisioned as more than just a theoretical activity, but is an actual ability to exercise personal choice. The method of scenario-based deliberation presented in chapter 10 and the discussion on ethical enactments in chapter 12 support agents in making sense of their situations and developing autonomy and control in the problematic situations they live.

Autonomy, as a dimension of human flourishing, is well recognized in philosophical principles of respect for individual rights, as well as in broader social and intellectual movements. The writings of philosophers who have made the search for self-fulfillment (e.g., the concept of praxis or formative action in Hegel and Marx) or self-determination and self-governance (Dahl and Shapiro 2015; Habermas 1997; Kant 1991) a cardinal goal of moral philosophy also speak of the pivotal importance of autonomy in understanding human flourishing. This was also the intent of the French Enlightenment era, which exemplifies how a movement of social and public education (e.g., *L'Encyclopédie*) favoring broader access to knowledge was envisioned as crucial to guide one's choices and to support one's self-governance. The development of autonomy was considered the epitome of human flourishing and the life worth living. In this light, an Enlightenment figure like Denis Diderot, whose style of writing (e.g., dialogues, invitations to readers to think independently) explicitly embodies the idea of the growth and autonomy of his readers (Diderot 1754/2005),[24] can be considered in some sense an ethicist. His ideas and actions promoted the flourishing of others within one of the most liberating historic moments of Western civilization.

Interestingly, contemporary psychology has expanded on insights in philosophy and shown how the fact that being autonomous, and more specifically, self-determined, is associated with important positive functioning and human flourishing as explored by self-determination theory (Deci and Ryan 1985; Ryan and Deci 2000, 2017). Self-determined choices and actions are autonomous and linked to deep beliefs and intrinsic motivations related to one's identity and the projection of oneself through time and across human relationships. The pioneers of this theory, Richard Ryan and Edward Deci, have brought to light, through multiple studies, that the attainment of intrinsically motivated goals is correlated with well-being, while the attainment of extrinsically motivated goals are not (Ryan and Deci 2000). Ryan and Deci describe intrinsic motivation as "the inherent tendency to seek out novelty and challenges, to extend and exercise one's capacities, to explore, and to learn" (Ryan and Deci 2000, p. 70). Accordingly, intrinsic motivation describes a "natural inclination toward assimilation, mastery, spontaneous interest, and exploration that is so essential to cognitive and social development and that represents a principal source of enjoyment and vitality through life" (Ryan and Deci 2000, p. 70). They stress that this phenomenon may be the most quintessential expression of positive human potential (Ryan and Deci 2000) and summarize some of their important findings about intrinsically motivated goals:

> Comparisons between people whose motivation is authentic (literally, self-authored or endorsed) and those who are merely externally controlled for an action typically reveal that the former, relative to the latter, have more interest, excitement, and confidence, which in turn is manifest both as enhanced performance, persistence, and creativity (Deci and Ryan 1991; Sheldon, Ryan, Rawsthorne, and Llardi 1997) and as heightened vitality (Nix, Ryan, Manly and Deci 1999), self-esteem (Deci and Ryan 1995), and general well-being (Ryan, Deci, and Grolnick 1995). This is so even when the people have the same level of perceived competence of self-efficacy for the activity. (Ryan and Deci 2000, p. 69)

Although self-determination theory describes autonomy, competence, and relatedness as key components of human flourishing, it does not provide as comprehensive an outlook as other integrative theories of human flourishing. Moreover, feminist scholars, especially those offering alternative accounts of autonomy, specifically relational autonomy, have argued convincingly that autonomy can collide with other aspects of human flourishing like positive relationships with others (Mackenzie and Stoljar 2000). Another limitation noted by Ryff and Singer is that "This aspect of well-being is undoubtedly the most western of all of the above dimensions" (Ryff and Singer 2008, p. 23); there has been a tendency to boil down human flourishing to autonomy in the Western world. Despite these limitations of the concept of autonomy, research suggests that the ability to self-determine, broadly construed, is nevertheless valued in different ways across different societies (Chirkov et al. 2003).

Deliberative Wisdom and the Incompleteness of Knowledge about Human Flourishing

In closing this review, there are two limitations that must be addressed on the topic of human flourishing. First, we must avoid envisioning current understandings of human flourishing as being somehow definitive such that they are viewed as complete or impartial. In this chapter, I rely heavily on research in psychology and Ryff's work on human flourishing. I am aware that there are competing theories of flourishing, and that Ryff's model has its shortcomings, even though there is also significant overlap with other important psychological theories (e.g., Mihaly Csikszentmihalyi's concept of flow [Csikszentmihalyi 1990], self-determination theory [Deci and Ryan 1985; Ryan and Deci 2017], flourishing [Seligman 2012]), as well as insights to be gleaned beyond psychology, notably in economics, management science, and human resources (Crespo and Mesurado 2015; Jacobs 2004). I am also aware that several other ethics scholars have offered alternate lists of core human functions, such as Nussbaum's ten central human functional capabilities (Nussbaum 2011) and Sabina Alkire's six dimensions of well-being (Alkire 2002). Unsurprisingly, there is significant overlap between these lists. Ryff's model has the advantage of having been built from qualitative research and eclectic intellectual insights in psychology and philosophy, then developed into a scale based on itemization, validation, and used across a wide spectrum of situations and human populations (Ryff 2014a). Nevertheless, it remains true that the science of human flourishing is still a burgeoning one.

One specific and important subissue related to the incompleteness of Ryff's account of human flourishing is that it does not integrate economic stability and income or health as key components of flourishing, for which others have argued (VanderWeele 2017). At this point, it is unclear whether the concept of human flourishing within an ethics theory should consider health and good economic conditions as genuine components of human flourishing or, rather, as conditions that impede or facilitate access to flourishing. Aristotle recognized that some amount of resources is likely required to support the flourishing life.[25] This point is also made forcefully by Sen. That said, there is also a sense in which socioeconomic status and health need to be distinguished from flourishing. Ryff's measure of flourishing has been shown to be correlated with better physical and psychological health (Ryff 2014a). But health is a complex concept. For example, some people with disabilities flourish and gain life meaning through their hardships, and their account and experience of health and flourishing may differ from those who do not have a disability (Ryff 2014b). Another example is the experience of poverty, which is not necessarily a barrier to positive relationships and developing a strong purpose in life. Being motivated by extrinsic considerations (e.g., wealth, reputation, success) can also be detrimental to the understanding of what really counts in living a good life (Racine et al. 2021;

Ryff 2014a). I do not argue that poverty or disease are favorable to human flourishing, but I do contend that we need to expand our understandings of human flourishing to accommodate the different forms of equilibria that people find in themselves and their surroundings to live their lives and contribute to the welfare of others.

A second limitation is related to common interpretations that take current understandings of human flourishing as gospel truth. Viewing ethical wisdom as nondeliberative wisdom, that is, knowledge that is established authoritatively and once and for all is antithetical to the proposed deliberative orientation. A top-down application of knowledge of human flourishing is also contrary to the ideas of open-endedness about the nature of human flourishing and the importance of learning with and from other human beings about the good life. It is the very intent of deliberative wisdom to cultivate an open mind about what we as individuals and others find to be an authentic life worth living and what activities and relationships contribute positively to this. In this sense, we must be careful about the biases induced by, for example, healthy or wealthy people who write about the good and meaningful life. The valuation of abilities and health has been criticized heavily by proponents of disability ethics when it takes the form of ableism. Being mindful of these and so many other biases invites us to be careful about the ideas we project and to be inclusive in the efforts that we undertake to imagine better lives for ourselves and others. Importantly, we must avoid the idea of mere "implementation" or "application" in this terrain and favor embodied accounts of human agency and collaborative methods of discussing the scenarios we wish to pursue. This is why I stress the participatory nature of ethics deliberation in chapter 10, as well as ethical enactment in chapter 12 as ways to situate ethical thinking. Deliberative wisdom represents a positive and open-minded orientation to learning about what matters to the good life; it is a form of wisdom that is more about learning than teaching, thus preempting the interpretation of ethical wisdom as a set of rigid doctrines.

Conclusion

We learn from the ancient Greek philosophers that what we designate now as "ethics" and "positive psychology" were once unified in the search of how to live the good and meaningful life. This was the crux of intense debate and exchange about the nature of eudaimonia as well as a series of existential experimentations carried out by these philosophers. Philosophy was a way of living (a living philosophy), a kind of teaching and learning put into practice. Interestingly, the pioneers of pragmatism like Peirce, James, Dewey, and Jane Addams, who urged and embodied a return to experience, developed a very similar living approach to knowledge. Their philosophy was a living and experiential one that was embodied in their lives and actions, including some of their courageous stances and activities to promote democracy, public education, minority rights, and social inclusion. Importantly, contemporary psychology offers further insights in the understanding of what contributes to the development of human beings.

In this chapter, I argued that human flourishing is a useful, substantive, and guiding focal point for ethics theory and practice. Ethics can be envisioned (or, more accurately, re-envisioned) as a science of human flourishing that directly concerns and speaks to our concrete existences. I designated the knowledge of what human flourishing is and of what experiences bring about human flourishing as ethical wisdom. This vision takes root in some of the ancient Western philosophical wisdom (as well as in other traditions [e.g., North American First Nations traditions, Eastern traditions] for which I am not qualified nor in a position to write about) and has surfaced in the writings of thinkers who have reflected on the meaning of the good and ethical life (e.g., Smith, Mill, Dewey) and of key figures in developmental and positive psychology (e.g., Adler, Rogers, Piaget, Maslow, Erikson, Lawrence Kohlberg, Gilligan). I attempted to show how contemporary insights can be mobilized toward the development of deliberative wisdom, a way of thinking of ethical knowledge that is participatory and dialogical. Accordingly, the science of ethical wisdom is fundamentally deliberative because human flourishing can only be understood with respect to the situations and existence of human beings. Our experiences reflect how we transact with the physical and social world and how our possibilities to do so are shaped by who we are. Accordingly, gaining knowledge about what flourishing is calls for an open and inquiry-like frame of mind—one open to new ideas and to learning about the world through the eyes of others and an evaluation of our own experiences and aspirations with them.

If this is granted, then our growing understanding of human flourishing does and will increasingly contribute to the development of ethics.[26] Indeed, if ethics is not only about knowing or teaching but also about learning, it would be surprising to claim that a defining and authoritative account of human flourishing has been established such that the rest is history. There are good reasons to think that any current understandings of human flourishing will and should be replaced (or refined and updated) with the advancement of knowledge. The first reason is that ongoing work on human flourishing will push the boundaries of what we understand to be in the better interest of human beings. Social debates, as they have done in the past, will continue to bring this question to the forefront, as will the evolution of economic, technological, and political aspects of human realities, calling on us to reconsider what we think human flourishing is and how it can guide our actions. As we do so, we may stumble across additional overgeneralizations about human flourishing and risk imposing a monolithic view on human ends of existence.

Another reason I say that human flourishing will continue to develop and evolve is that it is a personal orientation and goal. What matters for a specific person, what will bring them happiness and satisfaction, what will be consistent with the exercise of their autonomy, what activities and choices give meaning to their life, is determined on an individual basis and varies greatly between individuals with influence from the social environment (e.g., culture). Moral agents need to engage these questions for themselves and reflect on their own valuations, experiences, and struggles to find the unifying narrative and identity that helps make sense of their existence and guides them to new enriching experiences and commitments.

An ethical orientation toward human flourishing is an invitation to live one's life and to live it fully. Therein, a deliberative orientation is a method, an approach, by which experience remains at the forefront and existential learning is shared to serve ourselves and others. Instead of longing for established and fixed truths about human flourishing, we should welcome opportunities to learn about what makes human beings accept who they are, become positive forces for others, and contribute to the growth of capabilities in themselves and others. Knowledge of this kind not only needs to adapt to the existing and surrounding social and material conditions, it also needs to change as we learn about ourselves and others. Ethics, at least a genuinely scientific and philosophical kind of ethics, must espouse this kind of "love of wisdom," a kind of knowledge that grows in the experience we have of and reflection we cultivate on well-being and flourishing.

3 Situated Moral Experience

Chapter 1 provided a general situational account of moral agency in which the activities and abilities of the agent are envisioned as—and with respect to—transactions with the environment. The agent experiences these transactions, which together shape the existence of agents. The situated nature of these experiences is further explored in this chapter. I describe the basic concept of situation and explain how it provides an anchor to understand human behavior consistent with a situated account of moral and ethical agency. By grounding itself in established concepts and theory from psychology and social science, ethics theory can also pull from the immense body of knowledge generated in these fields, as well as their theoretical, methodological, and practical innovations. With these in mind, the following basic properties of situations are reviewed: (1) situations are made up of agent(s), environments, and their transactions; (2) situations are multidimensional (human, social, physical) units of (human) experience; (3) situations can change, as can our interpretation (meaning) and knowledge of them; and (4) the mechanisms of situational recall (episodic memory) play a key role in moral experience and ethical reasoning. General implications for ethics stemming from these properties of situations and from situational psychology and social science are described and discussed. The next four chapters explain the nature of morally problematic situations, but it is important to first understand what a situation is, more generally, to better appreciate the role and value of the concept of morally problematic situations proper.

> We are not determined by our experiences, but are self-determined by the meaning we give to them; and when we take particular experiences as the basis for our future life, we are almost certain to be misguided to some degree. Meanings are not determined by situations. We determine ourselves by the meanings we ascribe to situations.
>
> —Alfred Adler, *What Life Could Mean to You*, 1931

The basic concept of situation, within social psychology and social science, captures how human behavior and human attitudes can be understood as located or positional, that is, as shaped and influenced by, and meaningfully understood as responses to, environments in the broadest sense of the word. This contrasts with the understanding that individuals (e.g., their personality, their health, their performance, their well-being, their moralities) are separate or disconnected from their environments. Pragmatist theory and much of social science and social psychology recognize that agents are agents-in-context and, conversely, that psychosocial contexts include agents such that social phenomena cannot be understood without some basic assumptions about how agents think situationally.[1] Viewing moral agents and moral problems through a situational lens facilitates an understanding of moral situations in connection with the rich bodies of knowledge generated by psychology and sociology, notably on the nature of situated human behaviors, human relationships, human personality, and so on. It also opens up perspectives that can better explain how we understand or fail to understand situations based on capacities for situational intelligence and insight, such as competencies to envision and generate behaviors in situ, abilities to encode and categorize situations, as well as the self-regulatory mechanisms that allow agents to plan and act within and across situations with a sense of coherence (Mischel 1977). Understanding moral problems is thus not only a theoretical task *sensu stricto*, but a task of understanding a distinctively human reality in its context, with the help of knowledge generated by the humanities, the social sciences, and to some extent the biological sciences—as long as no positivist foundationalism is claimed by the latter (Racine 2005, 2008b, 2008c, 2013; Racine and Zimmerman 2012).[2]

The concept of situation is central for scholarly traditions in contemporary psychology (e.g., situational analysis) and sociology (e.g., symbolic interactionism) (Shalin 1991), where situations are cornerstones of scientific explanations of human behaviors and attitudes. Situations have an important status because they capture a fundamental insight into understanding human behavior and psychology, that is, that human attitudes and behaviors are contextualized, and that contexts are lived and experienced by agents through the interactions or the transactions between agents and contexts, or as Dewey puts it, between organism and environment. Situations can even be plausibly envisioned as the building blocks of human existence (Rauthmann, Sherman, and Funder 2015) via our recollection of episodes of existence through memory, notably episodic memory (Conway 2009), which constitute our selves. Psychological and social theories of situations may be less extensive and developed than, say, definitions and theories of personality (specifically those focusing on the agent) or those about impersonal contextual factors (theories that focus instead on context or environment). However, research in social psychology has reinvigorated a situationalist tradition of work in psychology, notably under the influence of Michael Argyle, Adrian Furnham, and Jean Ann Graham (Furnham, Graham, and Argyle 1981). This has also happened under the auspices of ecological psychology and sociology—notably the tradition of symbolic interactionism influenced by Mead and, to some extent, Dewey, that grant particular importance to situations (Blumer 1969).[3] These contributions have resurfaced because of the shortcomings of approaches that focus more on the individual agent (e.g., traditional psychology of personality) or on context (e.g., traditional sociology), and the need to pull from the strengths of both traditions. Moreover, the importance of interpretation—of the meaning-making capacities of human beings—and of the impact of those meanings on behavior have created a need for more sophisticated understandings of the complexity of human attitudes and behaviors to grasp how human beings make sense of situations. Hermeneutics, qualitative research, the sociology of practice, and pragmatism (e.g., through Pierce's semiotics) (Yakin and Totu 2014; Hoffmeyer 2011) all converge on the importance of interpretation and meaning making. They join recent theories on episodic memory, the memory of situations, in which conceptual knowledge and interpretation represent a handle on our recollection of memories (Conway 2009; Schacter, Addis, and Buckner 2007).[4] The concept of situation thus helps capture this complexity because each person has a perspective on the situation and these perspectives are part of the situation itself.

In this chapter, I pull from this insight and paint a broad picture of what situations are, and then describe four important features of situated moral experience (chapters 4–7), namely that (1) situations are constituted by agent(s), environment, and their transactions; (2) situations are multidimensional units of human experience; (3) situations can change, as can our interpretation and knowledge of them; and (4) the mechanisms of situational recall play a key role in moral experience and ethical reasoning.

Situations Are Constituted by Agent(s), Environment, and Their Transactions

The concept of "situation" explicitly captures the situated (in situ) or positional nature of lived human experiences; it emphasizes how human experiences are spatially and physically rooted, as its Latin etymology states.

Situations have a basic triadic nature composed, at the most basic level, of (1) agent(s), (2) environment(s), and (3) their transactions. Environments here include other agents (social context), as well as the physical context. In other words, situations are more encompassing units than environments, because a situation describes how an agent transacts with the environment, and this transaction is central to or constitutive of lived experience.[5] Indeed, if we start from an embodied and transactional ontology as described in chapter 1, we need to capture accordingly how human thinking and acting is nested in environments. Moreover, we need to account for situations as experienced and give way to interpretations that are the basis for building meaning and the intentional and projective nature of human thinking. The experience of a situation cannot be "neutral" or impersonal, as would constitute "a view from nowhere" (Nagel 1986). It is rooted in a particular perspective, and the value of the concept of situation is to make this positioned and action-oriented nature of human thinking about moral experience even more explicit than more impersonal and seemingly neutral terminology (e.g., "cases," "events").

The term "agent(s)" is denoted in singular and plural because not all situations are interpersonal, or at least they do not always directly involve multiple agents. A morally problematic situation can be triggered by someone learning something that does not involve another agent directly, but rather indirectly or perhaps very remotely as, for example, the media revealing information that reshapes one's understanding of a situation or changes in the natural environment (e.g., impact of global warming on the increase of catastrophic natural events), which would prompt self-reflection about habits and values. The stakes of nonhumans can also be at play in a situation, although human behavior is likely at cause, as when dealing with the welfare of animals or the environment.[6] As stated previously (see chapter 1), I use the terminology of agents (instead of individuals) because it deliberately stresses how people are or can be active in their own existences. Ethics is the project of empowering agents to take the reins of their moral life and their moral journey such that they live fulfilling and meaningful lives. The project of ethics commands that we assume a certain instrumentalist epistemological orientation consistent with this goal (see chapter 1).

"Context(s)" is also used in singular and plural form to grasp that what is designated as a situation can involve a number of other similar contexts wherein common and connected feature(s) appear(s) to the agent(s). This is an admittedly open concept of context that allows defining what contexts are based on how they are experienced and the evolving understanding we have of them. Take for example the case of a senior business manager whose behavior entails questionable practices of hiding important and consequential information from clients. At first, the manager's coworkers might be simply puzzled and have a hard time making sense of these behaviors, wondering if what they have seen is actually true. It may take a few instances of this likely covert behavior, as well as discussions among worried colleagues, for a more definitive interpretation of this behavior to emerge clearly as morally troublesome. For simplicity, we often speak of situations in the singular form, but quite often, there are a number of related situations in which moral concerns emerge. Importantly, this pragmatist account of context resembles the notion of "field" introduced by Lewin to consider contexts or environments as those that are lived or experienced by the agent (Lewin 1939).

The transactions between agent and context designate the importance of their co-constitutive interactions within a situated account of human experience. "Transactions" does not merely designate that the environment influences the agents or that agents influence their environments. It also captures that agents *transact* with their environment—they are in their environments and evolve in response to their environments and environments evolve as a function or agents; agents exist and act through their environments. As human beings have the capacity to interpret their lived experiences, they generate meanings that are tied to the lived situated experience. In other words, descriptions and understandings of situations are not ex situ psychological descriptions of human behavior that isolate a person's thought processes, attitudes, or behaviors from context (including other agents); they are not impersonal or third person accounts of human behavior. These understandings and interpretations are constitutive of the moral agent who always finds themself within a given context when a moral problem (or other kind of problem situation) occurs. Context can itself be part of the problem (e.g., lack of health resources creating a situation of injustice and unfairness for an individual or group of individuals), as can the agent's attributes, that is, their attitude or behavior with respect to the context (e.g., not following justifiable principles for resource allocation). Considering situations as fundamental units in understanding and resolving human (moral) problems implies that both agent(s) and context(s) need to be factored into the equation, and that, importantly, the meaning attributed to certain events be integrated into the understanding of situations.

Take for example the reaction of a very simple behavior like smiling, which would seem to lead to an unambiguous interpretation. Such a response can actually be interpreted and lived in many different ways depending on the meaning attributed to the smile by those observing or receiving it, even though it may be one of the most basic human behaviors (Darwin 1872). A smile can be a sign of harmless amusement. It can also reflect mockery and arrogance toward another person. It can be a sign of nervousness or over-confidence. Human beings have to interpret such signs to figure out what a given situation is. The meaning of behavioral and emotional cues such as smiling is amenable to cultural and personal variability such that there is no one universal meaning of a smile; entertaining such an idea would lead to wrongheaded interpretations of human behavior and potentially inappropriate behaviors (Krys et al. 2016). Even when cultural standards may be clear, interpretation of behavior still takes place guided by a cultural code about its common meaning. This insight about the importance of interpretation to meaning is attributable to Peirce—a pioneer in the field of semiotics—who noted that our experiences are always already constituted within the realm of signs, and that we make sense of experiences using signs (Crick and Bodie 2016). Importantly, the nature of the transactions between agents and contexts can generate greater complexity such that the "situation" can actually be a series of successive transactions between agents. For example, it is not uncommon for situations to involve multiple behaviors and layers of interpretations about the intents of other agents such that massive misunderstandings can be built from a few initial misunderstood words or reactions (Kihlstrom 2013).

To illustrate the situational lens and the agent-context-transaction triad, table 3.1 describes how the various components are involved in different strategies for understanding situations and how they contribute to a more comprehensive understanding of situations when they are all considered, especially how the agents make meaning out of the situation they are in. More comprehensive situational descriptions are the kind of descriptions needed to understand more holistically the lived experience of agents facing morally problematic situations to grasp how meaning making—essential to ethics—is embedded in human situations (Emmons 2003). Such descriptions and understandings set the stage for ethics to help moral agents use intellectual and material resources to understand, imagine, and enact in concrete lived situations.

Table 3.1

Comparison of different descriptive strategies for understanding situations*

Imagine the behavior of a nonprofit nongovernmental organization (NGO) worker fundamentally motivated by a commitment to helping others, who would also seem to be inclined to work more to earn a higher wage in a work environment where the incentive structure values and rewards extra time allocated to work.	
Description leaving out context	Focused on an agent's individual properties or capacities (e.g., their work ethic and other personality traits) but leaving out how the context (e.g., the institutional work setting) values their efforts.
Description leaving out agent	Focused on incentive structures, social factors influencing human behavior (e.g., wages) but leaving out the values held by the agent and their fundamental motivations.
Description leaving out transaction and meaning of context to the agent	Focused on beliefs of the agent (e.g., motivation for economic gain and greater wealth) and contexts (e.g., wages) but leaving aside the agent's interpretation of what these things mean to them and what they value in the grander scheme of things.
More comprehensive situational description	Attempts to integrate agential properties or capacities (e.g., a worker's work ethic), the context in which the worker evolves (e.g., payment structure), and the meaning that the worker attributes to the different components of the lived situation (e.g., their values and the meaning they ascribe to a situation, such as their views on earning more money or on the help they can provide to others through NGO work).

*This example is not an example of a morally problematic situation but simply an example of different explanatory strategies for human situations.

In sum, situations are broad, holistic units of analysis that capture the subjective first-person dimensions, the more objective third-person dimensions, and the social relationships that shape agents and contexts (the intersubjective) in the form of transactions with environments. The concept of situation stresses that agents hold interpretations (e.g., meanings, narratives, and stories) about the nature of situations, and as will later be discussed, about what is problematic within certain situations. There are several implications of this first feature of situated moral experience. The first implication is a methodological corollary of the basic triadic nature of situations. To understand situations well, from a research point of view and eventually from a practical standpoint too, there needs to be attention paid to understanding not only the individual agents and contexts involved, seen in isolation, but also to their connections and how they come together to form a particular situation lived in a certain way by an individual or a group. Although there is now a substantial amount of empirical ethics research, and more broadly of various scientific understandings on human morality (Borry, Schotsmans, and Dierickx 2005; Wangmo et al. 2018), it is far from clear that the findings of these studies give good situational understandings of moral problems. For example, in bioethics, almost two decades after a commonly cited "empirical turn," empirical studies remain mostly quantitative (largely surveys). When qualitative methods—methods that are particularly good at grasping context and meaning of life experiences—are used, it is mostly in the form of interview studies with limited use of context-oriented methods like participant observation (Wangmo et al. 2018).[7] These results are surprising considering how the use of *triangulation*, namely the use of multiple methods to understand a given reality, especially its context, is a mainstay in research and more broadly in social science.

The second implication is that, based on the definition of situations provided above, situations are an agent's experience of a life event described in light of the context in which they occur. Such descriptions of life events capture the agent, the context of the agent, and the transaction between them. However, it is important to acknowledge that there is no universal "weight" attributed to the different components of a situation. Psychologists have long debated—at least explicitly since the 1960s—whether personal characteristics, environmental characteristics, or their interactions best explain human behaviors and attitudes, without being able to come to a conclusion

(Kihlstrom 2013). Thus, a sliding-scale approach seems to be in order based on this work such that agents, contexts, and transactions can hold varying levels of importance given that the transaction is not mere interaction but an active process of the agent transacting (exchanging) with the environment.

Situations Are Multidimensional (Personal, Social, Physical) Units of Human Experience

As episodes of embodied and embedded lived experience, situations contain more than conscious awareness reveals, and more than what a single person experiences at a given time or can grasp, pay attention to, or remember (James 1890/1910). This is due to the necessary selectiveness of human attention as described by James and Dewey (Fesmire 2003; Pappas 2008). The inherent limitations of how we attend to and understand situations means that situations are susceptible to being reinterpreted, rewritten, or even re-experienced—given the capacity for imaginative reconstruction—in different ways, such that they become, in principle, an inexhaustible source of reflection on what is hidden and what is revealed to the agent in situ. This kind of insight about the limits of knowing led Kant to postulate the noumenon, and Sigmund Freud, the unconscious (Fesmire 2003).

There has been a revival of studies on the nonconscious dimensions of human behavior and psychology within the concept of habits, previously described by James, Dewey, and others, which reflects automatic behavior as also described by the psychiatrist and neuropathologist Theodor Meynert and later by Freud, who was his pupil (Arminjon, Ansermet, and Magistretti 2010). This old idea, which dates back at least to pre-Freudian psychology, psychiatry, and neuroscience (Hagner 2001), has resurfaced with greater attention to the implicit and often rather automatic aspects of human psychology and social behavior (Tversky and Kahneman 1974). The more recent evidence about automatic processing (e.g., Bargh and Chartrand 1999) has helped reestablish the idea of nonconscious attitudes (e.g., implicit racial attitudes) and cognitive biases within more scientific and experimental paradigms beyond Dewey's philosophical account of the nonconscious aspects of habits.

Understandings of situations should strive to reach a holistic comprehension of the situation, at least as a heuristic or aspirational goal. Seizing the entirety of an experience is, practically speaking, impossible and beyond human capability in an absolute sense. However, it remains that recognizing this aspirational ideal and the limitations inherent to it through the (position-based limitations of our understandings or knowledge-based limitations that call for the adoption of fallibilism) invites certain practical attitudes such as openness to learning, humility, and reflexivity, as well as certain methodological orientations favoring multiperspective understandings of situations. Thus, there is a need to recognize the unknown but no need to postulate unknowable essences since we do not really what they are and, also, our epistemic limitations evolve.

The implications of this second feature of situated moral experience include the importance of recognizing the value of multidimensional (rich, holistic) descriptions and experiential perspectives of moral situations, recognizing the existence of "noncognitive" aspects of the experience of situations and of some resolutions of morally problematic situations, as well as adopting fallibilism and humility with respect to understanding moral situations such that this understanding can always be, in principle, challenged and enriched. The multidimensional nature of lived situations also calls for interdisciplinary, multiperspectival strategies that yield descriptions of moral situations that can challenge established understandings, either in theory, practice, or both. By "established understandings" here I mean unquestioned, habitual ways of viewing situations, namely how agents think of themselves and how they think of their relationships with others and their physical environment. The commitment to holistic, multidimensional understandings of morally problematic situations implies that humility must remain present throughout the engagement with a morally problematic situation. Indeed, individual experiences and knowledge of these experiences are always limited because they are situated. Yet, the reality of human agency is at the same time constituted by experience and therefore the limited understanding of experience is a limited understanding of human reality and notably, the reality of other human beings. Opening up understandings of situations is thus a way to expand our understanding of human reality. In this effort, as a broad rendition of the interpersonal nature of contexts, the concept of situation should also make room for issues that are potentially caused by noncognitive (or not obviously cognitive) aspects of situations, such as physical aspects of contexts that can raise important ethical questions (e.g., for people who do not have physical abilities to adjust to certain physical contexts).

Situations Can Change, as Can Our Interpretation and Knowledge of Them

As dynamic units of lived experiences, situations change over time (Rauthmann and Sherman 2016) notably because all three components of situations can change (agents, contexts, transactions between agents and contexts). Let me further elaborate on this point. Because different aspects of situations can change (agents, context, transactions), the lived experiences attached to similar contexts can change (e.g., from being positive or negative) over time, for the same person, within a given culture, or across cultures and human groups. Abrupt changes can occur (e.g., the state of a severely ill patient may decline, changing the understanding of what the issue is and what the most appropriate therapeutic options are, and eventually what is in the best interest of the patient). Changes can also occur over longer periods. For example, an isolated situation can become more prevalent, developing into a matter of public concern because of, for example, public reporting of a problematic situation generating greater awareness. Conversely, a frequently encountered situation may become much less common and diminish in importance in response to, for example, an appropriate corrective intervention. The exact processes involved in the evolving nature of situations may defy classification[8] and explanations, given the inexhaustibility of situations. However, it remains that this is the reality in which ethics needs to operate.

Situations can involve multiple people and are shaped by broader social factors, all of which are also susceptible to change, whether in the short-term or long-term. Put simply, situations evolve, as do our interpretations of them. This may appear to be a truism, since if situations did not change—or perhaps even more importantly, if our understandings of them did not change—ethics would have no role in generating wisdom in furthering our understandings of situations and proposing new approaches. Otherwise, customary morality would suffice *ad vitam aeternam,* and ethics would not need to engage in inquiries and deliberation.

Implications for this third feature of situated moral experience include the need for practices that allow agents to reflect on the evolving understandings of moral situations because situations themselves evolve. This is a trait of some deliberative and participatory methods supporting case discussions such as moral case deliberation (Stolper, Molewijk, and Widdershoven 2016; Widdershoven, Abma, and Molewijk 2009). We also need to understand ethics and the role of ethicists in terms of companionship in addressing changing situations and not being antagonistic to change in contrast to human morality, which can veer into dogmatism. We therefore need to encourage epistemic and positional humility about ethics knowledge and ethicists (for like-minded proposals, see Inguaggiato et al. [2019a] and Walker [1993]). Ethics and ethicists are sometimes criticized for not naming uncontestable foundational principles according to which something like a new business practice or a new technology would be ethically right or wrong. It appears as if ethics and ethicists—for these critics—cannot simply and accurately state what is right and wrong. And if the opinion and analyses of ethics and ethicists changes, some would contend, then even worse for the credibility of ethics. But such views, although understandable as first reactions given the prevalent standard apprehensions of ethics, rely on profound misunderstandings about the nature of ethics and about morally problematic situations to which ethics responds. It is in the very nature of ethics—in contrast to morality—to provide novel, flexible, and intelligent adaptations to change, which can mean rejecting a given change since adaptation here does not simply mean compliance with change.

The Mechanisms of Situational Recall (Episodic Memory) Play a Key Role in Moral Experience and Ethical Reasoning

Situations are reported based on what is (the present) and what was experienced (the past), although they certainly engage visions about what we desire and hope (the future). Furthermore, memory of experience continues to be reconstructed across existence and its meaning reinterpreted in light of new experiences and thoughts (Glannon 2019). Therefore, mechanisms by which we remember or recall lived moral experiences have a fundamental role in subsequent ethical deliberations about situations; a phenomenon that has recently been described as the prospective brain (Schacter, Addis, and Buckner 2007) and was foreshadowed by Pierce's, James's, and Dewey's allusions to how ideas are embodied and projective (for further explanations see Gallagher [2014] and Martela [2015]).

Contemporary psychology refers to the memory of situations as episodic memory, which differs from semantic memory (the memory of facts), both of which are part of declarative memory. (Declarative or explicit memory is contrasted with nondeclarative or implicit memory, which deals instead with memory of processes and different basic forms of learning [Glannon 2019]). Episodic memory is often described as the memory of events, notably autobiographical events, a form of memory that that enable us to remember times, places, associated emotions, and other contextual information (e.g., the who, what, when, where, and why). As a unit of recall, though situations may involve some degree of semantic memory, they quite clearly involve episodic memory. It is also clear that implicit memory shapes our situated experiences without our awareness (Schacter, Chiao, and Mitchell 2003).[9] Appreciating the role of episodic memory helps understand individual variability in the ability to make sense of the past and entertain scenarios for the future, since future-orientedness (see chapter 10)—an orientation that is crucial to ethics as further developed in the current theory—can be impaired by various conditions that affect, notably, episodic memory, planning, and selective attention (Hirsch et al. 2006, 2007; Johnson, Blum, and Cheng 2014; Murphy et al. 2019). However, it suffices for us at this time to review a few key features of episodic memory (and memories). As a useful review, Martin Conway proposes nine properties of episodic memory, many of which are important in understanding the experience and recall of situations and situated experience which is our focus here (Conway 2009).

First, episodic memories are summaries of recorded experience and are therefore shorthand for the experience itself without corresponding to it completely (Conway 2009). Different circumstances (e.g., trauma) can modulate recall of episodic memory (e.g., emphasize or warp certain details) or impact one's ability to relive the situation. Even just recalling a memory can lead to reconsolidation, potentially changing the memory. Second, episodic memories can be inhibited or activated and are accessible through different cues or primes (e.g., features of a given environment, presence of a given person). The patterns by which these memories are inhibited and activated persist throughout time. Third, imagery, especially visual imagery, dominates episodic memories. Visual cues therefore play an important role in activating episodic memories and in recalling one's experience of situations. For example, damage to areas of the brain responsible for vision can cause amnesia (Conway 2009). This is an important feature of situated moral experience to keep in mind when reconstructing the situation through the lenses of different stakeholders. Fourth, episodic memories always seem to feature a vantage point, or what is called an "observer" or a "field perspective" in some circles of psychology. The perspective of the observer is preserved in memory, and the engagement of the observer's perspective implies that there is a construction process and an incorporation of other knowledge into memory, notably knowledge of oneself (Conway 2009). Fifth, a bit like situations themselves, episodic memories represent only brief portions of experience. It has been proposed that "the boundaries of episodic memories are marked at the opening boundary by information about actions and at the closing boundary by facts that are often details about the outcomes of actions" (Conway 2009, p. 2306). This view is consistent with theories stating that episodic memories (and autobiographical knowledge) are goal-related (Conway 2009), an idea congruent with the pragmatist view that ideas are themselves projective (Martela 2015). Because they provide information about goals as they are remembered in the context of a person's existence where memory provides a resource to adapt to new environments (Glannon 2019), "episodic memory also forms the basis for future goals and plan-implementation" (Conway 2009, p. 2306). The nature of the relationship between the ability to remember and the ability to plan in the future needs to be further investigated, but there appear to be significant effects of memory impairments on the ability to plan and envision new situations (Szpunar et al. 2013). This is a crucial point for ethics as a future-oriented activity informed by past moral experience, since the ability to overcome past or current difficult moral experience is contingent on the ability to develop scenarios for the future. In other words, ethics supports the ability to transition from past experience to future experience, and this is why learning—and what Peirce calls the "will to learn"—is such an important aspect of ethics, because learning represents the ability to use memories of experiences to adapt to new situations. A contiguous mechanism of remembering-imaging has actually been proposed by some whereby "specific memories of recent experiences and anticipated future experiences constitute a window of episodic consciousness that functions to keep us tightly connected with our current goals and plans" (Conway 2009, p. 2307). The temporal order—a sixth property of episodic memory—is constitutive of these memories and may underlie other abilities like planning and goal pursuit and, when it fails, a disorientation within time. Episodic memories relate to events that have taken place at a certain time or moment in life. They also have a temporal sequence with an episode starting with certain events and leading to certain outcomes. This is why accounts of situations and human existence in which situations take place can be narrated, consistent with the previous emphasis on narratives in moral meaning making (see chapter 1). Seventh, few episodic memories are remembered for long, but priming can lead to remembering events to which access is lost. Eighth, specificity of memory can be understood in terms of how memories relate to oneself and autobiographical knowledge and goals. For example, patients with depression who have overly general goals (e.g., lacking specificity and accuracy in envisioning future scenarios) tend also to have overly general memories; they lack access to episodic memories (Conway 2009). In other words, episodic memories are consolidated within human existence where the self integrates past, current, and future experiences (Crick and Bodie 2016). The final characteristic of episodic memories is that they can vary in complexity (Conway 2009). Simple episodic elements (event-specific experiential components such as visual images) make up simple episodic memories, which then make up complex episodic memory. Complex episodic memories are combinations of different simple episodic elements under a conceptual frame (e.g., the ability to group different episodic element under the description of an event or a situation). A conceptual framework is "a type of *interpretation* of the EEs [episodic events] that provides them with a personal meaning" (Conway 2009, p. 2309), again reinforcing how meaning-making narratives are connected to episodic memories. Finally, episodic memories—to be accessible for recall—are integrated into autobiographical knowledge. As I explained in chapter 1, narratives are accounts of human existence in which episodic memories (i.e., of situations) are integrated.

The implications of this fourth feature of the situated nature of moral experience (i.e., the mechanisms of situational recall [episodic memory] play a key role in moral experience and ethical reasoning) include a recognition of the connections between past, present, and future in understanding and resolving morally problematic situations, which aligns with narrative methods and methods that tap into imaginative capabilities to support ethical deliberation.[10] It also aligns with certain specific methods (e.g., the Doucet scenario-based method [Baertschi 1998; Dion-Labrie 2009], as well as Dewey's account of deliberation as dramatic rehearsal [Caspary 1991; Ralston 2010]), which envision ethical reasoning as the generation and enactment of scenarios (see proposal for scenario-based deliberation in chapter 10). This fourth aspect also entails that we can identify and recognize the limitations of memory and situational recall given the biases and shortcomings of human memory (e.g., Schacter, Chiao, and Mitchell 2003). Indeed, our ability to recall situations, including problematic moral situations, relies heavily on episodic memory and the relationship we have with this kind of memory. Episodic memory is memory of experience and notably—but unsurprisingly, given a situated account of agency (chapter 1)—the kinds of experiences that are goal-oriented and associated with self-knowledge (or autobiographical knowledge). Recent research on memory confirms these deep connections that James proposed between memory and self-identity: "Not only is our sense of self based on memories of past experiences based on memories of past experiences, as Mill contented, but our retrieval, recollection, and reconstruction of the past is, reciprocally, influenced by the self," because memories contribute to the "narrative that comprises one's sense of self" (Schacter, Chiao, and Mitchell 2003, p. 227).[11] The fundamental role of episodic memory in goal-oriented behavior—including future goal-oriented behavior—and its contribution to the autobiographical self helps explain why the future-oriented function of ethics is reliant on memories of past problematic situations and is also tied to concepts of the ethical self and personal growth, as these concepts are embedded in projections of ourselves as ethical and adaptive agents. As a science and practice of guiding action and reformulating problematic situations in light of human flourishing, ethics as deliberative wisdom relies heavily on resources offered by episodic memory and the ability to recall lived experience to then project futures through scenarios via processes of ethical imagination and ethical deliberation.

Conclusion

Understanding human existence and experience as situated calls for a robust concept of situation. The concept of situation, as a tool, espouses the transactional structure of our mental states about past and present experiences, allowing us to recollect lived experiences and eventually project future experience as scenarios, as future agent-context transactions (see chapter 10). The concept of situation helps connect important insights about human behavior and psychology, namely that (1) situations are made up of agent(s), environment(s), and their transactions; (2) situations are multidimensional (human, social, physical) units of experience; (3) situations can change, as can our interpretation and knowledge of them; and (4) the mechanisms of situational recall (episodic memory) play a key role in moral experience and ethical reasoning. On the latter point, lived problematic situations are in a sense those that are recallable while taking heed of all the limitations inherent to processes of memory recall. Perhaps on a more speculative note, the concept of situation may be useful, meaningful, and actionable precisely because the human mind and its recollection of lived experience through episodic memory maps well onto the idea that the human brain is a projective, endogenously idea-generating organ as described in some neuroscience and psychological theories (Changeux 1983; Edelman 1989; Schacter, Addis, and Buckner 2007). These theories advocate for a broader understanding of the fundamental units of experience in connection with the temporal and projective nature of human thinking and what some have described as "memory for the future" (Ingvar 1985). There are indeed striking features of episodic memory that are broadly applicable to situations as units of analysis of experience. However, the notion of morally problematic situations is more specific and characterized instead by the moral nature of the situation at hand. Further clarifying this point is precisely the object of the following four chapters, in which I pull from the literature in cognitive science and psychology on situations, and bridge this with ethics theory regarding what makes a situation morally problematic.

4 Morally Problematic Situations: Primary Objects of Moral Experience and Ethics

The theory of deliberative wisdom takes problematic moral experiences in their situated form as the point from which ethics begins and to which it ultimately returns. Moral problems are situated human social problems that have the added specificity of calling into question views and valuations about human well-being and human flourishing. As I present in a typology of moral matters, a moral matter can take the form of, for example, moral issues, moral questions, or moral challenges but takes its most definitive form as that of a morally problematic situation. At stake therein such problems are also, more profoundly, epistemological issues about the nature of human agency and the relationship between agents and their lived worlds. Attempting to bring greater clarity to this simple yet crucial notion of morally problematic situations involves building on both theories of human behavior and general concepts of situations from psychology and sociology to understand what a situation is (see chapter 3). Insights from ethics theories that help us understand what is morally problematic in such situations are needed as well, though there has been a tendency to leave out situational aspects of human behavior to focus on interpersonal relationships rather than holistic situational and contextual understandings of situations. To my knowledge, a synthesis about the nature of moral situations as presented in ethics, philosophy, social science, and psychology has not been undertaken since Dewey promoted the concept of morally problematic situations more than a century ago. In this chapter, I explain this first basic aspect of morally problematic situations and pull from recent literature to substantiate it.

> A moral situation is one in which judgment and choice are required antecedently to overt action. The practical meaning of the situation—that is to say the action needed to satisfy it—is not self-evident. It has to be searched for. There are conflicting desires and alternate apparent goods. What is needed is to find the right course of action, the right good. Hence, inquiry is exacted: observation of the detailed makeup of the situation; analysis into its diverse factors; clarification of what is obscure; discounting of the more insistent and vivid traits; tracing the consequences of the various modes of action that suggest themselves; regarding

> the decision reached as hypothetical and tentative until the anticipated or supposed consequences that led to its adoption have been squared with actual consequences. This inquiry is intelligence.
>
> —John Dewey

In chapter 3, I reviewed some of the key features of situations from the standpoint of psychology, cognitive science, and social science, such as their triadic, experiential, and dynamic structure that maps to episodic memory. Knowing about general features of situations helps us gain situational intelligence and insight. However, it does not really speak to the nature and peculiarity of morally problematic situations. Thus, in this chapter, I turn my attention to features of morally problematic situations proper and provide an account of morally problematic situations following the fundamental insights of Dewey, who specifically used this term to reflect an instrumentalist epistemology and the primary role of (moral) experience.[1] Building from this, I engage with contemporary literature in ethics, philosophy, social science, and psychology to specify four major features of morally problematic situations or, more precisely, of how we can better understand moral matters as derivatives of morally problematic situations: (1) morally problematic situations are the primary object of moral experience and ethics (this chapter); (2) the moral dimension of moral problems is best seen as being of an open-ended nature (chapter 5); (3) situations are experienced as morally problematic (chapter 6); and (4) moral situations are actionable and participatory units in understanding human experience (chapter 7). Before describing these four features and discussing their implications in this and subsequent chapters, I provide an account of morally problematic situations and explain the relationship of this account to a transactional ontology and to other forms of moral matters (e.g., moral issues, moral questions). The concept of morally problematic situations designates a reality but it also—when accounted for as in this book—is a specific tool to understand reality from an ethical perspective.

A Proposed Account of Morally Problematic Situations

In this chapter, I propose that morally problematic situations proper are the primary unit of analysis of moral experience and of ethics insight and moral problem resolution. By "primary" here, I want to quickly clarify that I by no means want to state that morally problematic situations are the *only* unit of analysis in ethics, but rather that they come first in being the most vivid and experientially rich form of moral experience. I come back to this issue in this chapter based on previous work on the plurality of terms used to describe moral matters (for the initial proposal, see Bogossian, Gorter, and Racine [2018]).

Building on long-standing, as well as contemporary work in social psychology and sociology that emphasize the agential (more subjective) and contextual (more objective) aspects of situations (Gross 2009; Mead 1934/2015; Mischel 1977), a situation can be understood as designating an episode of lived experience described in ways in which the agent and the contextual factors surrounding the agent are explicitly taken into account, including the relationships between the former two and the meaning of the lived experience of the agent. A situation is located (situated) in time and space and can be recalled/relived (with the help of episodic memory) or projected (with the help of imagination) as a scenario (or an action-scenario). A situation is morally problematic inasmuch as it raises questions or issues (see explanation in the next section) specifically about the flourishing it induces or prevents in human beings. It is also the case that overcoming morally problematic situations can be a source of growth but that the initial experience of these situations is in the problematic form.

This definition may be abstract, so let me provide an example. Imagine a 17-year-old confronted by a policy that restricts their personal autonomy, such as the legal recognition of the ability to give consent in a medical setting starting at the age of 18 years. Autonomy is one of the key dimensions of a flourishing life for most everyone (Ryff 1989; Ryff and Singer 2008) although the significance of independence—with which autonomy might be easily conflated—is less widely shared across cultures (see the background and data in Chirkov et al. [2003]). The person in this example is confronted with a morally problematic experience insofar as they are alienated from their own ability to decide for themselves and enact their choices (Racine et al. 2014a; Racine and Dubljević 2016). Those surrounding the person (e.g., parents, policymakers) can take a paternalistic attitude and impose this policy blindly, believing that it always fulfills its intended purpose of protecting youth from rash decisions. But they can also, or instead, look into ways to empower this individual by promoting their ability to exercise their decision-making and communication skills within the constraints of legal and medical rulings.

The initial experience of the situation as morally problematic is therefore not definitive; a situation like this can actually yield an opportunity for change and growth of the person, as well as those surrounding them as facilitators of their flourishing. The concept of the morally problematic situation, and situated views on agency more generally, provides—when instrumentalized—a framework to identify agential abilities (e.g., those of the youth at stake, those of healthcare professionals) and related obstacles (e.g., individual professional training and skills), as well as the context in which obstacles surface (e.g., institutional policies) and the lived experience of the agents involved (e.g., experience of suboptimal autonomy, desire to be recognized as an agent and exercise autonomy through choices and decisions that concern oneself). Thus, there is nothing particularly complicated in viewing moral problems as situations; this lens simply invites the recognition of the complexity of a lived life episode and its connections to context and to other people. It is a safeguard against reducing all moral matters to individual problems of agents (e.g., faulting the youth for not being fully autonomous) or to contextual problems (e.g., only faulting current policies and practices without trying to understand the role that agents can play therein).

As a word of caution, it is worth addressing head-on two preliminary reservations regarding the concept of morally problematic situations. First, it is important to recognize that adopting the framework of situations as units of analysis and stressing the importance of situational ethical insight as the understanding of moral problems in situated ways does not make the theory of deliberative wisdom a situationalist theory of ethics (e.g., Fletcher 1966), a theory in which ethical responses are relative to different situations or cases without any further critical reflection or attempt to look beyond each unique situation. This would imply that what works in a given situation is only understandable and justifiable within that situation. Situations are not isolated units of analysis within which valuations are protected from scrutiny. They are a starting point, not the end. With the help of imagination (Alexander 1993; Fesmire 2003), what is understood as a situation can be assessed by appealing to resources outside of the situation, either in space (e.g., "What would happen elsewhere in our society or in other societies?") or in time (e.g., "Is it a situation we want to replicate in the future? Is it consistent with historical behaviors?"). Imagination can help us delocalize or "de-situate" ourselves, to then re-situate by having a more holistic understanding of current situations. Thus, the commitment to understanding moral problems situationally does not entail a situation-bound normative gaze. As a unit of analysis, it is simply a concrete, embodied tool that taps into human experience and from which other moral matters (e.g., moral problems, moral challenges, moral issues) are derived (see Bogossian, Gorter, and Racine [2018] on this point).

Second, from a more descriptive and explanatory standpoint, the theory of deliberative wisdom does not imply a radical situationalist understanding of human psychology wherein psychological traits overpower situations or their characteristics (Upton 2009). A considerable amount of work and evidence, including famous social psychology experiments, such as Milgram's Obedience Experiment (Milgram 1963), Zimbardo's Stanford Prison Experiment (Zimbardo 2007), and the Princeton Theological Seminar Experiment of Darley and Batson (Darley and Batson 1973), have shown how moral behavior can be imperiled by the features of certain contexts. Although these may seem to undermine the positive role that the moral agent can exercise within a situation (Upton 2009), we must be careful not to overgeneralize such experimental findings to everyday experience.[2] In fact, we can accommodate for a sliding scale (alluded to in chapter 3 and explained in more detail below) of emphasis between the agent(s) and the context(s), including situations in which the agential power of moral agents could be quite limited, consistent with the "situational strength hypothesis" (Mischel 1977). According to this latter hypothesis, "The influence of person characteristics differs depending on whether situations are weak or strong, being higher in weak situations" (Knoll et al. 2016, p. 66). In some cases, knowing that contextual factors shape situations could lead to interventions targeting the contextual factors because interventions on the agential level would simply not address the root of the problem. Thus, we can imagine a sliding scale between factors that will guide where actions should be focused, depending on the nature of the situation.

Morally Problematic Situations as the Primary Object of Moral Experience and Ethics

In a broad sense, ethics is concerned with the full range of moral experiences and moral matters (e.g., moral questions, moral issues, moral problems; see below for further explanation on the range of moral matters), but the most definitive and existentially poignant form of moral experience takes the shape of morally problematic situations. In previous work, I have suggested that the language used to describe moral experiences can be analyzed in terms of the existence (abstract or concrete) and the temporality (realized or anticipated) of moral matters[3] (see figure 4.1, which provides a progressive view on different "kinds" of moral matters) (Bogossian, Gorter, and Racine 2018). To the original published table (Bogossian, Gorter, and Racine 2018), I have added corresponding illustrative tasks for ethics scholarship that take into account the nature of the type of moral matter at hand. For example, it is inappropriate to develop a detailed and contextualized response to what is still a somewhat amorphous moral issue or moral concern. Likewise, it is inappropriate to respond to a morally problematic situation by focusing solely on clarifying the problem at hand, which is a response better suited to more undefined moral issues and moral concerns. Importantly, ethics is not only concerned with morally problematic situations, that is, problems with which we are immediately and concretely confronted. Ethics is also concerned with anticipated problematic situations or moral problems, that is, moral issues and moral challenges. In this context, the concept of morally problematic situation can be an instrument and a benchmark for what a moral issue or a moral challenge could become.

This typology helps bring clarity to debates about the anticipatory nature of ethics. In recent years, many have faulted ethics for being overly concerned with problems that do not actually exist (e.g., in the context of a specialized field like neuroethics, see Fins [2008], Hansson [2020], Racine and Sample [2018], and Wexler and Sullivan [2021]). Part of this critique could be true and gets to the difficulty of doing anticipatory ethics well (Maslen 2015; Racine 2008b). But it is far from my intention to condemn more speculative and anticipatory ethical analysis. To mitigate worries about unbridled speculation, speculative ethics can be meaningfully informed by empirical observations to help create bridges between current realities and actual or possible moral experience. (For a proposal for methodological guideposts to develop "grounded speculation," see Racine et al. [2014b] and Racine and Cascio [2020] for a discussion about how pragmatist ethics tackles the relationships between current and anticipated experiences.) Moreover, clarifying upfront, at least roughly, whether what is dealt with is more in the realm of a moral situation or a moral issue can help avoid confusion.

Keeping in mind the situated account of moral agency described in chapter 1, it is important to stress that ethics is also about moral existence as

Anticipatory and Abstract ⟷ Realized and Concrete

	Moral issues	Moral concerns	Moral questions	Moral challenges	Moral problems	Morally problematic situations
Description	A potential matter that needs discussion	A worrisome matter that needs discussion and attention	A worrisome matter that needs discussion, attention and be answered	A worrisome matter that needs discussion, attention and be answered and surmounted	A worrisome matter that needs discussion, attention and be answered and overcome via a response	A worrisome matter that needs discussion, attention and be answered and overcome via a concrete response taking into account specific agents and context involved
Example	Confidentiality	Confidentiality in discussing sensitive information with parents	Should confidential information be discussed with parents?	How can I overcome the challenge of discussing confidential information with parents?	How can I resolve the problem of sharing confidential information with parents?	How can I resolve the problem of sharing confidential information with parents in this concrete situation composed of specific agents and a context?
Task of ethics	Bring clarity to the nature of issues, identify and validate issues	Bring clarity to the nature of concerns, identify and validate concerns based on impact to those involved	Bring clarity to the nature of questions, identify and validate questions based on impact to those involved, identify possible options	Bring clarity to the nature of challenges, identify and validate challenges based on impact to those involved, identify possible options identify possible strategies	Bring clarity to the nature of problems, identify and validate problems based on impact to those involved, identify possible options identify possible strategies, identify resources needed to respond to the problem	Bring clarity to the nature of problematic situations, identify and validate problematic situations based on impact to those involved, identify possible options identify possible strategies, identify resources needed to respond to the problematic situation, develop a contextualized action plan and the evaluation of its enactment

Figure 4.1

Typology of moral matters. This typology is theoretical and based on ideal types as proposed by Max Weber—it is difficult to distinguish clearly and authoritatively in which "category" a given example fits, especially since the categories are defined along a spectrum (Bogossian, Gorter, and Racine 2018). In previous work, I did not distinguish clearly between morality and ethics. This figure has been updated, and a column added to clarify relevant and legitimate tasks of ethics in keeping with the nature of the object it is dealing with.

a whole, not only about morally problematic situations, but that existence needs to be thought of in connection to situated experiences and their contexts. For example, my own moral life is in some respect very different from that of my father, given our different sensibilities, upbringings, and so on. It is also very different because of the kinds of situations we have each had to deal with. Our existences are different because existences, by nature, are unique and singular. They are different because of our respective growth and struggling with ourselves in the times we lived in. The more circumstances change between generations, the greater the need for creative reconstruction of valuations and personal reflection by moral agents. This also allows greater possibility for growth because of the need to adapt in order to live a fulfilled life. Growing up as a child in the 1970s and 1980s (instead of the 1930s and 1940s as did my father) means that my relationships with women were different from those of my father, relationships with people of other cultures, other religions, other languages were also freed from much of the prejudice that existed during his time. Ways of raising children and what becoming a good father meant changed drastically. The meaning of being a good worker and a good boss changed, too. Given all these changes, it would indeed be quite clumsy and inappropriate to simply import his views to the problems I deal with. Vice versa, judging his ways of thinking and behaving by today's standards or those I grew up with would be tremendously unfair, given that my own views and behaviors are largely the product of my time; they are far from being my own sole achievement. Time and space change moral existence and the nature of morally problematic situations in ways so profound that we must never lose sight of the connection between specific morally problematic situations and the more global aspects of moral existence.

My working assumption is that our existences as moral agents are fundamentally constituted by situated experiences and built upon them (Hunt and Carnevale 2011). Generally speaking, as described in chapters 1 and 3, situated experiences are transactions between us and the surrounding world; they are the units representing the transactional realities of which our existences as a narrated set of transactions are made of. As a specific set of situated experiences, morally problematic situations trigger the need to revisit how situations are dealt with from an existential standpoint because of stakes raised about human welfare and human flourishing. They spark questions about our behaviors, about what it means to be, to follow my example, a good father while avoiding traditional paternalism, a good life companion, a good and productive worker who sustains the welfare of others, and so on. Thus, morally problematic situations sometimes challenge our more stable patterns of self-interpretation (e.g., our cardinal values, our identity) that guide our existence and related behaviors (our habits, our personality) (Gross 2009; Hitlin 2003). Such an understanding of our existence can be nurtured by our experiences and our concrete and specific transactions with the world.

Methodological and Practical Implications

There are several methodological and practical implications of the fact that morally problematic situations are the primary object of moral experience and ethics. First, the plurality of moral matters (e.g., moral issues, moral challenges, moral problems) needs to be reckoned with, and a generous understanding of what they mean and entail reached. There is no need for ethics to settle for the consideration of only a certain category of realized or anticipated moral matters.

Second, morally problematic situations can serve as a categorial reference point. Several categories of moral matters exist, but morally problematic experiences provide a reference point in terms of their actual existence (concrete or abstract) and their temporality (present or anticipated). Clearly, it could be unwise to wait until, for example, anticipated moral issues become morally problematic situations to grant attention to them and address them. At the same time, the concept of morally problematic situations provides a benchmark to assess whether a moral issue (or another type of moral matter) will become or is becoming a "real" morally problematic situation. Knowing or sorting out what is dealt with (e.g., a moral issue, a moral question, a morally problematic situation) can help ensure the right kind of response and perhaps the appropriate level of attention, because what remains an issue may be less preoccupying than what is a lived experience of a problem situation.

Third, morally problematic situations—as episodes of situated moral experience—are ideally grasped using thick and situational approaches. The concept of situation reflects an effort to understand the valuations and experiences embedded in morally problematic situations as they are lived and experienced by those involved. It is a conceptual tool inviting the exploration of thick aspects of moral problems (Geertz 1973; Williams 1985) such as how certain questions and problems relate to a person's substantive worldview, that is, their views about their own welfare and purpose in life. Most of the time, our habitual ways of understanding situations may be sufficient such that this process goes unnoticed and does not cause problems. However, when a situation reveals itself as problematic, we should be open to revisiting our understandings of the lived human experiences in question and open up to the values and worldviews at stake. Practically speaking, a situational lens helps in understanding and engaging with a morally problematic situation through a process of inquiry as explained by Dewey in the epigraph of this chapter.

Fourth, empirical research helps understand and address morally problematic situations because it allows better understanding the lived experience and the context in which these surface. Although there have been ongoing debates about the contribution of empirical research or empirical ethics to normative ethics (Davies, Ives, and Dunn 2015)—what I re-describe elsewhere as the future-oriented outlook of ethics (Racine and Cascio 2020)—this is approached very differently within the theory of deliberative wisdom because there is simply no way of dealing intelligently with a problem if the actual problem is not understood, to paraphrase Dewey. To be sure, empirical research does not in itself solve problems; but the knowledge gained from it informs us about past and current experiences to support the imagination of what we desire for the future and whether and how we can enact projections of the future. Empirical research also helps to discern what kind of moral matter we are dealing with (e.g., moral issue, morally problematic situation).

Fifth, situational resolution strategies complement agential strategies. For example, a problem situation of justice and access may be resolved by providing more resources to those in need (i.e., focus on the context). If the agents providing the services hold unfair attitudes and behaviors, their attitudes may need to be problematized and education provided (i.e., focus on the agent[s]).

Conclusion

Morally problematic situations are the primary object of moral experience and ethics. They represent the most concrete and experienced form of moral matter. Theoretically, this means that morally problematic situations can serve, for example, as a categorial reference point to understand other manifestations of moral matters, and methodologically, that they must be understood as situations proper using thick approaches. Practically speaking, this means, for example, that situational approaches and varied emphasis on components of situations should be employed to tackle moral matters. This by no means implies that less immediate or concrete moral concerns such as ethical issues and ethical questions are not of interest to ethics. Those less concrete and less immediate moral matters are relevant to ethics, especially those that may become morally problematic situations affecting the lives of people. Accordingly, ongoing debates about the (overly) speculative nature of ethics are somewhat misdirected since some amount of anticipation is inherent to ethics and to its ability to prevent moral problems unless ethics becomes a short-sighted discipline. The usefulness of ethics in this case is only to put out fires without addressing the source of the problem and preventing future fires.

5 Morally Problematic Situations: Their Open-Ended Nature

There have been long-standing attempts in ethics theory, as well as in the empirical study of human morality, to settle the debate on what the "domain of morality" really is. Is human morality all about empathy or justice? Is it chiefly about respecting one's duties and obligations or about maximizing utility and happiness? Though attempts have been made to define the moral domain in terms of specific and definitive values, principles, or character traits, this domain should be envisioned as open-ended in nature given the multiplicity of valuations, the changing nature of valuations, and the evolving frameworks used to make sense of them. This chapter explains how such an open-ended view on human morality is a second feature of morally problematic situations—and of moral matters more generally—and can be nurtured to support ethics' task of reflection on paths to human flourishing within the wide-ranging pool of insights and moral opinions and practices. I explain how the fact that moral matters are rooted in human existence, notably through our valuations about better and worse human conduct, breeds extensive moral pluralism. This generates a vexing problem for those wanting to study moral behavior or moral attitudes, since relying on a rigid definition of the moral domain risks perpetuating a narrow view of human morality based on a single person or group's understanding of this aspect of human life. Vagueness about the moral domain is thus unavoidable and should be welcomed to reflect the fluid and evolving boundaries of the understandings of human moralities. There are both theoretical and methodological implications stemming from the open-ended nature of morally problematic situations, as well as practical insights on how to conduct research and provide ethics services.

> Those who cannot change their minds cannot change anything.
> —George Bernard Shaw

Ethics scholars have long attempted to define the precise nature of human morality and the kinds of matters considered to be *moral* matters (or moral questions, moral issues, and so on), as alluded to in the previous chapter.

Indeed, what is more frustrating for an ethicist or a moral psychologist than to be unable to precisely define their object of inquiry? In ethics, different shorthand strategies have been used to capture the essence of a morally problematic situation based on the insights articulated by major normative ethics theories (e.g., the moral nature of agents in virtue theories, deeds and obligations in deontological theory, and consequences in consequentialist theory; for a presentation of these strategies, see Dubljević and Racine [2014]), as well as more contextual theories (e.g., in business ethics, environmental ethics, healthcare ethics). Those active in undertaking empirical studies of human morality notably in psychology, as well as anthropology, but also within ethics itself (e.g., for a view from empirical ethics researchers, see Hunt and Carnevale [2011]) or acting as practicing ethicists (e.g., on the views of clinical ethicists, see Racine [2008a]) have tended to cultivate broader notions of morality, particularly because of their confrontation with the limits of definitions of human moralities provided by some influential and formal ethics theory (e.g., Mahowald 1994). Indeed, ethics generally proposes to delimit the boundaries of human morality based on critical analysis and deliberation; to propose a vision of what morality should be. This task—and its relationship to the boundaries of the moral domain or the domain of morality (Dubljević and Racine 2014; Wren 1990), that is, what we consider to be the set of attitudes and behaviors that are directly related to morality—is different from that of the empirical study of human morality, which tries to understand the fluid boundaries of moralities and the perplexities they run into. Between both perspectives, there is ongoing debate and clear tensions, and through the eyes of pragmatists, most seem to rely on potential misunderstandings about the nature of ethics and its relationship to the empirical study of human morality (Racine and Cascio 2020). Indeed, in the quest to find a few clear ethical principles that would guide the totality of human morality, some normative ethicists have tended to generate and deductively apply their preferred account of ethics to their understanding of human moralities (for critical analysis, see Dewey [1966], Dubljević and Racine [2014], and Hoffmaster [2018]).[1]

One of the most illustrative examples of the tension between ethics and the empirical study of human morality is the struggle of Lawrence Kohlberg, a pioneer of the study of moral psychology, with the definition of the domain of morality. Influenced by Jean Piaget, who was himself influenced by Dewey (Tanner 2016), Kohlberg clearly espoused a progressive and developmental orientation in his theory of moral development (Kohlberg 1968). He divided the development of human morality into three levels (preconventional, conventional, and postconventional), each of which is further divided into two stages.[2] The final level (postconventional) and stage (search for universal principles) of moral development designate an orientation in which moral thinking and actions are justified by appealing to abstract principles and values. Kohlberg was influenced by Kant's moral philosophy (Campbell and Christopher 1996; Gilligan 2001; Habermas 1991), which values autonomy and the capacity for universalization (in the four formulations of the categorical imperative) to weed out unethical intents and actions and test for the worthiness of the latter. Unfortunately, Kohlberg conflated the empirical study of human morality with ethics and projected a preferred ethics theory, namely Kantian ethics, onto his empirical investigations. The normative overtones of Kohlberg's empirical work suggesting the moral superiority of impartial, third person, ethics (Darwall 1983) is embodied in the appeal to universal principles of stage six, but ended-up hindering the credibility of his empirical work. Carol Gilligan, who worked alongside Kohlberg at Harvard University, pointed out that his moral perspective failed to attribute adequate value to human relationships (Gilligan 2001), suggesting significant biases (e.g., gender biases) in the alleged impartiality and superiority of this final stage of moral development (stage 6, universal principles).

The shortcomings of Kohlberg's work were understandable in the early days of moral psychology, but they must now be avoided, and clearer theoretical and methodological standards are being adopted in the empirical study of morality (for a proposal for standards in the neuroscience of morality; see Racine et al. [2017]). Importantly, Kohlberg espoused a more encompassing and comprehensive stance than subsequent narrower descriptive and explanatory trends in empirical studies of morality as exhibited in subsequent moral psychology and studies of moral development (e.g., Rest 1986). Kohlberg's work provided an initial bridge between the empirical study of morality and ethics proper, a bridge that has to some extent collapsed given the current sharp divide between the fields of ethics and moral psychology (Flanagan 1991, 2009). As I will argue, one of the underlying issues in how the moral domain is envisioned is that it is often anchored in conventional or common morality (e.g., what a majority thinks is moral/immoral). Critically examining whether this common morality is actually a useful tool in supporting the flourishing of individuals is the task of ethics proper. Ethics needs morality. It needs a pool of moral traditions and habits to pull from. Accordingly, ethics can benefit from the scientific study of human moralities because such studies offer insights into the realities that ethics tries to shape (Doris et al. 2020). However, there is still a regrettably staunch disciplinary split between ethicists and moral psychologists who barely interact with each other or read and cite each other's work (for a discussion on this, see Flanagan's work [Flanagan 1991, 2009]).

In the following pages, I detail some of the more specific implications stemming from the deliberate acceptance of the open-endedness of the moral domain as a feature of the concept of morally problematic situations. This open-endedness is broken up into several related claims which help understand how, for example, the moral domain is socially constructed, and how moral problematization is a social process but that despite this open-endedness, not every matter is a moral matter and not all negative situations give rise to ethical questioning and problematization. I also explain that the moral domain is inherently plural and that therefore, agreement on human morality is something to be (re)constructed, not imposed. Also, despite the unavoidability of a certain kind of reasoned-out vagueness about the nature and boundaries of the moral domain, moral matters can be made sense of in terms of their acceptability and their praiseworthiness, in light of human flourishing, keeping in mind the overview of these concepts provided in chapter 2 and chapter 4. Further, because the nature of a moral matter and nonmoral realities evolve over time, moral problematization is fallible. But, this being said, there must be real doubt or a genuine question for a moral matter to truly exist. I also explain some of main methodological and practical implications of this account of the vagueness of morally problematic situations such that it helps explain reconcile the empirical study of human morality with ethics.

The Moral Domain Is Socially Constructed, and Moral Problematization Is a Social Process

The moral domain is socially constructed and, likewise, the moral problematization of a given situation is not entirely, or perhaps not even chiefly a rational or purely logical process. It is a matter of social interaction and represents a socially constructed reality. Moral matters can be neglected, reframed, or ignored for a number of reasons: self-interest, fear of consequences, guilt, shame, denial, or other factors (see chapter 9 on barriers to moral awareness). Some moral matters, those common and pertinent in our society, have become recognizable almost intuitively, such as "breaches of confidentiality," or "conflicts of interest." Such heuristic-like labeled interpretations are useful, although they should always be critically examined to avoid reducing the situation to a premature or oversimplified understanding based on a single interpretation (Callahan 1973).

The fluidity and open-endedness of descriptions of moral situations may seem to engender daunting difficulties in defining morally problematic situations. Indeed, the moral domain is contested territory, especially for those who have stakes in avoiding a problem or for those who have an interest in maintaining the *status quo*. However, it is important to keep our minds open to different interpretations to truly reflect the nature of moral problematization as a social process. Stifling our morality such that it does not take into account its socially constructed nature prevents ethical progress because it precludes the recognition of changes in human morality and forecloses the ability of ethics to reorient morality toward flourishing. Thus, understanding the socially constructed dimensions of morally problematic situations is key to all subsequent progress in working through them more comprehensively and ultimately resolving them. Although this fluidity of moral boundaries appears at first to be a weakness, it is a theoretical and practical strength as exhibited by contemporary ethics consultation processes that build on robust yet open dialogical processes to apprehend cases (e.g., Stolper, Molewijk, and Widdershoven 2016; Weidema et al. 2013; Widdershoven, Abma, and Molewijk 2009).

Not Every Matter Is a Moral Matter and Not All Negative Situations Give Rise to Ethical Questioning and Problematization

Valuations about human beings and their welfare are implicit in our attitudes and behaviors because we think and act upon interests guided by goals and objectives, which are premised on certain such valuations. Stating this seems to imply that all of human existence is about morality and that all matters are moral matters. There is a sense in which this is true because valuations and values are everywhere such that morality is implicitly everywhere (see chapter 1), but not all conflicts of human and social life—for example, feelings of uneasiness—experienced by an agent point to the existence of a *moral* matter necessarily. A conflict can occur for a number of reasons without the term "moral" being the most apt in describing its nature. For example, colleagues can disagree about the effectiveness of a marketing strategy; parents can disagree about what the best school program for their child is; friends can discuss which physical activity is most fun or which wine is best, and so on. Although a generous, open-ended stance needs to be cultivated about what a moral matter is given the open-ended nature of human morality, some problems are not primarily moral matters because they do not chiefly concern—or are not best described as dealing with—harm to the interest of persons or the promotion of well-being and human flourishing. It is the task of ethics to identify how a given situation raises concerns with respect to important dimensions of human flourishing and therefore represents a genuine ethical problem with real implications for human flourishing. Clearly, the difference and transition is not clear-cut from a matter rooted in common habits and morality that is experienced as being morally problematic to a matter that is the object of ethics inquiry and becomes the object of discussion, reflection, and of reconstruction of purpose.

When discussing the social construction of moral matters, it is important to keep in mind that there is also a tendency to envision negative or difficult outcomes as moral ones, as being by nature the object of ethics. For example, when new technologies yield new risks, they are often perceived as raising moral matters (e.g., for a critical analysis see Carter, Bartlett, and Hall [2009]). There may even be a tendency to exaggerate the novelty and severity of questions, issues, and challenges related to such situations to bring in the attention of ethics (Hedgecoe 2010; Sample and Racine 2021). Sometimes what is of moral concern is even equated with risk (Nelkin 2001) and the control thereof via risk disclosure and informed consent procedures (Corrigan 2003). Physical or psychological risks related to interventions or actions can represent harms, and harms are relevant to ethics, but harms—even small harms, costs, or opportunity costs—are not always ethically important. Moreover, even ethical actions will involve risks such that the avoidance of risks is not in itself a guide for action. Therefore, the "negative" aspects of decisions such as physical and psychological risks of harm need to be seen in light of a balance with benefits and the goods pursued in actions.

What is of ethical concern is that harms caused by an action be justified in light of what that action contributes to well-being and flourishing (of oneself and others). For example, the situation of a car that breaks down on the side of the road may not raise any particular moral matter even though it may be frustrating or problematic. The fact that a mechanic may have inadequately repaired one's vehicle may also not necessarily lead one to interpret the situation as morally problematic because there may be multiple possible causes for this unfortunate event that do not invoke a genuine conflict of values or valuations other than frustration with one's goals. However, if the mechanic had mischievous intent (e.g., changed a wrong part or carried out the task carelessly to maximize their own profit, to the detriment of others), the owner of the car might start invoking their moral sense to view the situation as morally problematic due to the involvement of an untruthful and ill-intentioned agent. In this variation of the situation, the mechanic is deliberately impeding upon a relationship built on trust and expectations of professionalism; they are manipulating their client in dishonest and indefensible ways. The harm becomes a matter of morality and eventually of ethics, because at the core of the situation, views about human welfare are at stake.

The Moral Domain Is Inherently Plural

Every human being experiences harms and benefits in at least slightly different ways. Even individuals who share similar moral beliefs and habits through intersubjective experiences and views about what is proper to human welfare and flourishing (e.g., by being close friends, family members, colleagues in the same profession) will vary in their opinions, in what they value, and in what they consider to be moral or immoral in certain situations—though they may agree on a number of prohibited, permitted, and desirable behaviors. This is caused by the fact that each agent transacts differently with environments; each experience of the world is thus slightly different. This means that there is no single, universal, monolithic human morality, although ethics may strive to unite different human moralities and build from them to support a more robust orientation to human flourishing.

The last claim may sound relativistic, but it is not. There are some rather widely shared agreements on the prohibition of certain behaviors and attitudes (e.g., murder, theft, egoism), as well as agreements on the prescription of other behaviors and attitudes (e.g., caring, sharing, altruism, empathy) (Fesmire 2003), because most people, if not everyone, will recognize how those behaviors and attitudes help or hinder human welfare and flourishing. Agreement-in principle on some key guiding values and principles is even possible across humanity through joint declarations (e.g., United Nations 1948). However, it is commonly recognized that general ethical principles are not only incompletely enacted in practice but that they are also subject to, more philosophically, different interpretations about what they mean and express, what they imply, and how they relate to other rights such that the apparent consensus is and can only be partial.[3] There is a deeply interpretive aspect of laws and rules because we can hardly avoid taking into account surrounding political contexts and moral beliefs to make sense of them (Dworkin 1982). This does not mean, though, that adherence to and recognition of general ethical rules should be trivialized. General declarations of principles attempt to state the abstract values held by a majority, yet forming values is a deeply personal act rooted in daily experiences, so cannot be fully represented in universalized, decontextualized, dematerialized statements of value. What is most important is that the more personal and more universal views on valuations be put into dialogue such that bills of rights and declarations be effective tools in the enactment of rights and that the legal and political reflection on rights and values be informed by individual lives.

Despite appearances of a monolithic (and static) nature, morality reflects a moving body of valuations that are never exactly the same for anyone because moral commitments change, agents change, and social and physical contexts change. Finding coherence and systematicity in this area is part of the creation of a coherent self, a task that ethics supports insofar as helping one reflect on their values and developing a sense of coherence in light of moral meaning making is part of ethics (see chapter 1 for an account of situated moral existence). The path to self-coherence and moral maturity and development is not a given and is not predestined in anyone. Self-coherence of moral agents can dissolve when moral agents lack integrity (Jacobs 2004). In an ethical life, the process of our development of becoming who we are and who want to be as coherent agents also needs to be cross-checked by open questioning about the value of what we pursue for ourselves and others. In other words, ethics entails a certain cultivation of pluralism and experiences of diversity and divergence to transform them into moments of questioning and learning. The fact that the domain of morality is inherently plural means that human beings learn from each other about what is meaningful and valuable in life because general laws and principles are first and foremost generalizations about what we think is important. However, determining what will be important and what will guide our actions needs to build from these insights, sometimes take distance from them too as we envision paths to human flourishing through ethics inquiries.

Agreement on What an Ethical Human Morality Is, Constitutes Something to Be (Re)constructed, Not Imposed

If ethics is committed to the idea that human beings can develop authentic, reflective, and coherent moral lives, then within ethics itself there is something like an inherent commitment to a democratic way of life, rather than an authoritarian one (Pappas 2008). This often led Dewey to state that the method of ethics is the method of democracy (Gouinlock 2002). What is valuable, what is conducive to human flourishing, can be amenable to discussion and eventually to some agreement. But more fundamentally, what is considered to be valuable is an object of dialogue and mutual learning. Ethics, as a scientific discipline, critically examines how practices and behaviors—or sets of them—contribute to human flourishing (or not) in given contexts. Ethics puts moral claims and habits to the test and invites learning from experience. It therefore entails a commitment to the open-ended nature of morality and fosters ethics as a learning process, rather than a moralizing and teaching process. This is why ethics can lead to transformations and evolutions of human morality via education, organizational changes, awareness-raising, and so on. The more intuitively defined contours of human morality are bound to evolve and change in light of critical reflection, experience, the sharing of experiences, and so on. Accordingly, the domains of morality and ethics differ. The first is based on received wisdom about what is valuable in terms of well-being and flourishing, that is, common morality, while the second vets whether this received wisdom is effective wisdom. From the standpoint of ethics, but not of morality itself, what is within the purview of an ethical human morality is something to be (re)constructed, not imposed. Ethics reflects when the democratic spirit infuses and frees moral life such that each agent is granted the ability to become an authentic individual motivated by their own views about what matters, rather than being told by others what counts and what should be sought or avoided (Racine 2022).

Vagueness About the Nature and Boundaries of the Moral Domain Is Unavoidable

As mentioned previously, there has historically been a temptation to establish clear and rational boundaries of human morality. Beyond some general working definition of and rough agreements on this fundamental dimension of human existence, we should recognize that morality is rooted in practices and conventions such that no completely objective, impartial, or uncontested boundary for what counts as moral can be established, either throughout time or across human populations. One can quickly become worried here that the inherent vagueness about the boundaries of human morality implies that no delimitations exist or that there is no meaning in trying to agree on what is within or outside the confines of human morality and of the acceptable and inacceptable. Such seems the conclusion when a stringent positivist and rationalist stance is taken to this issue because no true objectivity can be attained. The subjectivist and relativist who rejects objectivity is also baffled because they cannot make sense of morality without appealing to the intersubjective aspects of human morality, whether it is the language we use, the norms we desire, or the futures we envision. However, ethics attempts to bring reason to the moral life without claiming perfect rationality or objectivity. It tries to sort the wheat from the chaff to determine which problems are problems that really impact human flourishing and which solutions actually advance it. Accordingly, the vagueness of terminology used to describe moral matters (or issues, challenges, etc.) as they are lived is deliberate and actually useful despite the apparent drawbacks of this ambiguity. In trying to define the nature of the moral domain, as expressed by the logician Carveth Read, "it is better to be vaguely right than exactly wrong" (Read 1898, p. 351). However, when attempting to resolve these situations, greater conceptual clarity is often useful within the boundaries of what can be reasonably achieved, but it must be tempered. Conceptual clarity and intellectual rigor need to be conjugated with the goals of empowering people to live flourishing lives. If the former two canons dominate the latter, we risk mistaking our means for our ends.[4] In ethics, this means that the logical, structured, evidence-informed discourse of experts needs to connect to human morality in a way that can adapt to and provide changes and evolution of human morality. It also means that the level of accuracy about the description of moral matters needs to be tempered with the willingness to reflect the existential and lived nature of problems as they are lived and as our experience of them changes. If ethics is severely disconnected from quotidian human behavior and attitudes such that it is meaningless to nonexperts, it can hardly be an effective tool for change (Fiester 2015).

Moral Matters Raise Questions About Acceptability and Praiseworthiness in Light of Human Flourishing

It is clear to anyone having experienced difficult moral situations that in such problems, there is an uncomfortable doubt with respect to whether a given attitude or behavior is conducive to human flourishing. However, moral matters—as a category—challenge the quest for clear, rational, and indubitable boundaries of human morality and of moral matters as concrete units of analysis. First, moral matters take various forms (see chapter 4 for a typology of moral matters) occurring in different contexts, such as questions of the coherence of one's attitudes, behaviors, and beliefs with respect to one's own well-being and flourishing and that of others. This questioning of our own attitudes can also be prompted by a myriad of factors: one's own sense of morality (thoughts, emotions, habits), other people's reactions, one's sense of moral coherence, that is, of having and pursuing a coherent moral life and narrative, or new knowledge that makes us more sensitive to aspects of a situation we have previously lived or currently experience. Second, moral matters are not a natural kind; they are not problems that exist independently of human beings and human interests. Rather, they represent a way of experiencing a situation in light of its impact on human well-being and flourishing.

In chapter 2 and elsewhere (Racine 2010), I have argued that ethics can be divided into two large sets of overlapping concerns that reflect two major complementary orientations. Ethical acceptability captures the negative obligation of not doing harm to others, whereas ethical praiseworthiness captures the positive goal of pursuing the good and alludes to obligations to others. These two categories are not mutually exclusive and should be viewed as instruments which help make sense of why, when, or how we find a situation morally problematic or a given action ethically acceptable and praiseworthy. Accordingly, despite debates framing these two orientations as opposites, dialogue between these two perspectives can enrich our understanding of moral matters because of their complementarity (Racine 2010). Table 5.1 describes how different situations can evoke moral concerns based on the different components of human flourishing (e.g., self-acceptance, positive relationships with others) based on an understanding of human flourishing I explained in chapter 2 (Ryff 1989; Ryff and Singer 2008) (for basic explanations of these dimensions, please see table 2.1 in chapter 2). It also explains

Table 5.1
Dimensions of human flourishing and their relationships with ethical acceptability and ethical praiseworthiness*

Dimensions of human flourishing**	Moral aspects of situations	Ethical acceptability	Ethical praiseworthiness
Self-acceptance:	Situations in which moral agents are not considered in their full complexity and situations that promote reductionist understandings of moral experience.	Does the situation induce tensions within oneself? Does it lead to a disquieting and lasting estrangement from one's values (e.g., moral distress versus moral growth)?	Does the situation offer a sense of completeness, grace, and equilibrium? Does it promote a balanced and reasonable esteem of oneself as a moral agent?
Positive relationships:	Situations in which relationships with others are potentially jeopardized and in which they are a key component of well-being.	Does the situation obstruct or harm meaningful relationships between people? Does it jeopardize a potential sense of community?	Does the situation promote meaningful relationships? Does the situation promote a sense of community and belonging?
Personal growth:	Situations in which the person seems alienated such that their growth and potential are undermined.	Does the situation diminish people or what they stand for? Does the situation involve reduction of opportunities and possibilities? Does the situation involve taking away opportunities for growth and learning by restricting them to the hands of a few?	Does the situation help people grow and learn? Does the situation lead to more opportunities and possibilities? Does the situation involve a sharing of opportunities for growth and learning?

Purpose in life:	Situations without guiding values or meaning or situations that disrupt personal narratives in ways that call for attention to values and principles guiding one's life or one's community.	Does a situation breach one's sense of moral coherence or the coherence of a person with their values? Does the situation ask someone to do something that goes against their basic values?	Does the situation promote reflection on the goals that one pursues and give meaning to their existence (self-reflexive practice)? Does the situation promote the authenticity of persons and of their decisions and actions?
Environmental mastery:	Situations in which the outcomes are negative, and the results are counter to the interests of persons, situations in which personal needs are at stake.	Is the situation preventing people from acting upon their situation? Is the situation one of alienation?	Is the situation promoting the ability of those involved to act upon their situation? Is the situation one of empowerment and meaningful transformation?
Autonomy:	Situations in which the autonomy of a person is at stake.	Does this situation disrespect the autonomy (choices, preferences) of a person(s)? Does this situation jeopardize the autonomy (choices, preferences) of a person(s)?	Does this situation promote the autonomy and empower the individual? Does it maximize autonomy even if the exercise of autonomy challenges established norms or customs?

* Because ethical acceptability is concerned with minimizing harm, it is expressed as infringements while ethical praiseworthiness is illustrated as a pursuit of ideals in more aspirational forms.

** Taken from Ryff and Singer (2008). See chapter 2 for explanations of these dimensions of human flourishing and connections to ethics literature. Chapter 11 further elaborates on ethical acceptability and ethical praiseworthiness in the context of ethical deliberations.

how these components can be analyzed in terms of their acceptability and their praiseworthiness. This exercise is not final and should be considered an illustration of how the open-ended nature of morally problematic situations can be linked to human flourishing in workable ways.

Despite the current proposal, it is clear that Western scholarship is limited in thinking about human flourishing and that other aspects of human flourishing can be conceptualized differently (e.g., in specific cultures), and different dimensions of flourishing can have different levels of emphasis. For example, the Assembly of First Nations Quebec-Labrador has developed, in concertation with its ten constitutive First Nations, a model of better-being (*mieux-être*) that incudes notions such as the power to act (*pouvoir d'agir*) and the social network (*réseau social*) that correspond to notions such as mastery of the environment (although the formulation of this ability as Western-style mastery over the natural world would be problematic for First Nations) and positive relationships. However, this model of well-being for First Nations communities in Quebec-Labrador also includes the land (*territoire*); this addition expresses a very deep connection to the natural environment (Sioui and McHugh 2021). This aspect is not found in common Western models of flourishing. However, there are authors (e.g., Aldo Leopold) who have proposed that it should be included in Western models of flourishing, in the form of a "land ethics" that reflects the change in "the role of Homo sapiens from conqueror of the land-community to plain member and citizen of it. It implies respect for his fellow-members, and also respect for the community as such" (Leopold 1949). It remains that even if certain features of human flourishing can be extrapolated from models and common views, there is a need to understand flourishing individually and contextually. Models and theories about human flourishing should be used most prudently and as useful but nondefinitive heuristics where each and every one's participation helps sustain the concrete life of this idea in persons and in society.

The Nature of a Moral Matter Evolves over Time

The nature of moral matters evolves over time. Some valuations are experienced as nonmoral preferences, such as is what tastes good or which physical activity or hobby is most pleasant to pursue. On the other hand, valuations that were seemingly nonmoral in nature in the past (e.g., attitudes toward animals, the environment, gender categories, human sexuality, nutrition) have become moral because their implications for the value and meaning of human life through human existence have been recognized.[5] Various patterns and configurations are possible. For example, some valuations and preferences remain primarily moral or nonmoral, whereas some may start off moral or have moral components but may ultimately be or become saliently nonmoral (or the reverse). Unsurprisingly, justifications for common moral positions or for the moral valuation of certain experiences in our everyday lives may be lacking (e.g., when we experience moral dumbfoundedness). Dubbing something as moral or immoral may constitute shorthand that ultimately delegates to established authorities or to common beliefs what is moral or immoral. This is not surprising given that valuations are deeply entrenched in our habits and are not usually actively reflected upon until questioned in light of situations that lead to their problematization. However, when such beliefs change, so too can our valuations and our habits.

A clear illustration of this aspect of the evolution of moral matters is the massive moral and social transformation that occurred in the 1950s and 1960s regarding religious authority, women's rights, ethnic minority rights, children's rights, and sexual minorities' rights (Kurtz 2003). The many things that were once considered to be moral based on tradition and custom (e.g., respect for patriarchal authority, limitations to women's autonomy, condemnation of homosexuality, racial segregation) became outrageously unethical in light of their detrimental impact on the people it affected. The moral and social transformation that began in the 1950s and 1960s, and even before as seen in the late eighteenth and early nineteenth centuries, is ongoing, with similar contemporary movements in several Western societies (e.g., recent MeToo Movement denouncing sexual misconduct, Black Lives Matter movement against systemic racism) in response to continued exposure and critical thinking about a number of moral and social experiences and embedded valuations about women, minorities (racial, sexual, etc.), and children's rights notably.

Moral Problematization Is Fallible

In approaching moral situations, it is important to stress that the initial recognition of the existence of a moral matter is subject to biases, denial, and misunderstandings. In other words, it is fallible. As I explain later in the book (see especially chapter 11), it is the task of ethics, particularly of ethics deliberation, to further clarify if a given moral situation is truly an ethical one. However, the fallibility of moral problematization processes is not disembodied or free-floating; it often reflects the struggle for recognition of people and of their interests to express their moral uneasiness (Honneth 1992; Taylor 1992). Importantly, it can be a courageous task to voice and bring recognition to morally problematic situations since the interests of those in power may be called into question. Recognizing this complexity helps explain the difficulty of formulating situations as problematic. Practices of moral problematization (e.g., whistle blowing, expressing moral distress) often involve raising questions about privilege and power when they intersect with valuations embedded in concrete situations.

There Must Be Real Doubt or a Genuine Question for a Moral Matter to Exist

Often, there is debate as to whether some behavior or attitude genuinely represents a moral matter or not; this carries over to positions held within moral situations because they relate to the importance of the stakes with respect to ethical acceptability and ethical praiseworthiness. In keeping with the pragmatist tradition, the experience of doubt or concern that represents a moral matter should be genuine such that there is an experienced morally problematic situation with a concrete action, behavior, attitude, or discourse (Misak 2013), not mere theoretical or gratuitous questioning (Fesmire 2003). As Peirce suggested (and has stated in the epigraph of chapter 1), what makes something genuine is the fact that there are real stakes and doubt as to how to move forward because one is confronted with competing values or with a situation that calls for a sacrifice of some equally praiseworthy outcomes.

Moral matters calling into question habitual morality are grounded in real-life settings and rooted in a genuine preoccupation with whether and how a situation or course of action (e.g., someone's behavior, attitude, discourse) is coherent with a flourishing life. However, it is sometimes difficult to sort out whether a problem really involves important stakes and for whom. Take for example the debates about the acceptability and desirability of public display of religious symbols in some Western democracies (Bilodeau et al. 2018; Kestler-D'Amours 2019). Some may have wondered whether there are real moral stakes (i.e., stakes which truly impact human freedoms) involved in this debate such that perhaps a problem has been fabricated, potentially as a wedge-issue to divide the public sphere. France has taken a staunch position prohibiting the wearing of religious symbols in public primary and high schools in order to, in their view, protect children from undue influence of religion and to foster gender equality because certain religious symbols may be understood to impose patriarchist dominance ("Loi encadrant le port de signes ou de tenues manifestant une appartenance religieuse dans les écoles, collèges et lycées publics" 2004). However, some religious groups (e.g., Muslims and especially women of Muslim faith) are disproportionately affected by such bans since they many adhere more to the idea that they are obligated to wear visible religious symbols than people of other religious traditions. Other countries, such as those influenced by English law and customs, tend to oppose such restrictions on the basis of freedom of religion. They may even think that this is a false problem, that is, that there are no moral matters at play in people wearing religious symbols of their choice such that policies such as that of France actually create a problem of stigmatization and discrimination.

In the state I live (Quebec), there is an ongoing debate and discussion (e.g., a governmental commission led by scholars Gérard Bouchard and Charles Taylor; see Bouchard and Taylor [2008] for more on this issue). A law promulgates restrictions of government workers in positions of authority (e.g., judges, police officers, and more controversially, school teachers) from wearing religious garments and symbols for representatives of the state ("Loi sur la laïcité" 2019). Those in favor of such restrictions advocate for (in the best cases) the construction of a postreligious political community where public spaces promote public discourse based on the use of rationality and intersubjective criteria to validate observations. Proponents entertain a view of a "good" or "better" community that removes potential obstacles posed by religious beliefs in a nondiscriminatory way (i.e., all religions are targeted—although perhaps not affected equally). They envision a political community built on freedom *from* religion and a commitment to constructing a community intersubjectively and positively based on a contractual model, not based on tradition or customs. However, those against these restrictions point out harms inflicted upon minorities who are more significantly affected and even targeted by such policy proposals, such as the increased risk of unleashing and encouraging xenophobic sentiments and conduct toward religious minorities. Since there is a lack of evidence of the threat of religious symbols in state functioning, those favoring secularism risk causing harm to religious minorities while not really tackling any substantive moral matter or a highly meaningful moral matter. This could be the case of a potential spurious moral problem—a situation which seemingly undermines important values—but which in the end is not a genuine ethical problem. There is an initial doubt, but subsequent inquiry may thus temper its impact and salience, at least when symbols are targeted.

Methodological and Practical Implications

The open-ended nature of morally problematic situations and of moral matters more generally, carries methodological and practical implications of which I will briefly underscore three important ones here with respect to empirical ethics research, the nonfoundational role of theories, and reflexivity and democracy.

The Open-Ended Nature of Morality Reconciles the Empirical Study of Human Morality with Ethics

The recognition of the processes implied in the construction of morally problematic situations and their intrinsically open-ended nature bears upon research strategies in ethics and their relationship to the empirical study of human morality.[6] The empirical study of human morality is (or should be, according to the current theory) open-ended with respect to the nature of the moral domain in order to understand human morality in the complexity of its manifestations and plurality. This goal is shared by several disciplines that attempt to understand, through different lenses, moral phenomena and the psychological, personal, relational, and social mechanisms they involve. In contrast, within ethics' use of empirical knowledge about human morality, there needs to be a focus, a point of integration, so that this knowledge can be used for the reconstruction of the meaning of specific actions toward human flourishing. Accordingly, empirical ethics is less a separate tradition of empirical research on human morality (like what constitute the anthropology, the sociology, and the psychology of morality) than a way of trying to bridge this kind of research with ethics theory and ethics practice (see the description of integrated ethics in van der Scheer and Widdershoven [2004]), which is profoundly tied to the idea of using knowledge to make practical changes. Accordingly, the use of empirical research to understand morally problematic situations in practice needs to take into account the specificity of ethics and its aim in gathering evidence about situated moral matters and support its reconstruction through deliberative and participatory methods. Thus, research aiming to understand morally problematic situations, their nature, and their manifestations shares much with work in many disciplines such as anthropology, sociology, political science, and interdisciplinary health and service research. The end of (empirical) ethics research is to understand the situation as a moral one, a goal shared with other humanities and social science disciplines. However, the normative component of ethics' goals (the focus on human flourishing) is often much less salient in these disciplines because of their commitment to take distance from judgments of values, which makes them appear more scientific, objective, and descriptive. Table 5.2 provides a quick overview of differences between the empirical study of human morality and empirical ethics.

Although there are clear similarities and likely much work done at the intersection of both ethics and the scientific study of human morality, it is important to recognize and appreciate their specificities in order to strengthen them both, as well as bring about greater recognition of their contributions. There is no need to choose or oppose these specific endeavors since they inform each other. Social scientists study human moralities more "neutrally", without engagement with the goals and needs of using that knowledge (what is essentially ethics) to provide a pool of insights that help examine, more critically, human morality and reconstruct it. Ethicists undertaking empirical research or using such research have the desire to reconstruct morality, realign it to solve real moral matters. They need to pull insights that help them understand situated moral matters and support deliberation about future scenarios. There are ways in which empirical research on human morality can be better tailored to more effectively support the task of ethics (e.g., by adopting methodological approaches that get to the situated nature of moral matters; by investigating how moral matters engage specific dimensions of human flourishing, such as autonomy and positive relationships with others; by setting the research process as a collaborative and participatory process of inquiry). However, proposing that all studies of human morality conform to the proposed use of the empirical research advanced by the theory of deliberative wisdom as an ethics theory would imply that this theory is settled and closed to learning and being challenged. It would condemn it to a certain kind of stagnation.

Table 5.2
Differences between the empirical study of human morality and empirical ethics

	Empirical study of human morality	Empirical ethics
Goal	Understand, describe, and explain the nature of human morality, moral psychology, moral behavior, notably in its plural forms.	Understand, describe, and explain the nature of human morality to support the reconstruction of meaningful and promising solutions to situations and problems.
Primarily prompted by	Practical and theoretical goals such as understanding the nature of moral emotions all the way to understanding how specific practical problems occur in specific contexts.	Mostly practical goals such that ethics inquiry is mobilized as a response to a morally problematic situation or another moral matter to which a course of action or scenario needs to be proposed.
Methods	Common quantitative and qualitative methods that allow describing, explaining human morality.	Mostly qualitative methods that support the construction of moral orientations such as deliberative and participatory methods, as well as implementation and evaluation methods allowing us to see if a given response-scenario helps solve a given moral matter.
Mainly conducted by	Psychologists, anthropologists, and sociologists who may not claim to be or want to be considered ethicists.	Ethicists, usually those having training in ethics, and some training in social science methods and philosophy, as well as social scientists with knowledge of ethics.

Ethics Theories Can Be Repurposed as Tools and Their Relationship to the Moral Domain Reconsidered

The proposed open-ended concept of moral situations does not grant any foundational status to different ways of making sense of the problematic nature of moral situations based on specific ethics theories, at least as far as understanding human morality goes. However, the understanding of morally problematic situations can be informed by the use of ethics theories, provided that theories and "foundational principles" be granted a more interpretative role than typically seen in ethics (Mahowald 1994). For example, ethics theories advocating for the importance of autonomy or human dignity can be used to make sense of important dimensions of human flourishing at stake in a given moral matter (see chapter 2 and table 5.1 where ethics theories are used to unravel dimensions of human flourishing). This expressive role of ethics theories is in tension with the common view that (normative) ethics theories claim to have the ability to provide a "foundation" for ethical reasoning or judgments and provide an objective way of identifying the domain of morality, what real moral matters are, and what a resolution to a moral matter should look like (see critique in Sherwin [1999]). The repurposing of ethics theories as interpretive or expressive rather than chiefly foundational instruments is supported by the fact that valuations are the starting point of human morality—and of ethics—and principles are ways of expressing the adhesion and commitment to these values (Czeżowski 1953).[7] Principles are not the anchors; experienced valuations are the anchors, and principles are the ropes that express our strong ties to these anchors. Anchors can also prevent movement and when needed, it is important to set sail to other destinations and allow our valuations to shift and our principles, too, provided that this is for the sake of well-being and human flourishing. However, when critically assessing moral situations to see whether they raise real issues for human flourishing, ethical theorizing takes on another role—beyond the understanding of the morally problematic situation—to guide and support the repurposing of our morality toward goals of human flourishing.[8]

Reflexivity and Democracy Support Open-Endedness about Moral Situations and Morality

The open-ended nature of morality inherent to ethics as described in this book prompts to reflect on what our implicit or explicit stances are toward what counts as belonging to the moral domain or not. Simply recognizing that different views exist is an important first step. This does not preclude deciding which view is most defensible; in this book, I do claim that a view that bestows upon its bearer growth and flourishing is most promising in opposition to views that diminish and discourage the search for additional layers of meaning in experience (Pekarsky 1990) (see chapter 2 on wisdom and growth). Contrary to what is often encountered in habituated morality, ethics in this book is open to evaluating moral stances based on their actual contributions—not solely on their power of persuasion, authority, custom, familiarity, and so on. Humility is thus the name of the game when it comes to claiming whether a specific situation is problematic or that a given response-scenario is actually preferable because only experience can really tell. Moreover, actively listening and learning from situations and those involved promotes and supports reflexivity; listening to and learning from different views is a way to re-examine one's own perspectives and test their ethical applicability or validity. This kind of learning orientation is crucial for ethical growth.

The humility implied by the fallibilism and experimentalism of moral habits is also highly consistent with a democratic worldview because in such a political system, opinions have to be vetted through the tribunal of discussion and experience. Additionally, a democratic worldview is most able to promote human flourishing of all (Helliwell et al. 2021) and expresses the important idea that individuals have equal worth. A democratic mode of social organization offers more opportunities for individuals to fulfill themselves because there is an increased chance that their experiences gain worth and consideration in political decisions in comparison to other forms of political organization based on greater concentration of political and economic power. Democracy also offers a richer social and public life than less open political regimes, thus providing a set of moral insights to pull from and support ethics as the pursuit of a flourishing life, and likewise, ethics supports democracy. They are in many ways synonymous and both rely on some open-endedness of experience (Pappas 2008).

Conclusion

The moral dimension of moral matters is best seen as being of an open-ended nature. This may sound like a very unsettling and defeatist statement to make. Paradoxically, understanding and tackling morally problematic situations using set and rigid views about what constitutes the moral domain leads to a number of theoretical and methodological problems, as well as practical problems, notably issues related to the imposition of one's view unto others. In response to these problems, I shared several ways that strategies for understanding moralities and ways of approaching the task of ethics to cultivate pluralism through an open-ended account of moral matters, not necessarily for its own sake, but as a first and inescapable starting point in thinking about the better ways of moving forward. The guiding light in this process remains the core aspects of human flourishing, but these aspects need to be understood contextually and in ways that reflect the flourishing of specific people, with their own unique experiences and narratives. Concepts of human flourishing should be used as facilitators of self-expression, not imposed dogma or received views about the best life to live. Otherwise, the idea of flourishing is perverted and becomes simply another way of defining the contours of the moral domain and of imposing itself as a form of ethics. It is thus in keeping with this participatory account of human flourishing that the very nature of moral situations be understood open-endedly.

6 Morally Problematic Situations: Their Experiential Nature

Morally problematic situations are not problematic in themselves, independent of agents and contexts. Moral situations are experienced as being problematic or not. This is the third aspect of morally problematic situations. It implies that the problematic nature of a situation is rooted in lived experience, which is a contextualized and temporally situated transaction. This is in contrast to theories that have tried to ground morality in certain specific and well-defined values and rational principles and sometimes overrationalize the practices embedded in implicit social attitudes. This means that understandings of human morality differ between human groups with similar or shared values—although the actual homogeneity within human groups should not be overestimated. Even between individuals, including those who are part of groups sharing traditions (religious, political, linguistic, etc.), there is diversity of moral beliefs, attitudes, and practices due to differences in experience. The experiential nature of moral existence and moral matters encountered in this respect is the starting point of subsequent ethical deliberation and ethical enactment. Lived experience of morally problematic situations takes many shapes and forms based on understandings of the world and self-interpretation and self-understanding. Being sensitive to and mindful about how these situations are experienced; the ordinary, nontechnical language used to describe them; the emotions associated with them; the impact of these situations on a person's well-being; and the connections of the experience of a problem to broader historical and structural ideological forces, such as colonialism and capitalism, are important to consider in theory, methodology, and practice. When ethics theories and methodologies lose sight of the connection between problems and existential experience, they risk falling prey to short-sightedness, unhelpful abstraction, and irrelevance. Experience is that from which ethics must start and to which it must return in order to meaningfully tackle moral matters and provide accompaniment on the path of seeking enriched experience and human flourishing. Empirical research in ethics has an inestimable contribution to highlight the experiential features of situations.

> Life is not a problem to be solved, but a reality to be experienced.
>
> —Soren Kierkegaard

Morally problematic situations are precisely such because they are *experienced* as morally problematic at a particular time and place by those in the situation.[1] Likewise, in psychology, it is sometimes said that a situation must be experienced for it to actually exist (Rauthmann, Sherman, and Funder 2015). A situation is indeed, also according to pragmatist theory, a unit of experience and analysis in which there is a transaction between the organism and its environment (Brinkmann 2011). When human beings transact with their social and physical environments,[2] there is a recognized, subjective, "lived experience" that includes the agent's view on the world (Fesmire 2003, p. 35; Foucart 2012/2013). The inclusion of the subjective aspect of human reality—in contrast to its exclusion in more behaviorist schemes (Skinner 1938; Watson 1913)—is a well-established theoretical commitment of contemporary social psychology that has proven its explanatory value (Kihlstrom 2013). However, more situated and transactional schemes of explanations have often taken a back seat in ethics because of mainstream ethics' ongoing commitment to impartial and disembodied reason. The advent of contemporary relational and feminist ethics, as well as more narrative forms of ethics inspired by phenomenology and hermeneutics, has challenged this commitment.

In this chapter, I explain that experience is central to the concept of morally problematic situations. Given the transactional nature of situations, the experience of situations diverges between stakeholders, such that taking into account these diverging experiences is an integral part of situational insight. Furthermore, the centrality of experience means that ethics inquiry absolutely needs to integrate lived experience. Likewise, because valuations are at the core of what makes a situation problematic, valuations are part of ethical deliberation processes. However, not all valuations are defensible and dedicating attention to understanding moral experiences does not justify those experiences because some moral experiences are ethically problematic. In fact, ethics does not pursue at all costs the defense of everyone's individual moral experiences; instead, it invites us to cultivate and understand experiences in order to enrich our moral lives. These clarifications about the experiential nature of moral matters carry methodological and practical implications such that empirical ethics research methods need to integrate experience and that they must do so holistically in order to seize narratives using nonalienating terminology. I explain how understanding the experiential and evaluative dimensions of morally problematic situations is an important task of ethics inquiry and that the process of doing so is an opportunity for co-learning and growth.

The Centrality of Experience in Morally Problematic Situations

The centrality of experience is a crucial insight from classic pragmatist philosophy that deserves dedicated attention in the holistic form from which it was envisioned, supplemented by insight from modern psychology and ethics theory (Brinkmann 2011).[3] Applying this insight to the domain of ethics means that morally problematic situations are not perspective neutral and do not exist independently of individuals experiencing them as morally problematic. The reasoning here could appear circular or relativistic: a situation is morally problematic because it is experienced by at least one stakeholder such that if no one experiences the situation as problematic, then there is no problematic situation. The short answer, strictly speaking, is yes. Agents can neglect the existence of problems related to human values, and they can also actively deny them. They can be insensitive to them or simply unaware. A morally problematic situation only arises when a moral agent experiences and eventually recognizes it as such whether they are the protagonist in the situation or more peripheral to the situation. There are also initial perceptions of a situation as morally problematic that, upon greater reflection, will evolve (see the previous chapter).

Experience is central to ethics and the theory proposed in this book recognizes and cultivates the moral dimension of human existence because of its value and contribution to human flourishing (chapters 1 and 2). Cultivating human morality with the help of ethics is a choice, one that is deliberate from a theoretical, methodological, and practical standpoint. Humans can deny the existence of morally problematic situations, they can discount other people's experience of situations as problematic, and so on. Human beings can also neglect morality just as they can neglect health. However, the proposed theory advances that we do so at our own peril; we risk impoverishing our ability to sustain productive and meaningful human relationships, leading to unproductive and unhealthy tensions and moral distress. As part of this broader pursuit, ethics is often viewed as starting when the problem is identified rather than as being more immersed in the experiential process of the situation's morally problematic nature and the complex social and psychological processes by which this experience is or becomes problematic.

The recognition and understanding of how moral experience gives rise to and can be approached through ethical deliberation is an important part of ethics that has often been ignored.

The literature shows that an agent's experience of moral matters can take various forms, such as feeling uneasy about a situation, being troubled by thoughts or emotions, receiving negative feedback from another person about one's behavior, witnessing the breach of what is an esteemed principle, or witnessing another's troubling behavior (Racine 2007). The clinicians and bioethicists Matthew Hunt and Franco Carnevale from McGill University have defined moral experiences as "encompassing a person's sense that values that he or she deem important are being realized or thwarted in everyday life. This includes a person's interpretations of a lived encounter, or a set of lived encounters, that fall on spectrums of right-wrong, good-bad or just-unjust" (Hunt and Carnevale 2011). This inclusive definition—useful in its ability to encompass the nature of human reality—resonates deeply with pragmatism's recognition that inquiries about problematic situations are triggered by unsettling perplexities and doubts; such situations are, as Dewey writes, "disturbed, troubled, ambiguous, confused, full of conflicting tendencies, obscure, etc.," and provoke the need for intelligence and reflective behavior based on inquiries (Dewey [1938/1986] as cited in Fesmire [2003, p. 35]). Supporting this description, literature on ethics consultations in healthcare settings suggests that the need for ethics consultation and the need for ethics support is often rooted in uneasiness, conflict, and other strong negative emotions (DuVal et al. 2001; Racine 2007; Tapper 2013).

These findings about the experience of conflicts and strong emotions at the basis of requests for ethics consultation is not surprising given that moral matters disturb the equilibrium of valuations related to human flourishing, notably the conditions that make our actions acceptable (or inacceptable) and praiseworthy (or blameworthy) for ourselves and for others, to build on Hunt and Carnevale's definition. Raising doubts and questions about what is esteemed or despised represents a disruption in our relationships with the world, within ourselves, and with others. Ultimately, it involves asking questions about our own existential position. Furthermore, moral matters—when they are legitimate problems raising such substantive questions—are not quickly resolved using the intuitive processes of human morality. It is the very nature of genuine moral matters to outpace our capacity for quick and automated evaluations and judgments. These kinds of problems involve subjective outlooks on and experiences of what the situation is based on the agent's self-understanding and judgments about what is at stake (i.e., their valuations and values),[4] as well as their biases and prereflective attitudes and behaviors. Moral matters touch on values that make up our identities and support our narrative self-constructions as moral agents (Hitlin 2003). Moral problems are therefore a kind of problem that generates some degree of existential anguish because they evoke our freedom to do otherwise (Sartre 1946/1996). Yet, this anguish is the price to pay for moments of growth (when in response to perplexities) following periods of what James called "incubation" and "fermentation" where "the results hatch out, or burst into flower" (James [1902/1985] as cited in Fesmire [2003, p. 36]). The ability to welcome and deal with existential disruption can be cultivated and accepted as being part of our existence; tools can be developed to understand this disruption and support people in making sense of it. Ethics is one such tool according to the current theory.

The need to capture the experienced aspect of morally problematic situations may further stress that the understanding of situations requires not only third-person perspectives (e.g., objective perspectives offered by experts or those who are seemingly detached from the situation), but also first-person perspectives of the moral agent confronted with moral issues, moral questions, or other types of moral matters (see typology presented in chapter 4). Furthermore, this requirement reflects that experiences of morally problematic situations take many forms and are not confined to a dilemma, value conflict, or the like precisely because the disruption of the equilibrium of valuations and values can be experienced in multiple ways and in countless contexts. We have yet to develop a robust typology of these perplexities; we lack a precise and thorough understanding of how, from a developmental perspective, they contribute to a flourishing life, although there is definitely progress being made in the understanding of the relationship between values and self-identity (Hitlin 2003). Any such typology could not only be about a given valued aspect of human existence (e.g., autonomy, positive relationships), but also about the psychological problematization and relational process with respect to that valued aspect of human existence.

Situations Are Experienced by Various Stakeholders

Because human existence is made up of transactions between agents and their environments, and given that each agent is unique, the experiences and valuations of different individuals are unique in that they constitute a certain reality, a part of reality. Hence, it is proper for human lives and existences to generate different experiences and thus to yield different ways of making sense of similar experiences. This pluralism is not an anomaly and results from the many mechanisms involved in the construction of human social realities (Kihlstrom 2013) and the complexity of life trajectories and the narratives we use to make sense of them (Polkinghorne 1988).

The individual nature of experience does not entail that human beings cannot share experiences and valuations about what kinds of experiences are pleasant or unpleasant, desirable, or undesirable. On these matters, we can agree or disagree as we can understand the experiences of others. But we can also fail to comprehend other's perspectives and be unable to engage in dialogue on this issue because of the embodied and entrenched nature of human experience. In order for ethics to support the construction of the mutual understanding necessary to learn from each other's experiences and grow as human beings, attention is therefore needed to understand situations as they are experienced by different stakeholders involved in or concerned by a given situation. Understanding how each person experiences reality, positively and negatively, creates a space in which to build shared futures as discussed in the process of ethics deliberation and imagination (see chapter 10 and also the excellent paper by Margaret Urban Walker [1993] on this point).

Understanding Morally Problematic Situations Is a Key Task of Ethical Inquiry

The identification of a morally problematic situation hinges on a moral agent experiencing (including thinking, feeling, or reacting to) a situation as a challenge to fundamental dimensions of human flourishing (see chapter 2 and chapter 5) or experiencing a challenge in identifying and deciding which paths or scenarios will bring about human flourishing. This insight carries the methodological counterpart (see also final section of this chapter) of the necessity to understand moral experience in greater depth since it is the starting point of an inquiry process to tackle a moral matter (Hunt and Carnevale 2011). For example, understanding why a given treatment decision raises an important moral matter for someone requires understanding not only what the treatment is but also someone's valuation of the different courses of action in a given context. A person suffering from an advanced form of cancer may, for example, greatly value the possibility of extending their life to be able to participate in a meaningful life event (e.g., a child's wedding, the birth of a grandchild), while another person may attribute greater value to quality of life and forgo invasive treatments in their final days. Whether and how that experience is or becomes morally problematic can vary immensely depending on the courses of action and contexts, since quality or length of life—as illustrated by this brief example—matter insofar as they are embedded in a given life narrative and the meanings of different experiences. They are dependent on how a person relates to the experience as a transactional moment in their existence and what they value. Likewise, appreciating how different business decisions have different moral implications requires an ability to assess the valued (or apprehended) experiences implicated in different courses of action, otherwise very inhumane and sometimes business-jeopardizing decisions can be made, such as in the Enron scandal, and become part of a corporate or institutional culture (Sims and Brinkmann 2003). It is therefore a key task of ethical inquiry to understand moral experience and connect what are sometimes rather abstract controversies and disputes about what is moral to the actual lived experience of human beings. Ethical inquiry has exactly the function of seeking understandings of situations that explore whether and how human experiences evoke important moral matters. Although the focus is on the situations themselves, the very experience of those situations implicates moral agents and their existence as such. This should prevent situation-oriented research from falling into a narrow problem-solving approach wherein only problems matter, irrespective of the existential meaning and their place in one's existence. As explained in chapter 1, situations occur within our moral existences that we make sense of as narratives as an identity- and meaning-making activity.

Valuations Are Inherent to Moral Experiences and to Ethical Deliberation

Valuations are inherent to moral experiences, and they reveal who we are because they speak to what we consider important and worthy of pursuing and being interested in. Through valuations—as well as more consolidated values—we express our identities and the direction in which we want to project ourselves (Hitlin 2003; Knoll et al. 2016). Our identities as moral agents take the form of narratives that sew together moments of our existence to create a more coherent whole and provide a cohesive account of where we come from and where we are heading (Schechtman 1996; Schectman 2009), including why we value certain relationships, situations, choices, or activities. Ruptures in life narratives, such as those caused by upsetting events (e.g., major life changes, disease, economic distress), are difficult experiences because they call into question the specific abilities and general capacity we have to act upon the world, self-actualize, exercise our agency, and thus shape ourselves in becoming and maintaining that sense of who we are (Bandura [1977, 1978] as cited in Kihlstrom [2013, pp. 18, 33];Wagner and Northoff 2014). It is as if, in these moments, our existence ceases until we find ways of constructing a resolution and mending our existence, amounting to biographical ruptures (Locock and Ziébland 2015). A range of bad or difficult situations cause such ruptures. Some, perhaps many, of these are not really moral in any clear or direct way (e.g., losing one's job, falling ill). However, they can be or become morally problematic if they call on us to reflect on our moral existence, including what we value in life and what we want to accomplish as human beings. Not all bad situations are moral situations that require ethics deliberation. Although these situations can incite and sometimes require self-reflection, very often, we also need to "simply" carry on with our objectives and values and seek other opportunities despite challenges (e.g., finding other employment, trying to manage an illness).

Consistent with the idea that valuations are inherent to our experiences, it is useful to envision ethics as not implying any rupture between experiential understandings of situations and valuations for courses of action. In other words, there is no need to posit a staunch dichotomy between facts and values. Rather, these can be envisioned following a continuum of valued experiences (rather than mere facts) and experientially supported values (rather than disembodied values) (Norton 1996; Racine and Cascio 2020).[5] Accordingly, moral matters can be alternatively described as "valued experiences" when the starting point is in human experiences, or as "experientially supported values" when realities are described from the starting point of more stable valuations to which experiences correspond. For example, the participation of youth with chronic illnesses and/or disabilities in their healthcare decisions can be reported as a valuable and enriching experience that helps build moral and psychological character (a concrete valued experience)

(Racine et al. 2013). At the same time, active participation in such decisions can also be seen as the realization of values such as autonomy, which are important in adolescent medicine and transition care (an experientially supported value) (American Academy of Pediatrics, American Academy of Family Physicians, and American College of Physicians-American Society of Internal Medicine 2002; Canadian Association of Pediatric Health Centres [CAPHC] 2016; Racine et al. 2014a). The description here depends on whether we are starting from lived experience or from reflection on the value of such experiences less contextually and more abstractly.

Importantly, valuations are in no way definitive or static (see previous chapter on the evolving understanding of morally problematic situations). They can be reoriented in response to changes in experience or reinterpretations of situations. Following one of the examples above, perhaps my mind might change about whether, hypothetically as an advanced cancer patient, I would prefer a longer life over quality of life because events around me could change or my mind might change. For example, my children might convince me that it is more important to them to see me pass through my final days in peace rather than trying to "make it" to the birth of their child. They may question me and reassure me that this is more meaningful to them and in the end, to me. Accordingly, understandings of moral experience—to be truly helpful in ethics problem-solving—need to incorporate the evaluative dimension in which morality is rooted, including the changing appreciations and understanding of various experiences.

Understanding Moral Experience Does Not Justify It

The desire to understand moral experiences in full and to maintain a charitable attitude toward the moral experience and the experiencing person does not entail in any way that an alleged moral experience has any ultimate ethical justification. It may even be that a given moral experience needs to be reframed, reconsidered, reoriented, or problematized in other ways because it embodies a problematic attitude with respect to human flourishing. Consider how it was not so long ago, and still sometimes now, considered immoral in mainstream Western (and other) cultures to belong to what is now called the LGBTQ2S+ community. This was even reflected in psychiatric diagnoses that medicalized and problematized sexuality and gender identities (Davies and

Davies 2020; Drescher 2015). Not too long ago, homophobic attitudes would have been a moral norm reinforced by many persons and institutions with an appeal to morality.

Situations involving people with queer identities were initially experienced up to very recently in many Western nations as immoral and shocking by many since numerous people learned this attitude from their families, their schools, their churches, and their politicians. Yet, these attitudes do not withstand scrutiny or any test of ethical acceptability or ethical praiseworthiness (see chapter 11), especially given the directly resulting harms to those who identify as queer. These harms impact LGBTQ2S+ individuals on many levels, including the intimate relational level, which is of great importance to human flourishing, but also on the levels of other meaningful life activities such as holding work and raising a family, which for many contribute to the pursuit of a meaningful, autonomous, self-accepted life, as well as other aspects of flourishing. Accordingly, even though people have experienced and (some) continue to experience queerness as immoral, that certainly does not justify it as "truly immoral" or what I describe as unethical, that is, when moral attitudes and behaviors are reflected upon. To cultivate human experiences and orientations that help human beings flourish, we first need to understand human experience. Sometimes these experiences are heavily shaped by problematic sentiments and impressions that will be later be debunked (see chapter 11 for the process of examination of scenarios).

Keeping in mind the ongoing discussion about the ethical acceptability of state representatives wearing religious symbols discussed in the previous chapter, we must understand what is being asked of those who wear religious symbols and we must eventually address whether these symbols do perpetuate servile obedience to religious beliefs and the alienation of human freedom, as is claimed (e.g., in fostering gender inequity and sexism). To do so, we first need to understand what these symbols represent for those wearing them, namely how and why they are valued and how they are connected to certain practices and habits, to community and family life, to metaphysical beliefs, and to cultures and histories. We need a good experiential understanding of the stakes at hand. This does not mean that understanding amounts to justifying religious beliefs. Also, engaging in a dialogical process of mutual understanding does not lead to automatically adopting the liberal multicultural ethical and political model as preferable. The mere sanctification of religious beliefs and the avoidance of thought-provoking dialogue within some multicultural approaches is not necessarily the wisest or most ethically defensible public policy option. Religion—when acting as a closed, dogmatic, and infallible belief system—can be a major obstacle for growth in broadening worldviews and apprehending the world in its complexity and richness, growth that is essential in facilitating enriched experiences. At the same time, disregarding and despising religious experience is usually no way to start a productive conversation about life experiences and the meaning of human existence. However, questioning and changing habits and perspectives can be painful and raise questions about the foundations of religious beliefs or about the foundations and limits of ethics, too.[6]

Enriched Moral Experience Is the Goal of Ethics

Ethics attempts to guide human existence toward greater flourishing and enriched life experiences. Enriched experience here designates episodes during which we gain greater understanding, meaning, and appreciation of ourselves and others, or during which we expand our abilities as human beings (Frankl 1966; Pekarsky 1990). Enriched experience is the manifestation of the flourishing life, and it is not only accomplished by realizing actions but by evolving in one's existential self-understanding. This satisfaction of being true to oneself, of self-actualizing, and of finding positive meaning in experience is hard to put into words. Aristotle and other ancient Greek thinkers spoke of the good life, the life worth living, that is, eudaimonia; Dewey of growth; humanistic psychologists (e.g., Rogers, Maslow) of self-actualization; contemporary positive psychologists of flow (Csikszentmihalyi 1990) and human flourishing (Ryff 2014a); and contemporary ethicists of capabilities and development (Sen 1989). Importantly, starting with Aristotle—and this holds true for many thinkers of human flourishing—the good and flourishing life was envisioned not solely as a state, but as a process, as growth, as an "actualizing tendency" (Rogers 1961).

As the endpoint of ethics, human flourishing is by definition dynamic, changing, and evolving because it involves successfully overcoming morally problematic experiences. Accordingly, the agent's identity, which is grounded in values (Hitlin 2003), becomes tied—from the standpoint of ethics—to their ability to become a flexible and adaptive agent able to produce outcomes and conditions of human flourishing. This becomes a truly rewarding process—in terms of intrinsic motivation—because enriched experiences are produced and then fed back into one's self-perception, sense of self-worth, and self-esteem. (However, as I explain above, growth can also be painful because of the experience of the limitations of one's views and values). In other words, ethics is a way of thinking about how a person and their actions actually contribute to their welfare or that of others, because it puts one's own values to the test of experience, thus generating growth in meaning and capacity of the moral agent (Pekarsky 1990). This task is not only or primarily for the person studying ethics or the person leading an ethics deliberation or consultation. It is relevant for every person involved who can grow through understanding the experiential aspects of other agents. However, as I later explain in chapter 10, experiential understandings are only the beginning of a process of seeking intersubjective understandings of situations and engaging those concerned in coming to a shared understanding and awareness of a situation, as well as engaging those concerned in participatory and deliberative processes. Accordingly, understanding the experiential aspects of morally problematic situations is not an end in itself; it is an important step in fostering a broader understanding of moral situations and growing the insight needed to foresee possible response-scenarios that can squarely address the problematic nature of the situation.

Methodological and Practical Implications

Recognizing that situations are experienced as morally problematic and understanding what that experience is carries a number of practical and methodological implications.

First, empirical ethics research needs to grasp the experiential and evaluative component of morally problematic situations. In-depth empirical understandings of moral situations are desirable to support the work of ethics and ethicists, and the experiential component of the situation needs to be tackled whenever one needs to understand a morally problematic situation in order to be meaningfully involved in its resolution.

Second, the use of holistic and narrative understandings of experiences helps get at the meaning of lived experience (Polkinghorne 1988). Given the transactional nature of human experience and the experiential aspect of moral situations, grasping the existential meaning of an event in a person's life is crucial to properly understand a morally problematic situation. Although seemingly abstract or theoretical, existential meaning and meaningfulness designates the very real impact and implications of an event on one's life and worldview, that is, the significance of an event in terms of one's valuations and orientations, including the meaning of life and life activities. As such, any given event can generate profoundly different meanings for different persons and support the development of diverse and even opposing narratives. In understanding the experience of moral situations, approaches that accommodate for and interrogate the meaningfulness of life events, what their implications are in terms of a person's existence, as well as their past and future, are uniquely valuable. Sometimes, in-depth approaches (e.g., fully narrative approaches) are required to grasp how valuations are rooted in life trajectories and life-changing events and habits to understand how the implications of different courses of action are experienced (e.g., as conflict, anxiety, pain, hope, or enthusiasm). Such methods have been used, for example, to foster recollection of difficult cases experienced by healthcare teams and to reflect on them (Aiguier and Loute 2016; Charon and Montello 1999).

Third, the use of open-ended and nonalienating terminology in understanding experiences is of great importance. An important corollary of paying heed to the experiential aspects of morally problematic situations is to be sensitive to the nature and expression of non-expert views. Since experience is precisely such because of who we are, that is, experience is the experience of that person in particular, the experience of moral agents must be understood in a way that does not alienate them from the description of their experience. The late Daniel Callahan eloquently described at the beginning of "Bioethics as a Discipline" how the expert ethical discourse he learned as a philosopher was ineffective in formulating actual problems at stake and in turn resolving them.[7] In the past few years, Autumn Fiester has explained how ethical principles can be "weaponized" against patients because their moral matters become captured by principles they may not even understand or relate to (Fiester 2015). This caveat does not mean that ethics theory should be abandoned altogether, but rather that its connection to moral experience needs to be assured and be thoughtful. Just like good scientific theories in disciplines like physics—no matter how theoretical and counterintuitive they are—need to help, in the end, explain concrete observations (Einstein 1979/1997). Ethics theories should also empower moral agents in dealing more intelligently and more effectively with moral matters. This is, to paraphrase James, the cash value of our ideas (Cotkin 1985). Ethics needs to work with life as it is experienced such that this is the starting point of ethical inquiries. It must cultivate the ability to return to these experiences, ensure its ability to be understood and support people in making sense of what they live, and provide constructive ways of dealing with changes and moral difficulties. As a meaning-making activity, ethics cannot alienate or invalidate everyday experience; it must orient it in a way that is in a constructive. Language is of chief importance here because it is the medium through which meaning is shaped and expressed. In the words of Dewey, language is the "tool of tools"; it is "the cherishing mother of all significance" (Dewey 1925, p. 186).

Fourth, understanding the experiential and evaluative dimensions of morally problematic situations is an important task of ethics inquiry, and notably of ethicists—it is an important task to engage with for all those involved in a morally problematic situation. Within resolution strategies to tackle these situations, there should be an explicit and dedicated effort to understand the experiential and evaluative dimension of situations not only from the point of view of stakeholders involved, but also to foster an inclusive understanding of their experience of the situation. When no such understanding of the situation is achieved, or remains incomplete (e.g., when people are asked to formulate an ethical judgment on a situation they know very little about), judgments of the situation should be reserved and expressed cautiously.

Fifth, co-learning about moral experiences favors growth. In everyday life and pursuits of deliberative wisdom, the commitment to understanding and starting from the experiential nature of morally problematic situations means that the life and voices of those concerned by a morally problematic situation should always be actively sought and remain at the center of our thinking. This implication is paramount in situations where the persons involved in morally problematic situations, such as children (Montreuil and Carnevale 2016; Montreuil, Teachman, and Carnevale 2019), persons with disabilities (Cascio and Racine 2019; Cascio, Weiss, and Racine, 2020), the elderly (United Nations 2021; Westwood 2019), the impoverished (United Nations 2016), or the disempowered (Lynam and Cowley 2007; O'Donnell, O'Donovan, and Elmusharaf 2018), are marginalized in ways that typically prevent them from voicing their perspective. Moral progress and growth (see chapter 2) could in fact be considered proportional to the extent to which a conversation about situations is opened up to the experiences and voices of those who have not yet been able to participate in the formulation of a common understanding of a given situation.[8] For example, in healthcare and social service environments, those engaged in collaborative strategies (e.g., nurses, social workers) that prompt children to articulate their experience as moral agents (e.g., when children with mental health challenges are asked about their experiences and opinions) report that the child feels both empowered and valued, and moreover, that the clinicians themselves experience a sense of growth in their professional perspectives (Montreuil and Carnevale 2018; Montreuil, Teachman, and Carnevale 2019). This growth is consistent with the kind experienced by those who feel the motivation to help persons who are marginalized, disempowered, and disenfranchised (Kern et al. 2001). As I will explain later, participatory deliberative methods that empower those concerned to participate in thinking and acting upon a given ethical problem foster learning and growth (Aiguier and Loute 2016; Inguaggiato et al. 2019a), which are contributors to a flourishing life.

Conclusion

Situations are experienced as morally problematic (or alternatively as non-problematic). Stressing that morally problematic situations are rooted in experience helps explain how lived experience is connected to the meaning of human lives. To understand this implication fully calls for a recognition that ethics needs to incorporate experiential understandings and that appropriate methods and practices must follow suit to capture the diversity of narratives within a situation. Theoretically, the connections between experience and valuations means that our transactions with the world reflect our interests and embody our orientation toward the world and what matters. Problematic moral experiences reveal our existential equilibria by bringing to light what disrupts these equilibria and the valued experiences constitutive of these equilibria. Clearly, these experiences can vary between individuals based on a wide range of factors. In its commitment to a robust ethical perspective, deliberative wisdom calls for a growing understanding of human reality and that understanding each other's experiences of the world is a step in the creation of a shared world, provided that this kind of knowledge be used with an eye to the equal benefit and flourishing of all. In this sense, ethics enriches human experiences because it provides additional insights into what others value, and thus, what is valuable and deemed of interest through open-ended conversation and dialogue. Human beings learn what to care about by understanding what others care about, comparing this to their own previous experiences and understandings, and reflecting on what kind of orientation is preferable. Experience is thus not only the starting point of ethics, but also that to which it must return by providing its users an enhanced vision of connectedness and insight about human values and their existential importance.

7 Morally Problematic Situations: Actionable and Participatory Units of Situational Insight

In this chapter, I describe a fourth feature of morally problematic situations, namely that they represent an actionable and participatory unit of situational insight that aligns with pragmatist theory. Often, ethics theory implicitly or explicitly takes choices, decisions, and dilemmas as the primary units of ethics. These expressions are not always used consciously, but they often give a rationalistic and positivist overtone to ethics, consistent with the overarching rationalistic epistemology of mainstream ethics. The proposed theory of deliberative wisdom invites the recognition of the non-neutrality of these words and underlying conceptual frameworks. It also argues that frameworks that support moral agency are preferable to those that misleadingly neutralize it or mask agency under the guise of objectivity, impartiality, or academic rigor. Understanding moral matters in situated ways—it is argued in this chapter—offers tools for agents to act within and upon situations because it supports an agent-oriented perspective on social realities and human knowledge. In this sense, the concept of situation positions the action possibilities of moral agents within reach, that is, as actionable units of moral experience. Moral agents are invited to situate themselves and act upon their situation in a way that grants that these situations are socially constructed and not simply a matter of individual choice. Far from reifying or neutralizing the impact of the search for the explanation of moral behavior, the current theory fully recognizes that knowledge is power; it is the power to act upon one's situation, which is anything but neutral. Ethics is a form of engagement with the world and with moral behavior specifically, such that the very intention of understanding human behavior situationally is itself an instrument for supporting human moral agency.

> There are no hopeless situations; there are only men who have grown hopeless about them.
>
> —Clare Boothe Luce

There are important relationships between the agency and the identity of moral agents (Gross 2009; Hitlin 2003; Knoll et al. 2016; Wagner and

Northoff 2014). Values are reflected in the actions an agent carries out, and those actions contribute to forming the identity of the agent, and the meaning of their moral existence. Accordingly, knowledge about morality (e.g., what it is, how it works) is a potentially empowering instrument in service of human flourishing, because human flourishing is about developing one's own potential and contributing to the welfare of others. To flourish as ethical agents and moral agents (see chapter 1 for the distinction between moral agents and ethical agents), individuals have to be active participants in their own lives, in their relationships, and in the meanings they generate in the situations that constitute their existence.

The tendency contrary to a flourishing and active agent orientation is the withering and the alienation of human beings and their potential to enact themselves, which can lead to mental and physical illness as well as social upheaval and violence (Peterson, Selgiman, and Maier 1995; Seligman 1975). It is no accident that imprisonment and its near total deprivation of the means to develop one's potential and to interact with others is such a terrible sanction to human beings. To resist this, during his 27-year imprisonment for having sought justice and opportunities to flourish for all human beings, Nelson Mandela cultivated an interest in others' points of view and tried to mobilize his prison mates toward small but meaningful victories, including fighting for more humane treatment. In other words, he upheld the deep commitment to the social movement he led (Lang 2004; Wooldridge 2013). He maintained human connections and his agency as best as he could and cultivated a positive orientation in his situation. This example of extraordinary resilience illustrates why participation in the understanding and resolution of morally problematic situations is so important, such that the ethics processes be oriented toward flourishing and away from languishing. This is also why epistemological frameworks that take away or neglect agency and neutralize knowledge by conceiving of and presenting it as disconnected from human interests and action are harmful. These frameworks do not reflect how values are embedded in ways of knowing, and they impede connections between knowing and acting because they lose sight of or obfuscate the transactional nature of human knowledge generation and of its use (Sullivan 2001, 2017). Alienation is also a source of epistemic injustice because it encompasses situations where the meaning and orientation of one's existence is determined authoritatively by others (Sullivan 2017).

The ethics theory elaborated in this book relies on an instrumentalist epistemological analysis of its own goals and orientations in light of human flourishing (Racine et al. 2019) such that its value is to be assessed based on how it services this orientation.[1] Accordingly, the terms "agent" and "agency" are used for the reason that they help mobilize the resources of human existence toward its own development and growth since they stress that we are actors and not spectators within our own existence. Although this language may appear biased and tangential because it stresses the active nature of human existence, there is nothing fundamentally more neutral or more objective in describing human beings alternatively as "individuals," "persons," or in specific contexts as "consumers," "patients," or "clients." These terms coalesce into matters of social and political thinking and related worldviews and may alienate human potentialities by obviating the transactional nature of human existence and human knowledge and reducing these to nonparticipatory schemes. Describing and envisioning human existence using the concepts of persons or individuals—just like describing it as using the words agents and agency—can reflect an implicit or explicit anthropological view, that is, a view about what human beings are and ought to be. Thus, the use of these terms is not neutral; it is part of projects and paths to the future that consider human agency according to a perspective (e.g., the patient and the agent, the doctor; the passive consumer and the active producer) or obfuscate the active role one can play. Thus, philosophical anthropological views that neutralize human capabilities are, according to pragmatist ethics, a disservice to the pursuit of flourishing.[2] Indeed, these views can be mere words, but they can also exert significant power when they are embedded in social systems since this gives the impression that they explain social reality as it is. These views can be embedded in broader moral and political philosophies and imaginaries as explained in Sheila Jasanoff's work on the relationship between technological imaginaries and political regimes (Jasanoff and Kim 2015; Winner 1980).

This chapter explains that a fourth key feature of morally problematic situations is that that they represent an actionable and participatory unit of situational insight, which aligns with pragmatist theory and the orientation of deliberative wisdom toward participatory human flourishing. In the following pages, I further explain how situations are actionable units of human experience and that because of the power of knowledge, learning about situations empowers moral agents. Also, these units can serve as participatory and experiential benchmarks. This chapter also explains how an account of situations as actionable units supports the use of participatory resolution strategies. Granting that situations are actionable and participatory units of situational insight and understanding leads to a number of methodological and practical implications such as the adoption of co-understanding and co-reconstruction strategies, mutual learning, and the recognition that participatory ethical agency in the domain of morality is a central goal of ethics.

Situations Are Actionable Units of Human Experience

The concept of situation—in contrast to other fundamental and often more theoretical and rationalistic ways of approaching morally problematic situations—suggests that lived experience is within the purview of the agency of the moral agent. For example, we "act upon situations" or "act within situations," we "take action in situations," but the same cannot be said for "events," which we "attend" or "witness," which we can "participate in" but mostly as spectators. Events are organized by some, attended by others. In contrast, situations are episodes "we go through"; they are "resolved," and they are "handled."

Although these metaphors and semantic networks may seem innocuous, Johnson and Lakoff have shown how Western worldviews, including our moralities, are structured by such metaphors (Johnson 1994; Lakoff and Johnson 2003); however, these authors may have gone too far in seeing this as the primary way that meaning is generated. For example, I do not subscribe to their thesis that "our ordinary conceptual system, in terms of which we both think and act, is fundamentally metaphorical in nature" (Lakoff and Johnson 2003, p. 3), since meaning is also derived from use and intent, denotation and connotation, and not just from metaphor. However, situations represent a way of understanding the world operatively, wherein human action and initiative make sense because agency and understanding of the lived experiences of moral agents are at the very center of this concept. It is consistent with an agential epistemological approach such as pragmatism, "a philosophical tradition—that very broadly—understands knowing the world as inseparable from agency within it" (Legg and Hookway 2019). For pragmatism, this action-oriented account of human existence and the use of the concept of situations to understand moments of an existence made of transactions are in line with science as a general attempt to vet meanings based on their implications, assuming that the implications of our ideas are contingent on the actions we take to experiment with them within inquiries (Peirce 1878).

Likewise, the idea that situations are "resolved" alludes more precisely to the repositioning that occurs when a new situation is provoked in an attempt to respond to a moral matter. I call these repositionings "re-situations" (see also chapter 10 on ethical deliberation and imagination). Re-situating oneself reflects the kind of adjustment made possible by learning about how to re-envision a path to flourishing using the resources offered by one's context, notably in response to changes in context. This is far removed from the potential reduction of moral situations to the form of static and abstracted dilemmas where responses to situations are envisioned in the cognitive forms of "answers" to "questions" (Racine 2016). Like the concept of dilemma (Beauchamp and Childress 2001), the concept of choice (Jones 1991) captures a rather limited portion of the reality of moral situations. Although other accounts of human action could potentially describe the experience of morally problematic situations, the notion of situation is a robust concept that helps individuals and groups relate to specific periods of their lives and their interactions with others. It captures—if envisioned as described in this chapter—the complex nexus of factors that make up human contexts, provided that the description of the situation is fair and informed by available knowledge. This concept also offers leeway for agents to act upon their situation intelligently with others to reflect comprehension of the physical and social mechanisms that underlie human situations.

On a more speculative note, the advantage of the concept of situation as embedded in an agential and transactional ontology could be due to the way that cognitive processes are structured. For example, the mechanisms of episodic memory (see chapter 3 for more discussion on the relationship between situations and episodic memory) are an important delimitation of the way that human thinking relates to and processes information about situations. Moreover, the possibility of imagining future situations, which I call "action-scenarios," represents a way to engage in forward thinking that generates possible courses of action in the context of deliberation (this is a point explored in greater detail in chapter 10) and is tied to how imagination works.[3] Indeed, when existence and situations are projected (akin to Heideggerian terminology where *Dasein*, an individual, understands themself and their world in terms of possibilities and of projection, *Entwurf* [Philipse 2021; Rousse 2011]), they take the shape of narratives and scenarios for human beings (Heidegger 1927/1996).[4] Sartre expresses a very similar idea when he writes that "Man is nothing other than his project, he exists only insofar as he realizes himself; he is therefore nothing other than the sum of his acts, nothing else than his own life" (personal translation, Sartre 1946/1996). Accordingly, the notion of situation and the future-related concept of scenario could reflect the mechanisms of memory and imagination, which map onto how human existence is lived and experienced as a series of situated projections of one's being.

The concept of situation designates actionable units of human experience and thus helps human beings take heed of their power to act on events and transform what is seen as simply the course of things into one's own projection. But here I do write "helps" because the concept is no magic wand; it does not instantly transform states of subjugation into states of liberation (Racine 2022). Situations may truly become mere "events" because human agency is taken away or the human factor is obfuscated. Situations embedded in the broader context of neoliberalism and globalization can sometimes be described in impersonal, non-agential ways—especially by those who benefit from such orientations and policies—such that human agency seems to be explained away by macroeconomic and sociopolitical forces. These ideologies need to be criticized not only from the standpoint of their social and economic implications but first for their very view of human beings. The same is true of situations where people are victimized and their existence becomes generally defined by what the situations they undergo. Re-envisioning those situations as moments in which one has the potential to act, react, and do something is not easy but is part of the process of emancipation and moving beyond the past to see a more open and constructive future (Lašas et al. 2017).

Learning about Situations Empowers Moral Agents

Francis Bacon is credited for stating the knowledge is power (*scientia potentia est*), but there has long been a concern that knowing more about human agency—including situational and social aspects of human behavior and thinking—constitutes a threat to the view that the agent is a freely acting agent. So go the common antinomies and debates between the understanding of human beings, behaviors, and attitudes as being (pre)determined and the view that human beings are free and determine their behavior of their own volition. This is seized on explicitly in Kant's third antinomy about the need to postulate transcendental freedom and the postulation of a *noumenal* self in the form of uncaused causal power to make sense of human freedom within mechanistic causality (Kant 1781/1787/1996). But Kant's transcendental idealism and other idealist philosophies reiterating their commitment to the freedom of human beings have been a concern in psychology and social science since Watson's behaviorism (Watson 1913) and Skinner's proposal of radical behaviorism (Skinner 1938). Indeed, "behaviorally oriented psychologists (i.e., 'situationnists') treat and manipulate people as if they were externally controlled rather than free, self-determining beings who are responsible for their own actions and growth" (Mischel 1977, p. 338). Applying strict causal analyses to human behaviors is thus often viewed as a threat to human freedom and human agency (Sartre 1946/1996), and reinforcing deterministic discourse has clear psychological and behavioral effects (MacKenzie, Vohs, and Baumeister 2014; Rigoni et al. 2012; Stroessner and Green 2001; Vohs and Schooler 2008). It fosters a view that human beings are without freedom, which undermines agency. It also provides the means to manipulate human behavior in ways that discount human freedom and its importance (Levy 2007; Saigle, Dubljević, and Racine 2018). In sum, knowledge is often developed as being, in the end, a threat to human freedom because freedom is envisioned as "uncaused causation," thus the act of revealing how human behavior is caused seems to undermine the freedom of behavior.

From the pragmatist view of human beings as transacting agents, knowledge about human behavior stemming from psychology, economics, and other sciences—even if it may appear to undermine beliefs in human freedom—is not an inherent threat to human agency (Racine and Dubljević 2017, 2016). Freedom of agency is something that is experienced and not invalidated by third-person perspectives (for a discussion and a pragmatist account of freedom and voluntary action, see Racine [2017]). This insight about the nature of human freedom and voluntary behavior has often defied positivist stances on human freedom that tend to gloss over the partly subjective nature of human freedom: freedom is not a given that can be studied objectively as a fact; it is an experience and an ability acquired while acting upon the world and understanding the world through transactions with it. Accordingly, the impact of knowledge on free human agency depends on its use; it is not something inherent to knowledge itself.

What we learn about ourselves and other human beings can become tools to empower us to act upon and within situations (Racine 2017). For example, becoming conscious of one's implicit biases, better understanding how one can convince others through sound argumentation and pleasant or charming discourse, can be used to make us more active agents; but they can also be used in manipulative ways to take our agency away and that of others, too. These examples, which are root concerns regarding the use of nudges (Thaler and Sunstein 2008) and other strategies that appear to circumvent conscious and voluntary human behavior (Cohen 2013), shake up our views of what it means to be a person in the sense of being a reason-guided being (Bell, Dubljević, and Racine 2013; Levy 2007). However, learning about and searching for an understanding of situations and the transactions between agent(s) and contextual factors does not take away individuals' agency, but rather situates it and invites us to revise our understanding of what we mean by such general concepts as freedom and predetermination. The better an understanding of situations and their key components (agents, contexts, transactions) that we have, the more we can bring context-sensitive information and resources to help resolve a given problem. When discussing the impact of knowledge on agency, there can be no real meaning to human freedom than a freedom of thinking and acting based on knowledge itself such that knowledge is not opposed to freedom but is the ally in the ability to envision options for action and having the means to carry them out effectively. As Dewey foreshadowed, "What men have esteemed and fought for in the name of liberty is varied and complex—but [it] certainly has never been a metaphysical freedom of will" (Dewey 1922, p. 303). Indeed, humans value the ability to act effectively upon their concrete situations. To act, they need to understand their situation, such that knowledge is an instrument in enacting freely.

Knowledge empowers agents and provides ways of thinking differently about one's situation. There are limitations to this, though. The ability to act calls for resources (e.g., human, financial, intellectual) within and outside of certain situations. For example, material resources are sometimes needed to change contextual aspects (e.g., increasing accessibility for those with disabilities) in ways that call on agents to look beyond the current situation. Sometimes, human resources need to be summoned (e.g., an ethics consultant) to help structure the processes involved in understanding situations and engaging in deliberation and open discourse. In other words, the resolution of moral situations, as in the process of re-situation, is not necessarily self-contained (i.e., located solely within the situation itself) and may call for additional knowledge and resources beyond that of those directly involved in the situation. Moreover, in appreciating how knowledge empowers agents, we must realize that like other kinds of situations, moral situations are not impermeable, static, non-evolving units of experience. Accordingly, the ability for knowledge to empower is restrained by the changing or unique nature of situations.

Finally, it is also true that some dire situations, despite any commitment to upholding high expectations for human agency (see introductory citation from the writer and politician Clare Boothe Luce) may constrain everyone's will. The will is not a static essential component of human behavior but a self-interpretive concept that fluctuates according to one's understanding and experience of the situation, including the understanding of oneself (Racine 2017). It is in this sense that learning about the situated nature of behavior and agency, such as learning about the factors that constrain or facilitate action, empowers moral agents (Racine et al. 2021), because knowledge supports the imagination of possibilities therein and freedom is contingent on having choices. But simply having choices is insufficient; there also needs to be an ability to enact preferences. Knowledge about contexts and situated agency is thus useful to maximize the ability to undertake action and enact successfully (Racine and Dubljević 2017).

Moral Situations Are Participatory and Experiential Benchmarks

One of the benefits of building a theory of ethics on a strong participatory concept of moral situations is that it helps locate where a given problem stands with respect to its relationship to lived experience and concrete realities. In other words, the concept of morally problematic situations helps to scale down what might appear to be insurmountable problems to the scale of the actionable and the resolvable (or lack thereof). Indeed, the proposed concept of situation offers the potential to ask whether a given problem is an existing situation to be reckoned with, based on the lived experience of agents, or if it is rather an issue that has not yet been experienced but may lead to difficult situations (see chapter 4 regarding a typology of moral matters) (Bogossian, Gorter, and Racine 2018). It also offers a framework to assess the usefulness of ethics theories and concepts in understanding moral realities and making them actionable as a benchmark of the success of an ethics theory. For example, an ethics theory that does not connect to actual real-world experiences is unhelpful because it does not even allow first making sense of situations. Moreover, if it fails to grasp lived experience, it will likely fail in helping generate potential action-scenarios in response to problematic situations. Likewise, situations that have never happened or that have little chance of happening may not be as worthy of our attention unless they are perhaps so consequential and transformative that they justify prospective contemplation. (By writing this, I understand that much discussion could be needed to sort out whether a given issue is a real problem or not, hence the value of more speculative and anticipatory reflection and the value of the concept of situation therein. For a proposal for grounded speculation in ethics and examples of exercises, see Racine et al. [2014b] and Voarino, Dubljević, and Racine [2017]). Likewise, hypothetical scenarios that are considered potentially problematic but never actually materialize can be discarded as false positives.

There is always a certain amount of speculation involved in forward-looking exercises examining potential action-scenarios (e.g., the implementation of policies; the implications of technologies; the use of knowledge generated by scientific research). Importantly, some amount of forward-looking attention is necessary for ethics since it is concerned with future actions and thus needs to engage in some form of speculation. However, ethics—as proposed in this book—also builds from strong scientific understandings of past and current empirical evidence about past and current situations, which means that speculation about the future is ideally infused as much as possible with our knowledge about the world. Accordingly, concerns for the actionability of ethics theory and its empowering effects with respect to human agency do not dismiss the need for anticipatory analyses. Envisioning moral situations as participatory and experiential benchmarks provides a milestone for how useful and important certain problems are, or how important they could become, and how ethics theory is positioned with respect to the actionability of situations and its usefulness in benchmarking these situations.

An Agential Account of Moral Situations Inspires Participatory Resolution Strategies

Situations typically involve multiple stakeholders, the stakes of situations themselves being defined according to the different persons involved.

Stakeholders are not observers or bystanders. As the word implies, they are part of the situation and, when this is a problematic situation, they are directly concerned by the problem because they are constitutive of the situation and its problematic aspect. Thus, the concept of situation serves to remind us of the importance of engaging with the situation from the different points of view to understand its multifaceted and multiperspectival nature. Concretely, this means that those engaged in moral situations should be allowed to fully participate in the process of deliberation since they can be helpful in understanding the situation. Likewise, they may be able to participate in its resolution (in generating response-scenarios) and may even be able to learn and grow from this process as they enact their ethical agency. Importantly, "protecting" those deemed incapable of these tasks (e.g., as is sometimes done with children or the elderly) can impede successful understandings and resolutions and may hinder those involved gaining recognition as moral agents. Accordingly, the default should be inclusion or sound justification provided for exclusion (e.g., incapacity to understand the problem, risk of catastrophic reaction to bad news). In the end, a situation is said to be resolved only if the problematic experience is resolved through changes in the agent(s), the context(s), or transaction(s) between the two. Thus re-situation (resolution) entails that moral agents participate in efforts to understand and think about the situations they are in. Solutions from outside the situation (e.g., a solution proposed by someone not involved in the situation, drawing on resources from outside the situation) can sometimes be effective but also risk suggesting that the stakeholder's agency is insufficient to resolve the situation. This can sometimes reinforce broader contextual factors that perpetuate problematic situations by undermining the agential resources (e.g., autonomy, control over one's environment) conducive to human flourishing. Not all broader contextual forces can be dealt with at once, so moral agents often have to prioritize more immediate problems. They might also adopt a staged process in which some of the more immediate aspects are dealt with and then undertake a more serious effort to reflect later and in a better-suited structured form. Simultaneously, partial responses must be acknowledged for what they are to avoid any intellectual and ethical dishonesty (see discussion in chapter 12 on ethical enactments and discussions on moral residue).

Seeing moral situations as actionable and participatory units of situational insight and understanding human experience does not mean that situational insight into morally problematic situations is an ethical magic bullet. Rather, the situational insight reflected in agential epistemology invites the recognition that agents need to understand their positions and the problems they encounter as situational such that the means of responding to problematic situations thus become clearer and the actions to take within the situation become more effective and meaningful. Here, agency and participation are not equivalent to ethics but when action is guided toward human flourishing, participation becomes the promise of the enactment of one's potential as a moral agent—what Dewey calls "consummatory experiences" and what Maslow describes similarly as "peak experience" (Dennis and Powers 1974). This insight about the instrumental nature of situational understandings of morally problematic situations carries a number of more practical implications with respect to the use of participatory methods in ethics.

Methodological and Practical Implications

A first implication of the fourth feature of morally problematic situations is that it further underscores the multifaceted benefits of participatory methodologies and intervention methods and highlights their ethical merits. There is a very real ability of participatory methods to support the development and flourishing of agents as evidenced in literature on empowerment (Budig et al. 2018; Clark, Biggeri, and Frediani 2019; Perkins and Zimmerman 1995). Of additional interest here is that participatory research and participatory enactment of changes actually represent effective ways of bringing about change because change is most effective when agents or participants identify with that change, and they see change as being connected to their intrinsic motivations (Chirkov et al. 2003; Deci and Ryan 2000). Participatory methods (as a form qualitative research) typically recognize the centrality of meaning and of the personal point of view. As explained before, meaning and the process of meaning making are tied to the moral life (see chapter 1) because values are a key component of personal identity and well-being (Hitlin 2003), and the concept of morally problematic situation as an action-oriented concept emphasizes the connections between valuations, habits, and behaviors (Gross 2009) and the important role of narrative (Ricoeur 1992) in making sense of ethical meaning. Participatory methods also foster mutual learning between people in both understanding each other and working together and getting to know each other as part of a project, vision, and an agenda for change (Chevalier and Buckles 2013; Reason and Bradbury

2008; Whitehead and McNiff 2006; Whyte 1991). Moreover, when this is the case, the development of such shared visions for action is also an additional moment of mutual learning and co-creation because of the processes implied in trying to agree on different action-scenarios and of trying to identify the best action-scenarios and the underlying reasons why they represent better options. Finally, participatory approaches embody the value of processes of co-understanding and co-reconstruction of the problem situation from the point of view of those who are involved (Inguaggiato et al. 2019a; Inguaggiato et al. 2019b). Genuine co-construction invites self-reflection on the limitations of one's conceptual framework to grasp a problem situation based on others' understanding and experience. It promotes awareness about issues and challenges as lived situations and hence embodies situational awareness about oneself and others, as well as goals and contexts. These are important features in the pursuit of a deliberative form of wisdom (see chapter 2).

Second, when envisioned as participatory instruments of knowledge as described in this chapter, morally problematic situations also favor the use of participatory methods of evaluation per se. Keeping in mind Peirce (1878, 1877), the highest level of knowledge is achieved when we understand the implications of our ideas. But the outcomes of ideas need to be evaluated by those who are concerned by or involved in their enactment. As Einstein stated, "Everything that can be counted does not necessarily count; everything that counts cannot necessarily be counted" (Einstein 1979/1997). Participatory evaluation methods, because they typically engage stakeholders in setting out the goals that are meaningful to them and lend themselves to an assessment of outcome based on participatory methods, can help ascertain a connection with human flourishing such that we can know if something really counts for someone. In other words, participatory methods allow us to evaluate what something means and implies for human beings as agents who have a say and are involved. This incorporation of agency and learning capabilities of participants in participatory evaluation methods offers important safeguards against common pitfalls associated with the evaluation of outcomes detrimental to the well-being of participants. Indeed, participatory evaluation approaches allow assessing outcomes without either (1) reducing outcomes to predefined epistemological criteria that only partially capture the sought-after outcome, or worse, distorting the outcome sought because of the measurement tool used and the way a given outcome is operationalized; or (2) measuring outcomes that have poor connection with what is actually sought in the first place (real-world outcomes). Additionally, participatory methods of evaluation, because they engage stakeholders, help recognize that some solutions are only partial resolutions to moral matters; there is often moral residue at the outset of a moral problem resolution effort (Webster and Baylis 2000). Moral residue, following Webster and Baylis, represents the remaining distress or perplexity once the events generating the moral distress or perplexity have ended or a candidate solution has been enacted.

Third, the fourth feature of morally problematic situations helps explain how we can productively make sense of ongoing discussions about alleged dichotomies between is and ought or between facts and values (Racine and Cascio 2020). Let it be clear that describing the world as we desire it to be is not the same as describing the world as it is experienced. However, these two processes are often dichotomized (what ought to be and what is, the normative and the descriptive, values and facts). Using the concept of morally problematic situation, we can see them as two kinds of time-and-space relationships to the world, one more concerned with understanding our past and current moral practices (the description of situations), the other more concerned with projecting what we desire and arguing for this based on life scenarios and the imagination of possibilities (the projection of desired situations and experiences). The recognition of this nondichotomous relationship through the fourth feature of the concept of morally problematic situations can help instill productive synergies between experience and values; it is a tool to help movement from past and present into the future. In this sense, embodied understandings of situations help the enactment of valued experiences in ways that allow ethical agents to participate in their own moral lives.

Fourth, doctrines, theories, and ideologies that neutralize agency are not consistent with ethics when ethics is oriented toward human flourishing. Hence, the agential yet situated outlook described in this chapter and elsewhere (see also chapter 1) offers a critical perspective against nonparticipatory orientations. Nonparticipatory orientations are inconsistent with the very nature of ethics as a life project for human beings who can learn, experiment, grow, and live an open-ended life in which they decide how to best live. This does mean that our way of understanding situations is tilted toward desires to make a certain optimistic vision of situated agency come about. However, as I have argued above, although other (e.g., more positivist) frameworks may appear more objective, they can veil their own commitment to a certain conservative and static vision of human beings. These frameworks may even attempt to sever any relationship with human agency such that knowledge seems impartial to particular action-scenarios. In some situations, individuals may actually be staged or indoctrinated to consider themselves atomistic individuals based on, for example, the effects and conditions of neoliberal policies (Pitts-Taylor 2010; Racine et al. 2017b). This comes at the cost of diminishing agency and making human knowledge a potentially disempowering tool. In contrast, an agent-oriented epistemology explicitly recognizes the role of human actions in understanding the world and that knowledge and practice are not separated, as claimed by partisans of more positivistic (e.g., object–subject) epistemologies.

I realize that reformulating things this way is a bit of a Copernican revolution insofar as the exercise of description and explanation are not seen as primary or foundational but are rather recognized as always being instrumentalized, that is, in service of interests and values. Recognizing this clearly does not take away from efforts to be objective but re-situates these efforts in a process where knowledge is recognized as being embedded in goals, values, and interests. As I described earlier, this was very much the practice of ancient schools of philosophy where the aesthetic of an ideal life led to choosing a specific school of philosophy where one learned the teachings and practices to pursue that ideal (Hadot 2002). To understand our work and explain it is to already shape its course because it is partly an effort guided by views about the world and cannot be separated from it.

Conclusion

Moral situations are actionable and participatory units of situational insight and understanding of human experience. This point need not be overdone to become witch-hunting about allegedly disempowering words and terminology, thus leading to a crusade for linguistic purity to remove all non-agentive concepts. Nothing could be further from my intent. Rather, I explained how there are connections between worldviews in which we are spectators in various forms (e.g., individuals, patients) and specific moral-political philosophies that can take away agency, sometimes unintentionally, sometimes intentionally. When we envision ourselves as agents open to experience, we are not merely spectators, patients, or individuals. Rather, we are active citizens in a democracy, and the words we use—when accuracy, focus, and meaning really count—are more aligned with being agents, because envisioning ourselves as able to act is a first step in enacting our agency and developing our abilities. The point of this chapter was to stress that when agents face difficult circumstances, it is empowering to envision them as situated agents. Situations represent an approachable way of understanding and tackling life situations in ways that create room for agency. Theoretically, this insight reflects that knowledge of the psychological and social aspects of situations empowers agents. More practically and methodologically, an epistemology of situations and situated agency offers participatory and experiential benchmarks and supports participatory resolution strategies. In closing, in this chapter, which concludes a set of four chapters on the features of morally problematic situations, I also recognize the limitations of this account of morally problematic situations and the need to better connect ethics theory to increasingly advanced scholarship in the social sciences and psychology. I hope that the current chapter sparks interest in pushing the boundaries of situated moral understanding further and in bridging different bodies of knowledge to empower ethics theory to support situational insight.

8 Moral and Ethical Awareness

If ethics is a kind of inquiry, it must first start with a recognition of the existence of the kinds of matters about which it inquires. In previous chapters, I describe those moral matters as being situational in nature in their most concrete and temporally immediate form. In this chapter, I dedicate greater attention to the process by which we can become aware of moral matters, that is, moral awareness. Moral awareness is, in this sense, contiguous with subsequent attention to further understanding the nature of moral matters and engaging in open discussion to find creative solutions. Because it helps grasp moral matters, moral awareness defines the contours of the kinds of solutions that will be deemed fitting to the matter at hand. Further, because it is rooted in the moral agent and their experience of the world, moral awareness says something of the person who experiences the matter. Accordingly, moral awareness is often described as a first step for ethics and as a set of psychosocial mechanisms (e.g., emotions, relationships) by which the existence of a moral matter is recognized as such. Moreover, this process can be cultivated and guided such that the initial awareness of a matter is enriched by taking into account the perspectives of others. When this process of moral awareness is reflected upon to be guided in the interest of human flourishing, it becomes ethical awareness proper. Otherwise said, ethical awareness is the process of identifying the matters that genuinely affect human flourishing. In this chapter, I expose the concept of moral awareness, situate it within ethics theory and particularly within the theory of deliberative wisdom where awareness is viewed partly as an instrument that can be refined in the form of ethical awareness. Indeed, not all matters identified as moral matters withstand the test of affecting human flourishing, and some matters not initially seen as moral matters should be recognized as such. Refining one's moral awareness is contingent on the ability to be sensitive to what impacts other people's flourishing and welfare and therefore is also reliant on the recognition that the well-being of others is inseparably linked to one's own flourishing.

> No quality of human nature is more remarkable, both in itself and in its consequences, than that propensity we have to sympathize with others, and to receive by communication their inclinations and sentiments, however different from, or even contrary to our own.
>
> —David Hume

> As we have no immediate experience of what other men feel, we can form no idea of the manner in which they are affected, but by conceiving what we ourselves should feel in the like situation.
>
> —Adam Smith

Many readers will be familiar with the famous philosophical riddle, "when a tree falls in a lonely forest, and no one is nearby to hear it, does it make a sound?" This riddle may partly originate from the work of the idealist philosopher George Berkeley, who used similar formulations in discussion about the nature of perception and reality, and argued that to be, is to be perceived or to perceive (*esse est percipi* [*aut percipere*]) (Berkeley 1734/1957). This perplexing question, or at least its perplexing formulation, has vexed the greatest scientific minds. In a discussion with Albert Einstein, Abraham Pais, a quantum physicist and biographer of Einstein, formulated a similar thought experiment, here involving the existence of the moon irrespective of anyone looking at it. No matter how hard he would try, Einstein could not prove in their discussion that the moon existed independent of being observed (Pais 1982).

In my eyes, the relationship between the existence of a moral matter and its existence in the eye of the beholder is different than the formulation of the classic forest and tree riddle or its lunar reformulation by Pais. The third-person recognition of objects that exist independently of human beings is different than the recognition of realities that are constituted by human relationships and the engagement of human beings with the moral dimension of their existence. If one lacks the ability to recognize moral matters for what they are—whether it is through bluntness of character, lack of moral education, pathologies affecting moral sensitivity and awareness, pure mistake, or strategically minded self-deception—then, for that person, the moral matter will not exist. To put it otherwise, the existence of a functional sense of moral awareness, defined here as a capacity to identify and recognize the moral dimensions of an otherwise multifaceted situation and to assign importance to it, is a condition used to identify moral matters and eventually to engage in ethical deliberation. A person lacking moral awareness may be caught up later by the neglect of the morally problematic nature of an experience because their lack of awareness may cause others to be offended by their behavior and disturbed by their callousness. It can also be the case that ignoring the existence of moral matters generates significant moral distress for the person ignoring the matter, as well as others impacted by this neglect (Abbasi et al. 2014; Rathert, May, and Chung 2016). In other words, these kinds of matters are ignored at one's peril. Increasing the ability to identify and discuss these matters can thus reduce moral distress (Abbasi et al. 2014). Having a good mechanism of moral awareness ensures adaptation to new situations and eventually to a number of connected and related situations that call for more profound change in society or in oneself. As John Kekes writes, "Moral sensitivity is needed not to change oneself, but to remain open to the possibility of doing so, if it is warranted" (Kekes 1984, p. 14). In other words, moral awareness is necessary for growth, notably because "heightened moral sensitivity depends on the discovery of new moral possibilities" (Kekes 1984, p. 14).

If a person is truly lacking moral awareness, as psychopathy and severe forms of autism may cause (Bollard 2013), this can generate significant issues in the different contexts (e.g., education, work, family) where smooth social interactions are expected and some ability to recognize moral dimensions of situations is presumed. As Kekes writes, "The proper characterization of moral conflicts requires understanding the kind of tensions that exist in sensitive moral agents" (Kekes 1984, p. 4), and an understanding of these tensions requires due attention to the capacity that allows us to perceive them. Accordingly, moral awareness and the recognition of moral situations can be deemed a precondition for moral life upon which ethics, as a deliberate effort to resolve moral matters, is scaffolded. As Ann Tenbrunsel and David Messick write, "As with most embedded problems, the first step—recognizing and accepting the problem—is often the most difficult" (Tenbrunsel and Messick 2004, p. 235).

In this chapter, I explain the nature of moral awareness. Then, I discuss the important role of moral awareness in ethics theories. Finally, building on the distinction between morality and ethics, I explain how ethical awareness is a form of enriched moral awareness. The next chapter is dedicated to reviewing barriers to the recognition of moral matters.

The Nature of Moral Awareness

Moral awareness (or moral sensitivity) is defined variously as, for example, a caring response, a skill in identifying ethical dimensions of care (in healthcare professions), an ability to consider others' well-being, and a component of ethical care (Weaver 2007). In the literature where morality and ethics are not systematically distinguished as they are in this book, the concepts of moral/ethical sensitivity, moral/ethical sensibility, ethical intuition, moral awareness, and moral perception are often used in ways that correspond to the concept of moral awareness (Weaver 2007).[1] Many of these concepts and their theoretical accounts share the assumption that moral awareness is a first step or process for the subsequent engagement to deliberate on a moral situation and propose solutions to it (Butterfield, Trevino, and Weaver 2000; Rest 1982, 1986; Reynolds and Miller 2015; Weaver 2007). Scott Reynolds draws on the work of Kohlberg to express the thought, which aligns with pragmatism's description of the projective nature of ideas, that "moral awareness represents more than just the recognition of the moral matter, it also portends action, which necessitates a process of considering and reflecting on decision alternatives" (Reynolds 2008, p. 1036). Indeed, within the current theory, moral awareness is the springboard from which to refine the understanding of a moral matter and engage in deliberation to envision different courses of action. Within the initial grasp of the moral matter, the awareness of the matter, there is an implicit understanding and description of what the matter is.

Moral awareness is connected to other aspects or processes of (moral) psychology and behavior (Rest 1986). In fact, moral awareness can influence moral behavior (e.g., efforts to raise implicit moral awareness can increase honesty or reduce immoral behavior (e.g., cheating); higher moral awareness is inversely related to bullying and directly related to defender behavior) (Reynolds and Miller 2015). As I will explain, moral awareness is the beginning and certainly not the end of ethics inquiries, since our more spontaneous understandings of moral matters can be examined critically and refined. I leave aside, for now, the issue of what the domain of morality is, which I addressed at some length in earlier chapters (e.g., chapter 5). I come back to it when I discuss the nature of ethics deliberation (chapter 10). For now, it will suffice to keep in mind that the domain of morality is rather loosely defined and that mechanisms of moral awareness are shaped by customary identification and exposure to what is considered to be right or wrong, while the domain of ethics, including the process of ethical awareness, is an attempt to critically guide moral awareness toward the recognition of matters that have real stakes for human flourishing. These "real" stakes are those that are further validated beyond the more immediate recognition of the existence of a morally problematic situation. Further validation can mean many things, like ascertaining that the matter is not based on a misunderstanding, gathering evidence that the problematic situation is indeed real and problematic through the accounts and testimonies of other people concerned, and ensuring the problematic nature of the situation is based on open discussion with those involved.

One caveat about the distinctiveness of moral awareness with respect to other related aspects of human psychology and behavior, such as general social thinking and human relationships, needs to be articulated. The contours of moral awareness and sensitivity surely overlap with other constructs and capacities found under the banners of empathy, prosocial thinking, and so on. Furthermore, it may not be clear that what is commonly described as moral awareness and moral sensitivity stand separate from other cognitive systems or sets of behaviors from the standpoint of science (e.g., neuroscience, social psychology) or of everyday behavior (e.g., where lack of moral awareness may be loosely described as impoliteness, rudeness, untowardness, or inconsiderate behavior). Again, I will not take a firm stance on this issue, given the complexity of settling this matter. Furthermore, there have been numerous more or less fruitful attempts to find in the biological origins and psychological mechanisms of moral awareness (e.g., empathy), strong lessons for how we should use this capacity or define it in ways that map to a distinct psychological or social function or mechanism (e.g., altruism and prosocial behavior). A more productive endeavor is to identify common shortcomings of moral awareness and spell them out as fallacies of perception and recognition, much like vision can be investigated for its failures because of the shortcomings of the perceptive mechanism itself or its susceptibility to being tricked by a wide range of perceptual illusions. Doing this is a process of building an ethical awareness, that is, a cultivated and deliberate form of moral awareness following the distinction I introduced between morality and ethics (see introduction and chapter 1). I recognize that there is continuity between both phenomena. Yet, it is crucial to identify strategies with which human morality can be supported, enhanced, and augmented through individual effort (e.g., cultivating one's understanding of other human beings and their cultures) or social intelligence (e.g., learning from others and using cues from others to recognize the existence of a morally problematic situation), as well as reaching an understanding of how this moral awareness can be dampened or diminished.

Moral or ethical sensitivity is often used synonymously with moral awareness (Miller, Rodgers, and Bingham 2014), although "sensitivity" is used frequently used in the healthcare setting and may stress the receptive or reactive component of the capacity and less its prospective function of being on the lookout, expressed by the notion of "awareness" (Weaver 2007). "Moral or ethical sensibility" and "moral or ethical perception" are used to capture much of the same reality as moral awareness (Weaver 2007) and to designate the ability to "see" a morally problematic situation (Weaver 2007), but neither captures the attentiveness implied in "awareness" or the less cognitive and more affective aspects of moral awareness. "Moral disengagement" sometimes designates the opposite of moral awareness (Miller, Rodgers, and Bingham 2014), though I would prefer the more neutral concept of moral unawareness since "disengagement" could suggest an intention to be disengaged (which is sometimes true to the situation) while, in this chapter and book, the concept of moral awareness (and unawareness) also designates the nonconscious processes that can create blind spots in our moral visions (Weaver 2007).[2] Sometimes, authors (such as Hartman et al. [2020]) distinguish *moral* awareness (as a more spontaneous and intuitive capacity to recognize) and *ethical* awareness (as a more reflective and cognitively demanding task) as I do, but often the terms "moral" and "ethical" are used interchangeably (Weaver 2007).

The Important Role of Moral Awareness in Past and Present Ethics Theories

The description of the functionality of the biological, psychological, and social mechanisms of moral awareness is part of attempts to understand human morality—but it is the task of ethics to use this knowledge to repurpose and orient (when needed) this functionality toward human flourishing.[3] In other words, the functional description of what the biological and psychological equipment underlying moral awareness (e.g., emotional reactions related to moral awareness) should do and when and how it should be put to use is guided by human (ethical) aspirations. Understanding how these aspirations are possible or not, given our biological and psychosocial condition is one thing, but possibilities are evaluated in light of our aspirations and are not only constrained by biological and psychosocial realities. In the next chapter, I address barriers to this aspiration, but it is important to understand the functionality of moral awareness, which is defined by ethics theories.

Many celebrated moral philosophers acknowledge, in some form, the need to be sensitive to the existence of moral matters, but Hume is perhaps one of the marking philosophers—alongside Adam Smith, the philosopher and father of economics, and Francis Hutcheson, a founding figure of Scottish enlightenment—who elaborate on this important dimension of moral life. Hume goes to great lengths to argue against intellectualistic and rationalistic ideas about what allows individuals to identify a matter as a moral one, and argues that this capacity is part of human nature and human experience. He also emphasizes that the awareness of the existence of a moral matter is not simply a matter of fact (although it relates to facts and experiences) but requires some interpretation and experience in order to be reliable. He wrote, for example, that vice and virtue are "perceptions in the mind," much like sounds, colors, and temperature:

> Take any action allow'd to be vicious: Willful murder, for instance. Examine it in all lights, and see if you can find that matter of fact, or real existence, which you call vice. In which-ever way you take it, you find only certain passions, motives, volitions and thoughts. There is no other matter of fact in the case. The vice entirely escapes you, as long as you consider the object. You never can find it, till you turn your reflection into your own breast and find a sentiment of disapprobation, which arises in you, toward the action. Here is a matter of fact; but 'tis the object of feeling, not of reason. It lies in yourself, not in the object. So that when you pronounce any action of character to be vicious you mean nothing, but that from the constitution of your nature you have a feeling or sentiment of blame from the contemplation of it. (Hume 1739/1975, pp. 468–469)

The fact that the recognition of a matter as moral is contingent on one's moral experience and moral awareness as explained by Hume does not mean that the issue does not really exist, at least for others. Hume points out that "nothing can be more real, or concern us more, than our own sentiments of pleasure and uneasiness; and if these be favourable to virtue, and unfavourable to vice, no more can be requisite to the regulation of our conduct and behaviour" (Hume 1739/1975, p. 469). He describes the awareness of a moral matter as a "perception" but also goes as far as arguing that moral judgments are more analogous to feelings than ideas, a rather extreme point of view. Hume also tends to reduce the ability to distinguish between different vices and virtues, to "*particular* pains or pleasures" (Hume 1739/1975, p. 471). Again, this seems an unnecessary reduction of moral valuations to purely subjective and personal impressions that can be difficult to be reasoned, refined, or discussed. Therefore, Hume's suggestion that initial impressions about moral matters must be defined in relationship to the feelings induced by situations (or characters in his writings) is important (Hume 1739/1975), but it should not prevent us from trying to understand patterns and factors or causes that impede the functioning and growth of this capacity, since initial moral experience forms the terrain on which ethical conversations will be set.

In contemporary scholarship, the moral psychologist James Rest has perhaps emphasized more than any other figure the crucial role of moral awareness (Rest 1982, 1986). Indeed, Rest's famous four-component model of morality is built on the theory that there are four key components of human morality, one being moral sensitivity, delineated as the ability to identify and discern morally problematic situations.[4] Rest defines moral sensibility (moral awareness) as an interpretive process in which the individual recognizes the existence of a moral matter or that a particular ethical principle or value is at stake. This definition is useful because it stresses the mediation happening in the experience of the situation. In fact, James Rest describes this first step very appropriately as "interpreting the situation" (Rest 1982) and defines it as "the perception that something one might do or is doing can affect the welfare of someone else either directly or indirectly (by violating a general practice or commonly held social standard)" (Rest 1982, p. 29). He states how it is rather easy to be unaware of the moral aspects of our decisions and behaviors for different reasons, such as not knowing the full implications of one's behavior or choices, and the difficulty of predicting the outcomes of complex actions (e.g., a surgery) (Rest 1982).

To Rest's definition, Scott J. Reynolds, an important contemporary figure of moral psychology, adds that moral awareness can be defined as an individual's determination that a given situation has moral content and is fitting to be considered from a moral point of view (Reynolds 2006). In addition, Kekes specifies that "one crucial moral task is to perceive accurately the situation in which one is called upon to act. And accurate perception depends on the breadth and depth of one's appreciation of the available moral idioms: they are the dimensions of moral sensitivity" (Kekes 1984, p. 10). Kekes makes a point of recognizing that moral sensitivity is fundamental to moral life even though much focus in contemporary ethics is on action and decisions leading to actions. Indeed, it is important to bear in mind that in moral matters, the issue at stake is not necessarily the action or the choice itself but also the thinking underlying the action or choice, and the effort to "become aware of the considerations inclining one in different directions and in trying to sort out the elements composing one's mixed motives" toward this choice (Kekes 1984, p. 5). Kekes also describes moral sensitivity as an aliveness to moral life (Kekes 1984). In this sense, moral sensitivity and awareness is a condition for the adoption of a "moral point of view" (Baier 1958).

Interestingly, much research on moral awareness comes from contemporary psychology and nursing research. In these fields, the idea that moral matters are real-world matters, but whose existence is contingent on processes of interpretation and positionality within experience, is blatantly clear. Scholars of the mind and of behavior have quickly realized the importance of the psychosocial processes (e.g., cognitive, perceptive, emotional/affective, relational, and social) involved in the actual identification and recognition of matters as moral matters. Nurses and scholars of feminist ethics identified early on the importance of moral awareness in ethics, notably the lack of attention paid to a number of important moral matters in which they were involved in healthcare (Robertson 1996; Walker et al. 1991). They may have been more cognizant of social and psychological processes at stake in moral awareness because of the impact of hierarchical positions with respect to physicians (male-dominated, especially at the beginning of these writings) and related gender dynamics, as well as the work and relationships of nurses with patients.

Ethical Awareness as Enriched Moral Awareness

Keeping in mind the distinction introduced between morality and ethics in chapter 1, I would like to explain how moral awareness differs from ethical awareness, which is essentially an explanation of why moral awareness matters for ethics theory and the kind of role that it plays within the broader ecosystem of a pragmatist ethics theory like the theory of deliberative wisdom. This may sound trivial, but it also helps situate the contribution of the scientific study of morality (e.g., the scientific study of mechanisms of moral awareness in the form of empathy and others forms) to ethics theory and how further understanding mechanisms of human morality helps refine the awareness of morally problematic situations.

Moral awareness designates the everyday, implicit process by which we recognize the existence (or fail to recognize the existence) of moral matters.

It is a complex assemblage of skills and abilities (e.g., mind reading, theory of mind, empathy). Following the distinction I draw between morality and ethics (see introduction and chapter 1), moral awareness could be assimilated to a system 1 process, while ethical awareness could be designated as a system 2 process (Kahneman 2011) whereby processes of moral awareness are, for example, questioned, problematized, scrutinized, and deliberated on (Kahneman 2011). Thus, ethical awareness is a form of enriched moral awareness because it is a form of moral awareness guided by reflection on its impact on the well-being and flourishing of others. For instance, showing empathy and being concerned with how someone is suffering in a situation (e.g., how immunosuppressed patients are affected by the end of public health measures against COVID-19) occurs rather naturally once the stakes are minimally understood. However, determining what role this empathy plays in broader thinking about how to best accommodate the interests and positions of those who are at risk and anxious in the face of the end of restrictions is not necessarily something that can be done by relying on our quotidian habits and reflexes. For example, this pandemic has brought significant politicization of public health measures such that it makes open discussion about the impact of our valued practices more difficult. Many, such as the majority of people with well-functioning immune systems, may not be aware of the implications of abandoning the wearing of masks in public spaces for those who have compromised immune systems. A portion of the public may also feel antagonized by people wearing masks such that they become rude and unpleasant even toward immunocompromised people wearing masks. There will be a need to cultivate our moral awareness by understanding how we can guide our moral sensitivities with the interests of others (i.e., interests that may differ from ours) kept in mind.

Ethical awareness can be envisioned as a form of enhanced and intelligent moral awareness, that is, a capacity to be cultivated and that plays a key role in a deliberate problem-solving effort with human flourishing in sight. By becoming aware of the factors that limit our moral awareness and sensitivity, we can cultivate a richer initial understanding of moral situations and of those who are front-and-center in those situations. Research has shown that individuals can experience moral development and that enhanced moral awareness can be an integral part of this development (Killen and Smetana 2013; Rest 1986). As Kekes writes, keeping in mind here that what he describes as "the function of reflection in moral sensitivity" is similar to ethical awareness, "the function of reflection in moral sensitivity is to give breadth and depth to the employment of moral idioms, to see both that out moral idioms are the conventional products of the social context we happen to live in, and that underlying the various conventions there is an abiding concern with benefit and harm and with living a good life" (Kekes 1984, p. 15). Although Kekes may overstress the value of existing customs and conventions, the point is well taken that moral awareness is a process of acquiring moral insight relevant to actual behavior and social practices.

Despite stressing the importance of scrutinizing critically habits of moral awareness, I want to acknowledge that this can also lead to an intellectual bias whereby a more sophisticated or intellectualized form of ethical awareness—one that is described in abstract concepts and loaded with theory—may seem to offer a better tool for increasing moral awareness. However, it is not hard to understand that those who are ostensibly experts in discussing moral situations (e.g., ethicists) may not necessarily be advantageously equipped with a broader or richer moral awareness, that is, a functional moral awareness that works in everyday life and is an instrument to human flourishing. In fact, an expert understanding of the initial origins and triggers of moral situations could possibly impede—although this would warrant dedicated investigation—important dimensions of moral awareness if it becomes overly intellectualized and rigid (see for example the work of Arthur Caplan, Walker, Barry Hoffmaster, and Steven Fesmire on the rationalistic tendencies of ethics and the problems this has generated [Caplan 1980; Fesmire 2003; Hoffmaster and Hooker 2009; Walker 1993]). All of this is to say that strategies to reflect on and increase moral awareness should be examined critically. For example, the temptation to increase a certain manifestation of moral awareness, such as empathy, by undermining other forms of more cognitive moral awareness through the teaching of ethical reasoning skills (e.g., as in the behavioral ethics movement) could be unfounded because cognition plays an important role in behavior (Maxwell and Racine 2010). Moreover, the ability to recognize and be sensitive to moral matters does not guarantee a due process of sorting out the meaning of that initial experience in light of its impact on others to ensure, for example, justice and fairness (Decety and Cowell 2015).

This and the preceding chapters on morally problematic situations should make clear that ethics, as a deliberate moral matter problem-solving endeavor, is ineffective if a basic form of human awareness and human morality does not exist and, to some extent, is not shared to the point that moral matters are recognized as such by human beings and collectivities. Accordingly, research on the conditions and barriers of moral awareness is of great interest. Over the years, different strategies have been deployed to enhance moral awareness, notably in different professional contexts (Weaver 2007), and some of these are reviewed in the next chapter. There are also lessons that have been learned through the development of ethics pedagogy. For example, in clinical settings, teaching about ethical awareness by using case studies and vignettes, which are already formatted to trigger basic moral awareness, may not yield robust strategies to raise moral awareness because people are primed to see these matters as moral matters (Weaver 2007). Newer teaching methods that increase ethical awareness through media—for example, the use of more narratively rich materials such as literature and films—are promising because they expose learners to more complex and less formatted situations; however, they may need to be evaluated empirically (Kiosses, Karathanos, and Tatsioni 2016; Weaver 2007). In addition, the efficacy of the application of knowledge generated by such efforts could be counter-checked by broader factors that structure, for example, the economic valuation of behaviors, workplace cultures, practices in educational systems, and behaviors in healthcare institutions. Research on ethical awareness needs to integrate the cultural and social factors that increase or decrease sensitivity to certain moral matters or certain types of moral matters, and not only investigate the individual and their moral awareness. Although the recognition of a moral matter is only the start for ethics, it is nonetheless (as many have already stated) a crucial step.

Conclusion

Recognizing a moral matter as such is an incontrovertible first step for ethics. Ethics—and its orientation to human flourishing—explains why it is important to first recognize moral matters, notably because unrecognized moral matters will generate, sooner or later, other troubles such as moral distress or social unrest. In the end, these problems can only be avoided or neglected at one's own expense, but maybe even more importantly, to the detriment of others. There is also an important issue to the claim of ethics researchers to be the sole authority able to articulate moral matters through their research. Such dogmatic authoritarianism has no justification, but more importantly, as explained, ethics expertise does not necessarily impart robust moral awareness that is useful in everyday life. It is vital that open-mindedness be cultivated to ensure that blind spots in ethics research do not perpetuate further blind spots in our moral world (e.g., an issue described by a non-expert as a side effect of ethical research being dismissed by professional ethicists not willing to recognize this issue). With this in mind, the next chapter describes common barriers to moral awareness.

9 Barriers to Moral Awareness and Gateways to Ethical Awareness

Recognizing the importance of moral awareness, its role in moral life, and its refinement with the help of ethics theory, is easier said than done. Several forces play against the recognition of moral matters, because these matters are vexing and disruptive to our identity and psychological integrity. They bring us to reflect on ourselves as ethical agents, what we should be, and what we should desire to project through our actions and decisions. However, ethical progress relies on progress in refining this recognition and cultivating moral awareness, that is, ethical awareness that is informed by observation and reflection on the limitations and blind spots of our moral awareness in light of human flourishing. A life of flourishing is a life where we are able to articulate what it is that we do that contributes to our own well-being and development as well as the welfare of others. To do this, we need to be attuned to what the welfare of others is. In other words, we must be sensitive to what shakes others and what raises deep and disturbing issues regarding what matters to them. This does not mean that we need to become hypersensitive individuals and societies or that our only guide should be the avoidance of disruption to self-narratives. A life of flourishing requires open-mindedness and the acceptance of truth and change. It is a life that accepts that one's view of the world and of oneself is potentially inaccurate and in need of adjustment. Thus, moral awareness can err, and lead to the identification of false negatives. In other words, the guiding light should not only be our sense of being hurt or comforted; we should be guided by the higher pursuit of our development, and that of others, as ethical agents and flourishing persons. In this chapter, I explain common barriers to moral awareness and group them into affective, cognitive, relational, and social barriers. Each group is described in general terms, and some ethical gateways, that is, strategies to tackle these limitations, are described. The spirit guiding this chapter is that these barriers are not so much things we need to fight against, but rather things we must acknowledge and think about critically in order to grow as human beings. There is no such thing as the perfect moral agent; there are only moral agents willing to learn, that is, ethical agents. Learning is as close to cultivating excellence and perfecting themselves as human beings can come to.

> Perfection is not attainable, but if we chase perfection we can catch excellence.
> —Vince Lombardi[1]

The literature on moral awareness and other constructs that shed light on the capacity of recognizing a moral matter as such, is extensive. This chapter can only serve as a sketch of the barriers to such recognition and awareness. One guiding assumption of this chapter is that this "capacity" or "ability" is multifaceted and likely relies on a set of distinct behavioral and cognitive processes (some emotional and some more cognitive; some individual and some more social) of which we only have at this time a partial understanding (Rest 1982). Given that the focus of this book is the use of this capacity within an ethical framework and not a treatise on the basic underlying psychological processes, I focus on the actual barriers to moral awareness instead of trying to unpack their mechanisms. I recognize that these barriers, and overcoming them, may imply tacit assumptions about the workings of moral awareness and the efficacy of certain remediation and enrichment strategies, which I also review in this chapter as *gateways* to ethical awareness. Without claiming some clear delineation of or between these barriers, or making any claim about them constituting distinct phenomena or processes, a number of sets of barriers to moral awareness can be identified that partially map onto different dimensions of the concept identified in the literature (Miller, Rodgers, and Bingham 2014; Weaver 2007):

- ***Affective* barriers to moral awareness:** Obstacles created by the *inability to feel the existence* of moral tensions and key aspects of moral situations. These barriers are well evidenced in literature about moral callousness, as well as emotional empathy.
- ***Cognitive* barriers to moral awareness:** Obstacles created by the *inability to apprehend moral situations or certain aspects of moral situations* because of cognitive processes such as implicit associations or because moral content is only narrowly understood as relating, for example, to principle(s) or virtue. Literature on ethical predispositions, implicit biases in morally relevant judgments, and other cognitive limitations in apprehending moral situations documents the existence of such cognitive barriers.
- ***Relational (interpersonal)* barriers to moral awareness:** Obstacles created by the *inability to see the other person as a person*, take into account their interests, and relate to others as human beings. Literature on perspective-taking, altruism, and selfishness illustrates how different factors create obstacles to relating to other people and the impact of this lack of recognition.

- ***Social* barriers to moral awareness:** Obstacles created by *social and contextual factors that impede the ability to recognize the interests of others* related to moral awareness and the recognition of genuine moral situations. Literature on the impact of social factors (e.g., socioeconomic status) and how they can lead to neglect, avoidance, social distance, and so on, is relevant here.

Importantly, the sets of barriers reviewed in this chapter (affective, cognitive, relational, social) are not unchanging and natural kinds. Indeed, the shape that these barriers take varies across time and space. Moreover, even if I use "barrier" terminology, we should be prudent and thoughtful about what we consider to be our aims for action and the real meaning of these barriers (e.g., barrier for whom, under which circumstances). Barriers are not necessarily overcome by simply opening gates mechanically. Becoming more deeply aware of moral matters often implies a significant amount of personal reflection and ethical maturation, a process that is not always easy, comfortable, or pleasant, not to say anything of the social and institutional aspects at stake.

The following pages describe each of these sets of barriers and give examples of how shortcomings of moral awareness generate important issues.[2] Additionally, gateways—that is, strategies to address barriers and cultivate an ethical form of moral awareness—are presented. Readers should keep in mind that the construct of moral awareness is subject to different views and definitions (see previous chapter) and that the evidence base prevents the proposal of a comprehensive theory (Miller, Rodgers, and Bingham 2014; Reynolds 2006; Weaver 2007). There could also be arguments to the effect that a simple theoretical account of moral awareness could not be possible, given the multifaceted nature of the construct, with affective and cognitive dimensions that do not share a common social or psychological basis or mechanism (see previous chapter). Indeed, recognizing or being aware of moral matters may constitute a holistic process that is not necessarily broken up into discrete steps, but that makes sense within the reconstructive effort of ethics. However, it has the merit of mobilizing forces toward sought aims and prepares us as ethical agents to be motivated to act upon the awareness of the situation. In other words, as users of our own morality, ethics helps us adopt the stance that the sought outcomes and the debates about the nature of moral awareness need to be envisioned within the broader context of human flourishing.

Before going further, I would like to make a methodological comment about the barriers I have in mind. There are some serious (pathological) impairments to different aspects of moral awareness such as those resulting from severe forms of autism (e.g., affecting perspective-taking and theory of mind) (Baron-Cohen, Leslie, and Frith 1985) or psychopathy (e.g., including emotional callousness) (Blair 1995). Although it is sometimes estimated that understanding normal function (e.g., moral attitudes and behaviors) can be gleaned from examining dysfunctions, with which I do not disagree, I want to focus on the refinement and nurturing of moral awareness rather than reflecting on more serious situations where severe conditions could be at play, deeply affecting moral awareness abilities. That being said, I would not necessarily preclude that those with severe shortcomings in moral awareness cannot overcome some of their difficulties through learning, counseling, and targeted interventions to improve moral awareness. However, given the gaps in knowledge to establish this firmly and the attention that would be needed to carefully review the clinical literature, which has a much narrower focus, I feel justified in moving forward without presuming that clinical and non-clinical barriers to moral awareness should be closely associated at this time. Importantly, the fact that someone would entirely lack moral awareness for clinical reasons would not lead, in my eyes, to a judgment about the value of that person not being part of the community of humans, since we have obligations toward those who have difficulties, including difficulties related to moral agency such as moral awareness (Dubljević and Racine 2013).

Affective Barriers to Moral Awareness

One set of obstacles to moral awareness is created by the inability to feel the existence of moral tensions or be attentive to the affective cues that would indicate uneasiness, anxiety, discomfort, frustration, and the like in oneself or in others (Rest 1982; Weaver and Morse 2006). There are many ways to describe this phenomenon (e.g., callousness) (Lethbridge et al. 2017), but the inability to recognize a morally important matter in this affective way is commonly described as a "lack of empathy." There is literature, both conceptual and empirical, suggesting that empathy is associated with increased moral awareness or the capacity for moral awareness (Hughes and Dunn 2000; Simmons 2014). Although the mechanisms of how this works are debated, a role of emotions and empathy in allowing moral agents to be aware of the existence of morally problematic situations is often recognized (Charon 1993; Halpern 1993; MacLean 1967).[3] Even the more rationalistic ethicists—often considered to have cognitivist bents because of their emphasis on the importance of ethical principles—recognize the role of emotions and empathy in ethics. For example, the authors of the *Principles of Biomedical Ethics*, in an approach often unfairly considered to be entirely deductive and rationalistic, write that "In addition to expressing their feelings in their responses, agents also need to attend to the feelings of persons toward whom they act. Insight into the needs of others and considerate attentiveness to their circumstances often come from the emotions more than reason" (Beauchamp and Childress 2001, p. 373).

Emotional Empathy

Empathy has been described in many different ways to designate the ability to feel what it is like to be in a certain situation (sometimes described as *emotional* empathy) or to imagine what it is like (sometimes described as *cognitive* empathy, which reflects cognitive aspects of moral awareness in the typology I use) to be in a certain situation. Nancy Eisenberg and Janet Strayer write, "Empathy denotes the vicarious experiencing of an emotion that is congruent with, but not necessarily identical to the emotion of another individual" (Eisenberg and Strayer 1987, p. 146). They also note that "Social perspective taking entails cognitive understanding of the feelings and motives of others and, as such, is an instrumental skill. Empathy, on the other hand, entails emotional responsiveness to the feelings experienced by others and, as such, is an expressive experience" (Eisenberg and Strayer 1987, p. 246).

There is extensive literature on barriers to empathy and empathizing, which suggests that the origins of these barriers are manifold (e.g., acculturation, desensitization, emotional dampening, fatigue). For example, the notion of a "hidden curriculum" designates the habitual ideas, emotions, and behaviors that are not openly intended to be taught but are nonetheless conveyed or enforced more or less implicitly in the training of physicians and other healthcare professionals. This hidden curriculum could be at the basis of a notable decline in empathic function during their training (Hojat et al. 2009; Ward et al. 2012). This phenomenon could be caused by multiple factors, such as the encouragement of emotional detachment during training, lack of empathic role models, and time pressures (Martineau, Decety, and Racine 2020). Other affective factors leading to moral unawareness (reviewed in Weaver and Morse [2006]; see table 9 in Weaver and Morse's paper) are uncaring attitudes, insensitivity to harming others, desensitization caused by one's own suffering, and blindness toward the vulnerability of others. It is important to note, however, that empathy can be important in feeling morally problematic situations but, in and of itself, may be insufficient or even harmful if not redirected toward a constructive goal (Bloom 2016; Decety and Cowell 2014). Indeed, being overly sensitive can increase anxiety and the likelihood of professional challenges, including burnouts (Fenton 1988; Grant 2002; Smilansky 1996; Weaver 2007). Also, anxiety sometimes impedes action or proactivity in resolving or addressing a situation through phenomena such as avoidance (Flett, Blankstein, and Martin 1995). Furthermore, empathy can actually impede a more comprehensive awareness of the stakes at play because, for example, one can empathize more easily with those resembling oneself or those with whom one has relationships (Cialdini et al. 1997; Decety and Cowell 2014, 2015), leading to biases against those more distant from oneself. Thus, empathy is likely only one of the components of the affective aspects of moral awareness and of human morality more generally (Batson et al. 1995) since empathy may rely on more basic mechanisms of emotional perception.

Empathy and Emotional Perception

Although empathy is almost equated, by definition, with the recognition of the affective cues relevant to defining a morally problematic situation, it falls short of this equivalence. Indeed, the entire gamut of emotions can signal that there is a morally problematic situation since different situations can lead to entirely different reactions: for example, the anger experienced in being or seeing others in an unfair situation, the sadness of witnessing unresolved tensions of interests and values, or the shame experienced by not holding to one's own ethical standards. These experiences are all emotions leading to the awareness of a moral problem, which are shaped by the affective components they feature, and are likely not so well captured by typical understandings of empathy. While empathy could potentially be considered an ability that allows someone to understand emotions in others, it is most often conceptualized as an ability to understand another's perception of negative emotions and suffering. However, emotional empathy, perhaps understood in this extended sense, best approximates the affective aspects of moral awareness and, therefore, barriers to emotional empathy can be instructive.

This being said, emotional empathy implies other kinds of abilities, perhaps more basic ones, such as the ability to perceive emotions and recognize them. These abilities may seem rudimentary but should nevertheless be kept in mind as potential barriers, notably when some in the situation are less perceptually sensitive than others (or have difficulties processing information gleaned from the expression of emotions) or when cultural diversity could mean that expressions of emotion are interpreted differently. A number of conditions such as anxiety, depression, schizophrenia, autism spectrum disorders, and posttraumatic stress disorder have been associated with differences in emotional perception or the processing of emotional information (e.g., Rowland et al. 2012; Simcock et al. 2020). Thus, the inability of some individuals to perceive emotions should not be understood as a malicious lack of empathy, where the motivation of the agent could be at stake, but importantly as a more basic challenge in gleaning and processing emotional information. This basic process is analogous with the role played by "moral attentiveness"—which includes a reflective and perceptual counterpart—to capture the mere perception of moral situations (Reynolds 2008).

Gateways to Affective Aspects of Ethical Awareness

An array of strategies has been proposed to address moral unawareness based on affective and emotional factors (Martineau, Decety, and Racine 2020). In the clinical context, there are a great number of proposals to increase empathy in healthcare professionals (Kiosses, Karathanos, and Tatsioni 2016), and this surely represents the context in which most attention has been attributed to increasing empathy. Studies conducted include experiential interventions involving active engagement, such as role playing and group discussions (e.g., practical exercises using role play, communication skills training using workshops, empathy training using discussions and reading), as well as non-experiential interventions involving less active engagement (e.g., training using videos featuring vignettes and testimonials, training to facilitate responses to concerns of patients using a lecture and training modules) (Kiosses, Karathanos, and Tatsioni 2016). Importantly, the reasons why empathy is targeted in clinical contexts varies and are not confined to ethical concerns, although they are often very closely related to them. For example, empathy can be promoted as an effective tool to promote good clinical communication, but good communication skills are not always used ethically.

A review of randomized clinical trial studies found that a variety of strategies could be effective to promote empathy, although there remains a number of important methodological concerns such as the duration of the effects induced by the interventions (e.g., after more than a year) and the comparative effectiveness of different strategies. These differences are most notable in generating the desired effects since empathy is often a mediator for other outcomes such as increased quality of patient care and patient satisfaction or the fostering of a healthier work environment (Kiosses, Karathanos, and Tatsioni 2016). Some have proposed teaching about the neuroscientific underpinnings of emotion and empathy as a kind of educational intervention. Riess et al. (2012) investigated a set of three one-hour training modules on the neurobiology of emotions and empathy. The intervention resulted in improved empathy of physicians, as evaluated by patients, and improved ability of physicians to decode emotions contained in subtle facial expressions.

In business ethics, Alice Gaudine and Linda Thorne (2001) suggested that organizations should encourage their employees to recognize their emotions at work, what they signal, and their possible influence on their decision-making process. Likewise, Rommel Salvador and Robert Folger (2009), as well as Jennifer Mencl and Douglas May (2009), recommended using training programs or educational strategies to support managers in recognizing different types of information, including emotions, as part of ethical decision-making processes. Another type of strategy can consist of encouraging social proximity and relatedness between colleagues. For example, managers can be encouraged to get to know their employees personally in order to encourage empathic responses toward them (Mencl and May 2009). Additionally, some suggest that business school education, management programs, and executive coaching can and should be used to promote empathy in leaders (Holt and Marques 2012; Mahsud, Yukl, and Prussia 2010). This kind of initiative, as well as initiatives targeting business students, could counter the "cold blooded" culture in businesses. Another study proposed that business schools foster empathy in their curriculum, not in a theoretical way, but by providing students with multiple opportunities to develop ethical traits throughout their entire pedagogical experience (Brown, Kitson, and McKnight 2010).

In sum, affective barriers to moral awareness are numerous and cover the spectrum of causes that could lead someone to be emotionally insensitive to morally problematic situations. Lack of empathy is one way to capture this complex phenomenon, but it is certainly not the only angle to adopt. Others include literature on moral disengagement (Knoll et al. 2016) and ethical fading (Tenbrunsel and Messick 2004). For our purposes, we should remain attentive to how affective dimensions can shape our moral awareness and be self-reflexive and open to dialogue about these matters. We should also be attentive to scientific evidence about the value and limitations of affective sources of awareness since they could also induce bias toward those with whom one empathizes (and thus conflict with principles of justice and fairness) (Decety and Cowell 2015), or induce anxiety in the person who is sensitive to the situation of others. Importantly, the nature of a barrier (e.g., affective, cognitive) does not determine the nature of the intervention needed to bring remediation. For example, in an electroencephalography (EEG) study, Feng Sheng and Shihui Han (2012) showed that the racial bias induced by empathy (i.e., empathic neural responses to perceived pain are stronger to racial in-group members than to racial out-group members) can be decreased by the use of cognitive strategies such as instructing participants to pay attention to and observe others' feelings of pain, which increased neural responses to pain expressions for other-race individuals (Sheng and Han 2012). Thus, perspective-taking, an aspect of cognitive empathy, can be deployed as a strategy to elicit emotional empathy, concern for others, and altruistic motivation (Batson, Early, and Salvarani 1997; Decety and Cowell 2014). Accordingly, the best remediation strategy to make someone morally aware could be cognitive, even if the barrier appears to be affective in nature.

Cognitive Barriers to Moral Awareness

Cognitive barriers are obstacles created by the inability to apprehend moral situations or certain aspects of moral situations. They are caused by various processes involving thinking and consciousness, such as inattention, inability for perspective-taking, and challenges in information processing, notably information derived from ambiguous or complex social cues. Additional psychological factors that can impact moral awareness are difficulties with self-regulation (higher self-regulation is associated with higher moral awareness) and mindfulness, that is, being attentive to one's place in the present (Reynolds and Miller 2015).

Importantly, an unduly narrow conceptualization of the domain of ethics can hinder the recognition of moral situations if a strict definition of ethics is imposed on moral situations. For example, one more recently researched set of cognitive barriers is found in implicit biases (also called implicit associations), moral inattentiveness, and the difficulty in taking the perspective of another person, akin to cognitive empathy (Miller, Rodgers, and Bingham 2014). If ethics does not open up to the impact of such implicit biases, it will further its ongoing neglect of cognitive barriers to the recognition of certain moral matters and the mechanisms at stake.

Implicit Biases

At this time, cognitive and psychological science research has documented a series of cognitive biases or "implicit associations" that could impact how one is aware of and able to recognize a morally problematic situation (Bargh and Chartrand 1999). In this literature, implicit or automatic cognitive processing is described as nonconscious—which is not the same as unconscious—such that information processing is informed by past experience, but the influence of that past experience is unidentified or inaccurately identified (Reynolds, Leavitt, and DeCelles 2010). In this context, implicit[4] associations are just one of the automatic, nonconscious processes engaged when one encounters a morally problematic situation (Miller, Rodgers, and Bingham 2014), and they can be counter to one's explicitly held moral beliefs and attitudes (Haidt 2001), such as holding implicit racist attitudes while being committed to anti-racism. Importantly, beyond the scope of the current chapter, the literature distinguishes between different types of implicit associations such as implicit attitudes, implicit stereotypes, implicit self-esteem, and implicit self-concept (Reynolds, Leavitt, and DeCelles 2010).

Moral Inattentiveness

Another cognitive barrier may lie in a basic lack of attention to moral phenomena. Moral attentiveness, a concept proposed in the literature, refers to the "extent to which an individual chronically perceives and considers morality and moral elements in his or her experiences" (Reynolds 2008, p. 1028). Moral attentiveness is different from moral awareness but could overlap with it in the sense that awareness is situational, triggered by the features of the situation itself (Jones 1991) or "influenced by internal factors such as gender (Ameen et al. 1996), nationality (Blodgett et al. 2001), or education (Sparks and Hunt 1998)" (Miller, Rodgers, and Bingham 2014, 27). Moral attentiveness "pertains to the process by which an individual actively screens and considers stimuli related to morality" (Reynolds [2008, p. 1028] as cited in Miller, Rodgers, and Bingham [2014, p. 27]). Reynolds distinguishes between perceptual moral attentiveness, by which "information is automatically colored as it is encountered," and reflective moral attentiveness, "by which the individual uses morality to reflect on and examine experience" (Reynolds 2008, p. 1028). In the context of this book, I prefer to maintain the holistic understanding of the process of recognition of moral matters (beyond attention-related processes) and therefore use the terminology of moral awareness for reasons that are explained above, as well as that of ethical awareness for the more reflective processes regarding moral awareness. However, moral attentiveness may designate a specific and identifiable trait of individuals such that one's more stable disposition to attend to those situations (or lack thereof) may represent a form of cognitive barrier that needs to be addressed (Reynolds 2008). It could also be an area that is amenable to teaching and training. For example, Reynolds suggests that measures of moral attentiveness could be used (likely alongside other measures) with select employees or to focus training efforts on those least attentive to moral matters (Reynolds 2008).

Ambiguity of Moral Situations

Another cognitive barrier to moral awareness may reside in the ambiguity of moral situations, namely the difficulty of untangling the various issues at play in a given situation. From a cognitive standpoint, some moral situations are clear, but others are ambiguous such that it is hard for those exposed to them to understand exactly what is going on. Rest gives the historical 1964 case of Kitty Genovese, who was raped and murdered in New York City, just outside her apartment building. It was initially thought that thirty-eight fellow apartment dwellers looked on at her attack without responding, apparently because many were unsure of what was happening (e.g., perhaps it was a lovers' quarrel in which they did not want to interfere) (Rest 1982) and not necessarily because they lacked empathy. The ambiguity of such emergency situations can affect the response of bystanders and generate an interpretation of the situation as a morally problematic situation (Rest 1982). Since then, the accuracy of the original reports on this case have been called into question (Manning, Levine, and Collins 2007) but there remains evidence for what has now been called the "bystander effect" in an extensive body of research (Fischer et al. 2011). A similar point is made by Tenbrunsel and

Messick (2004), who advance that errors in perceptual causation are a chief source of ethical fading (see below for a description of this process).

Perspective-Taking

The inability to take the perspective of others, which can sometimes be caused by conditions such as autism (Baron-Cohen et al. 1985), can also be impeded by a number of cognitive factors (as well as affective, interpersonal, and social factors). It is now well recognized that there are various degrees of individual skills in perspective-taking and that a number of factors can influence this ability (Apperly 2012). Perspective-taking is influenced by domain-general factors (e.g., language, executive function) (Conway, Catmur, and Bird 2019); impairments to those abilities can then compromise perspective-taking. However, it is also possible that this ability functions somewhat like facial recognition insofar as different aspects of minds are easy or more challenging to understand by a given person based on that person's own background and psychology. There is evidence that different aspects of the mind (e.g., understanding that others have desires or have beliefs) vary across cultures such that the developmental sequence varies across cultures. Just to give one illustrative example, children from Iran and China acquire the understanding that seeing differs from knowing before they acquire the understanding that people have diverse beliefs, compatible with collectivist cultures, while this order is reversed in Australia and the United States (Shahaeian et al. 2011). These aspects of perspective-taking point to the need to be vigilant regarding the possible impression that lack of awareness of moral matters is necessarily voluntary or due to a general inability to understand other minds. In addition, to domain-general processes and factors, there are domain-specific processes and domain-specific difficulties such that perspective-taking in a particular context will be easier or more challenging depending on one's familiarity with the background context, as well as the relevant aspects of perspective-taking. For example, visual perspective-taking is different from other forms of perspective-taking (Schurz et al. 2015), such as perspective-taking in language, which can itself be divided into different aspects of language and semantics (e.g., spatial and action domains of linguistic perspective-taking) (Beveridge and Pickering 2013). Understanding in greater depth perspective-taking processes helps move away from strawman depictions of this capacity and helps generate strategies to expand abilities of perspective-taking.

Ethical Fading

Ethical fading is another kind of cognitive barrier to the recognition of moral matters because it involves transforming a morally problematic situation into a nonproblematic one to which we no longer need to be morally attentive. The concept of ethical fading is particularly useful to understand how, consciously or unconsciously, morally problematic situations are described as nonmoral situations from a more cognitive standpoint. This concept and related theorizing proposes four mechanisms through which the fading of the moral salience of a situation occurs: (1) language euphemisms (e.g., harm to others becomes "collateral damage"); (2) slippery slope of decision-making (e.g., numbing through repetitive exposure and progression toward banalization of problematic habits and practices); (3) errors in perceptual causation (e.g., not attributing causality to the proper agents); and (4) the constrained representation of self (our view of ourselves is not objectively accurate—it is shaped by the point of view of others and we only really experience our own existence, not that of others) (Tenbrunsel and Messick 2004). On this last point, self-deception can be instrumental to ethical fading (Tenbrunsel and Messick 2004).

Narrow Definitions of the Moral Domain

One's own views on what exactly the domain of morality is can shape one's awareness of situations triggered by aspects that do or do not fit with such views and thoughts. Here, I not only have in mind highly theoretical accounts of what the moral domain is but also implicit views held and nurtured, that is, embedded in practices that themselves embody yet-to-be theorized and articulated meanings of different sorts. In Kekes' terminology, "A person's sensitivity to a moral situation depends, in the first instance, on the moral idioms available to him. They are provided by his language, tradition, and culture" (Kekes 1984, p. 7). Hume and Dewey, two experience-oriented philosophers, clearly alluded to this interaction between one's views of morality and one's ability to recognize moral situations, with the latter directly criticizing unidimensional views of the moral domain because of the narrow views they embody (Dewey 1966).[5] Importantly, recent research has investigated what can shape one's moral awareness and how it can be considered a form of attention, directed by personal (predispositions) or contextual factors (Reynolds 2006). In other words, our moral perspective is influenced by what triggers our moral awareness.

Reynolds has explored this relationship through the Measure of Ethical Viewpoints Scale, a scale commonly used to measure the different "ethical predispositions" of teleological (utilitarianism) and deontological (formalism) moral decision-making (Brady and Wheeler 1996)[6] in samples of business students. Reynolds shows that one's initial moral disposition (or personally stable preferred style of moral decision-making) shapes their sensitivity to different features of a moral situation (Reynolds 2006). Importantly, higher levels of utilitarianism and higher levels of formalism are associated with greater moral awareness. However, those with higher scores for utilitarianism rate the moral salience of a situation higher when it contains harm (e.g., a delay in a shipment will lead to the worsening of the condition of a patient) than when it features a broken norm (e.g., shipping clerk has told the truth or not). Formalism is associated with greater awareness to both harms and rule breaking. This is interpreted by Reynolds as implying that deontological reasoning somehow subsumes both sensitivity to harm and to rule breaking, while this is not true of utilitarianism.[7] In my own collaborative research with colleagues Veljko Dubljević and Sebastian Sattler, we have developed and tested a specifically designed instrument (the Preferences for Precepts Implied in Moral Theories Instrument [Dubljević, Sattler, and Racine 2018a, 2018b]) and shown how moral judgments are actually structured by three types of moral insights (namely about agents, deeds, and consequences) aligned with three major ethics theories (virtue theory, deontology, and consequentialism) (Dubljević, Sattler, and Racine 2018a; Dubljević and Racine 2014). Our research shows how these three factors shape and predict moral judgment (Dubljević, Sattler, and Racine 2018a), notably in relation to how people affiliate with the moral precepts of these three theories (Dubljević, Sattler, and Racine 2018a).

Regardless of the precise nature of the relationship between ethical predispositions as measured by the Measure of Moral View Points Scale (Brady and Wheeler 1996) or the Preferences for Precepts Implied in Moral Theories Instrument (Dubljević, Sattler, and Racine 2018b), studies using these measurement tools point to important relationships between how we envision morality and our moral awareness of moral matters. At the very least, this cognitive phenomenon can be recognized such that the definition of the moral domain in relation to, for example, consequences or principles or human relationships is not in fact unbiased because they may structure but also limit moral awareness and subsequent processes in the broader conceptual and methodological ecology of the theory of deliberative wisdom.

Gateways to Cognitive Aspects of Ethical Awareness

There are allegedly several strategies that can be used to counteract different forms of cognitive barriers to moral awareness. Some concern the individual and involve nurturing greater self-awareness of one's own cognitive limitations and biases; others relate to groups of individuals who need to sensitize themselves to their lack of moral awareness (e.g., teams, families, or communities and societies).

Recognizing the existence of different cognitive barriers related to how we think represents an important step in tackling them. The literature in cognitive science, and in psychology in particular, offers a rich set of insights about how different effects such as framing (the effect of how information is presented, notably positively or negatively on choices, see Tversky and Kahneman [1981]) and implicit attitudes make us aware of certain moral situations but neglect others. For example, a proposal has been made to help judges recognize implicit biases toward racialized minorities (National Center for State Courts [Race & Ethnic Fairness in the Courts] 2012) and tackle these documented biases. However, remediating such biases has been found to be challenging. For example, increasing self-awareness of implicit biases may not necessarily lead to behavioral or attitudinal changes. Given the possibility of becoming aware of implicit associations, notably implicit prejudice and implicit attitudes, some have discussed whether (and to what extent) one can be held accountable for their implicit associations (Faucher 2016; FitzGerald and Hurst 2017; Levy 2017).

A recent review of interventions (FitzGerald et al. 2019) designed to reduce implicit bias concluded that some interventions, such as those promoting the engagement with another person's perspective, may not be so fruitful (Galinsky and Moskowitz 2000). Likewise, being exposed to one's implicit biases can actually have contrary effects and amplify biases (Payne, Lambert, and Jacoby 2002), while being simply aware of one's biases may induce false confidence that one can overcome them (Moskowitz and Li 2011). Likewise, being aware of framing effects does not mean that one does not fall prey to the biases induced by frames of reference. Being exposed to examples that contradict one's bias or prejudice could be a promising strategy, as could examples in which viewers experience high identification with those portrayed in the scenarios (Lai et al. 2014).

Regarding narrow definitions of the moral domain, I have argued before (see also chapter 4) that rich and inclusive understandings of the moral domain should be entertained to capture everyday views about ethics by using strategies where the limitations of one's own views are challenged and questioned (Racine 2008a). More specifically, in the context of elaborating on the concept of everyday ethics and its implications with Natalie Zizzo and Emily Bell (Zizzo, Bell, and Racine 2016), I suggested that different approaches to ethics be operationalized to offer guiding questions in understanding and tackling moral matters in ethics teaching and the facilitation of ethics discussions. For example, generating questions rooted in different ethics theories can help bring to light different dimensions of a situation, as also argued by Mary Mahowald (1994). This strategy can help entertain a comprehensive view of the moral domain. For example, we can ask whether a situation is problematic because there are undesirable outcomes based on consequentialism, or because the situation raises questions about someone as a moral agent (according to virtue ethics), and so on. Fostering a comprehensive view increases openness to what could represent a morally problematic situation beyond the precepts of a single theory, and importantly beyond a single person's understanding of the moral domain, which exists, by its very nature, socially and is socially shaped.

These explicit and elaborate strategies may not be practical in everyday life. They may require too much attention and cognitive resources for someone to actively mobilize the arsenal of strategies to overcome cognitive biases. However, this literature can help explain why someone could be oblivious to another's plight. Accordingly, when there is a signal that something is not going right and there is no agreement on the nature of that situation as morally problematic, the origins of the misunderstanding can perhaps be partly rooted in cognitive barriers. Here again, as for affective barriers, there are many reasons why our way of thinking about moral situations and being aware of them may be reduced because of our thinking habits or our lack of thinking. Importantly, as a word of caution, the real-world effect of implicit cognitive processes may be overestimated in experimental studies in the sense that individuals evolving in real-world environments may be less prone to error than in tightly controlled experimental setups, which tend to be used to exhibit cognitive barriers to moral awareness. For example, James Druckman argues that in real-world environments, there are resources (e.g., access to credible advice) that help individuals correct for framing effects and, therefore, such effects may be overestimated (Druckman 2001).

Relational (Interpersonal) Barriers to Moral Awareness

We can be unaware of moral matters because of the nature of our interactions with others, which are also shaped by affective, cognitive, and social factors. Such relational or interpersonal barriers to moral awareness impact the ability to see the other person as a person, to take their interests as genuine interests, and to relate to them as human beings. Encroachment on the recognition of human beings *qua* human beings, that is, with their own interests, autonomy, dignity, and entitlement to respect, is captured by the Golden Rule ("Do unto others as you would have them do unto you") and Kant's formulations of the categorical imperative.[8] Like I wrote earlier in this chapter, moral (un)awareness is not disconnected or separate from other aspects of human morality and social behavior. As these references suggest, a clear relational threat to moral sensitivity is the lack of recognition of the interests of others, and this threat is a fundamental threat to ethics itself. If other persons and their interests are not considered, the conflicts to which moral situations give rise will not even be seen, felt, perceived, or experienced. In fact, the moral point of view is often synonymous with taking a point of view that moves beyond one's own self-interests (Baier 1958, p. 189). However, some amount of self-interest may be contiguous with moral behavior, because "a person could not have a reason for doing anything whatsoever unless his behavior was designed to promote his own good" (Baier 1958, p. 187). Contemporary research has shed much light on egocentrism, selfishness, and barriers to other-oriented thinking, altruism, and prosocial behavior with useful observations to keep in mind to understand interpersonal barriers to moral awareness.

Egoism

Egoism designates a pronounced tendency to center on one's self and one's own interests beyond what is usually seen as appropriate or justified with respect to a given situation. It thus represents a possible relational barrier to moral awareness because moral awareness requires one to identify and consider the well-being of others involved. Some form of general self-centered and goal-directed behavior is a foundation for individual survival, hence the emphasis on the exaggerated or pronounced nature of this tendency in egoism. More sophisticated accounts of the relationship between egoism or selfishness and morality or ethics can be found in the literature (e.g., such as the account that the only ethical point of view is the egotistical point of view, discussed by Baier [1958]), but for our purpose here, basic descriptions of the phenomenon of egoism are sufficient (Stebbins 1981).

The tendency to focus on one's self and one's interests could occur incipiently such that the person has an egocentric vision of a reality that places the interests of others in the shadows. The American sociologist Charles Cooley, who foreshadowed George Herbert Mead's ideas about the self, considered selfishness as related to problems of inaccurate or inexistent role taking (Stebbins 1981). This is a form of egoism that can perhaps be corrected by greater attention to the interest of others since it does not necessarily involve ill-intent or bad faith. In contrast, some egoistic tendencies can be deliberate and become part of a strategy to use others as means to one's ends. This strand of egoism is more corrosive because it undermines the very nature of ethics as a guiding force toward human flourishing. It is a deeper or more engrained personality trait indicating that self-centeredness has become a mode of action, potentially rooted in personality. Remediation of such habits may thus be more difficult since the psychological reward systems of such a person are self-centered, while human flourishing and an ethical life commands levels of other-orientedness such that one asks what their impact is on the meaning of the lives of others (Ryff and Singer 2008). Remediation involves correcting psychological tendencies to align them with longer term goals and satisfying human relationships based on more intrinsic motivations (Deci and Ryan 1985).

Nonrecognition of Personhood

The lack of recognition of others as persons is a significant barrier to moral awareness. Indeed, if someone's person or their interests are not deemed to be equal to those of oneself or of another person, they can be neglected in different ways such that the concerns of that person are not considered, and moral challenges involving their interests will not be fully taken into account. A strong example I encountered in my own research was a situation involving young adults with cerebral palsy. Cerebral palsy is a chronic neurological condition typically diagnosed early in life that, due to varying levels of brain injury, can lead to wide-ranging levels of motor impairment, and in some situations, to cognitive impairments. I was interested in understanding the ethical challenges faced by some young adults with the condition, especially because a person close to me lives with the condition, and I have been exposed to the struggles it can impose.[9] In a qualitative interview-based study, my research group showed—supporting other studies—that the motor impairments, which can lead to slower speech and more difficult enunciation, were often conflated with cognitive impairments. As a result, these individuals were perceived by some clinicians to be unable to speak for themselves. Importantly, these young adults reported being extremely frustrated by this situation. They shared with my research team how they felt belittled by clinicians who would not talk to them or engage them in conversation but rather spoke to their parents instead. When clinicians did speak to them, they often infantilized them by overly simplifying their messages (Larivière-Bastien et al. 2011). This is, in my eyes, a clear example of the frustration created by the lack of recognition of personhood in another human being. Personhood is a complex concept, but it captures, in these cases, not being treated as an equal, not being respected, not being treated with dignity, not being considered capable or worthy, or being reduced to some specific aspect of oneself (e.g., disability), and so on. The dismissing of people, such as those with cerebral palsy due to perceived disability, by clinicians causes the latter to blind themselves to the experiences and difficulties lived by this population. In other words, not seeing others as genuine people suggests that we will neglect and be unaware of their challenges. Moreover, we are deprived of their problem-solving strategies and experiences that could contribute to the tackling of moral matters and enable them to thrive as human beings.

In bioethics, the principle of respect for persons carries the intent of correcting for such situations. In work with the anthropologist and bioethicist Ariel Cascio, following the concept of patient-centered care, I have argued for the "holistic respect of persons" in the context of research[10] because situations in which persons are misapprehended are unfortunately frequent. This requires remediation in the context of research on disability and beyond (Cascio and Racine 2018). In the context of cerebral palsy described above, the lack of recognition of personhood can take many shapes and forms, such as not paying attention to one's interlocutor, not considering their interests, leaving them out of a discussion, or not engaging with them. Sometimes, this recognition for personhood is in fact framed as a struggle for recognition (Pilapil 2012). Accordingly, demands for recognition, can be interpreted "as demands for inclusion into personhood through which the moral dimension of recognition struggles is properly understood" (Pilapil 2012, p. 39). On this topic, Appiah, cited by Renante Pilapil, writes:

> To be more precise, recognition is not simply about recognizing a person of a particular kind but more fundamentally recognizing their moral status. Conversely, to be misrecognized is to be morally wronged as a person. Misrecognition ignores, violates, or disrespects the status of persons as persons. That is to say, they are treated as mere objects, animals, or numbers. It is no wonder that when identity groups–say women, Blacks, or Muslims–fight for recognition, they make such a claim on the basis of their being a *moral equal of a person* of any other gender, race, and religion, especially white, male, or Christian. (Appiah [2005, p. 109] as cited in Pilapil [2012, p. 46]) (emphasis in the original)

Although this cannot be discussed here in any detail, one of the driving ideas underlying the importance of recognition is that the status of personhood is not to be found in some fixed or abstract essence (e.g., some key defining attributes), because the status of personhood is socially constructed through the recognition by others such that even individuals with minimal consciousness still need to be treated as persons (Blain-Moraes, Racine, and Mashour 2018; Demertzi et al. 2012; Sample et al. 2019) even though they may lack what some consider to be defining traits (or the essence) of being a person. In other words, recognizing people as persons who have interests, a life to live, and a story to make and to tell about their own lives allows us to recognize the problems they live as real problems. Thus, our ability to recognize and be aware of moral matters is contingent on our ability to recognize other people as persons, as full moral agents.

The inability to recognize persons holistically, as well as a self-centered perspective that prevents someone, naively or strategically, from seeing moral matters, are but some of the many factors that could create interpersonal blind spots in moral awareness. For example, altruism and prosocial (e.g., helping) behaviors can be considered important factors in generating responsive moral behavior and may be also related to moral awareness, although this would merit further investigation to see specifically whether, for example, higher levels of altruism and social behavior correlate with higher levels of moral awareness. Furthermore, natural affinities and resemblances between individuals may make it easier for someone to understand the interests and perspective of another person.

Gateways to Relational Aspects of Ethical Awareness

Being something of a personal disposition, moral awareness can be influenced by one's upbringing and psychological profile and varies between individuals, as shown in numerous studies of moral development (Limone and Toto 2022). Accordingly, strategies to increase moral awareness, notably their interpersonal aspects and related barriers, overlap with strategies of moral and ethics education and training about interpersonal relationships. They can include promoting cultural diversity such that one is confronted with other points of view and ways of knowing and being (e.g., in educational settings; see Dewey [1938/1986] and Sikandar [2016]). Stimulating efforts to value diversity in professional environments could also send signals that different points of view coexist (e.g., through institutional volunteering programs). Promoting diversity does not mean that all points of view are equal, but the recognition of diversity may support a context of openness and mutual learning between moral agents of different positionalities within, for example, a social or institutional context. It also means that tendencies that diminish interactions between groups in society (e.g., increase of socioeconomic disparities, private systems of education and healthcare) undermine the social mobility that helps provide a shared pool of insights that help establish and maintain the foundations of moral awareness through healthy human relationships and recognition of the worth of others. Although these issues related to social mobility are broad, they warrant attention as they may narrow our moral outlooks and our ability to see interpersonal barriers to the recognition of moral situations, thus pointing to another set of barriers next discussed: social barriers.

Cultivating a broader ethical awareness implies being open to being challenged about the biases that operate to narrow our moral visions in ways that exclude others. This recognition as moral agents or human beings is often one of the core demands of those affected. The example cited above from the research I have been engaged in concerning youth with cerebral palsy illustrates this point well, though there are countless examples of such situations. Clinicians, for example, can become oblivious to important dimensions of the personhood of some individuals in ways that became a modus operandi for certain clinicians and clinical environments. This can be the same for business managers who come to see others (e.g., employees) only as means toward (financial) goals (Craze 2020). Such situations speak to a context rooted in a broader social environment that undermines and impedes moral awareness systemically.

Social Barriers to Moral Awareness

There are obstacles to moral awareness created by social factors that diminish the ability to recognize the interest of others in ways that would trigger moral awareness and lead to the perception of the situation as a genuine moral situation (Reynolds and Miller 2015). This is well identified in the extensive literature on social distancing and social avoidance in various contexts such as racism (Bailey et al. 2017) and disability (Temple, Kelaher, and Williams 2018). It is also discussed in ethics literature. For example, Thomas Jones proposes that "moral intensity," a concept he uses to describe the salience of a moral situation, is influenced by six factors,[11] including some that appeal to social consensus on the evil or good nature of an act (Jones 1991). Although they overlap with interpersonal factors, social factors entail that broader forces (beyond interpersonal relationships) are at stake in shaping more or less directly one's moral awareness. For example, socioeconomic status, race and ethnicity, disability, gender, age, physical appearance, and language have all been recognized as having an impact, sometimes a very profound impact, on how we behave toward each other, including how we recognize the legitimacy of our interests and our worth as human beings. Here, the history of racism, sexism, ableism, and ageism—all of which are socially constructed phenomena—are clear examples of how social factors have created situations in which it becomes "natural" (or second nature) to view others as being less worthy or less deserving in ways that shape moral awareness. Scientific research has also shown that subtle embedded or hidden sociocultural factors (e.g., the hidden curriculum in medical training [Bandini et al. 2017]) can impact moral awareness (Reynolds and Miller 2015). Moreover, work climates or organizational cultures can shape human moral behavior (Reynolds and Miller 2015; Tenbrunsel and Smith-Crowe 2008). Sometimes, assumptions and biases relating to gender and race are implicit such that our behaviors are modified even against our own self-entertained view of ourselves and self-declared values (Reynolds, Leavitt, and DeCelles 2010). I here exemplify the phenomenon using areas where research has documented situations in which social factors impede or modulate moral awareness.

Framing Effects

Framing effects concern how a situation is understood and presented, sometimes described as "scripts" (Butterfield, Trevino, and Weaver 2000). The research on framing effects is vast. With respect to the study of morality, the research is telling as it shows how contextual cues can induce a number of biases in judgments and moral evaluations (Butterfield, Trevino, and Weaver 2000). In business ethics, the famous Pinto fire case illustrates a lack of moral awareness caused by frames. In a nutshell, Ford expediently developed an affordable car model in the 1970s (the Pinto, priced at $2,000 on the US market) to rival competing German and Japanese car makers. In doing so, Ford cut corners and failed to recognize that the crash tests showed that the gas tank setup of this model led to dangerous leaks and fires after rear-end collisions. When confronted with a decision to make a change to the design of the tank, the cost ($5 to $8 per car) was considered too high given the constraints on the total cost of the Pinto (Shaw 2005). Additionally, individuals working in Ford's recall department never interpreted or perceived the moral aspects of the problem that resulted in the death of hundreds of individuals once the vehicle went to market. One of the key players in the case, Denny Gioia (Ford's recall coordinator at the time), reported that ethics was never part of the available scripts to describe the situation such that it never came to mind because of the organizational context in which the employees were situated (Butterfield, Trevino, and Weaver 2000). Accordingly, framing—or failing to frame—an issue with moral language (or being sensitive to cues that reflect moral matters) can impact moral awareness, and such framing can be shaped by work culture and the broader social context in which it is rooted (Butterfield, Trevino, and Weaver 2000).

Butterfield and colleagues have shown, for example, that very subtle language changes in the description of a situation raises moral awareness (Butterfield, Trevino, and Weaver 2000). Their findings suggest again that commonly accepted frames influence awareness about moral situations especially given that moral situations are embedded in social cognition, that is, ways of thinking about social behaviors and social attitudes. Likewise, Tenbrunsel and Messick describe how framing situations with allusions to morality (e.g., as being about competitive behavior or simply as a business situation without allusion to its moral aspects) influences behavior (Tenbrunsel and Messick 2004). For example, their research has shown how the mere presence of a surveillance system in a work setting (versus a context without a surveillance system) changed the frame from a mostly ethical one (without surveillance) to a mostly business one (with the surveillance system) (Messick 1999; Tenbrunsel and Messick 2004). Accordingly, social cues can influence frames and significantly impact the construal of the situation as moral. In turn, recognizing the situation as moral or not impacts the ability to be aware of problematic features of the situation. There is thus an interdependence between social cues (found in frames) and moral awareness.

Social Consensus

The existence of a social consensus regarding a norm, defined based on Thomas Jones' work as "the degree of social agreement that a proposed act is evil or good" (Jones [1991, p. 375] as cited in Butterfield, Trevino, and Weaver [2000, p. 990]), can be another social factor that affects moral awareness. Although Jones considers this factor as being constituted by objective factors associated with the moral matter itself (e.g., a situation featuring reprehensive acts such as theft or murder), I side with Butterfield and colleagues who claim that moral awareness relative to alleged consensuses about morally good or wrong acts is modulated by social factors. Indeed, "it requires that the individual perceives that social consensus exists about how ethically problematic a particular action would be" (Butterfield, Trevino, and Weaver 2000, p. 990). The existence of social consensus about the evil or good nature of an act could reduce ambiguity for the moral agent, clarify the nature of the issue, increase the certainty about the moral nature of the situation, and potentially increase their sense of responsibility toward the act (Butterfield, Trevino, and Weaver 2000). However, in doing so, it can reduce active moral awareness. For example, viewing the context of business as being intrinsically moral, that is, morally defensible and acceptable (e.g., believing that the free market is intrinsically morally good), can diminish moral awareness with less critical engagement of moral agents (Reynolds, Leavitt, and DeCelles 2010). However, research has also found, for example, that contrary to what is often expected, viewing a business environment as highly competitive can actually increase awareness of moral matters (Butterfield, Trevino, and Weaver 2000; Reynolds, Leavitt, and DeCelles 2010), perhaps because such an understanding of the business environment can make one more sensitive and prepared to recognize how such a context can generate moral matters.

Socially enforced moral awareness can have truly tragic and substantive implications, when something considered to be normal and morally acceptable behavior is profoundly problematic from an ethical standpoint. For example, systemic racism against Indigenous people (First Nations, Inuit, Métis) is a historical and ongoing reality in Canada. It took many decades to be brought to the general public's attention and to really be seen as a genuine and substantive problem (Truth and Reconciliation Commission of Canada 2015). In recent years, Indigenous patients have been part of tragic cases showing the inhumane treatment they still suffer; even more disturbing, these kinds of treatments are not problematized in some circles. For example, Joyce Echaquan was a 37-year-old woman, a mother of seven children, and a member of the *Atikamekw* First Nation located in the northern territory of the Saint-Maurice River (*Tapiskwan sipi*). She was severely mistreated at the Centre Hospitalier de Lanaudière in Saint-Charles-Borromée (a.k.a. Joliette Hospital) and died on September 28, 2020, under horrific circumstances, while filming herself the entire episode in which clinical staff insulted her (Radio-Canada 2020). One of the orderlies involved in these acts went as far as claiming that a clinician's racist insults were benevolently intended (Nadeau 2021). Another similar devastating situation has also been brought to light: Brian Sinclair was a 45-year old Indigenous man who was transferred by a community healthcare center to the Winnipeg Health Sciences Centre to have a catheter changed (Browne et al. 2017). After waiting 24 hours in the emergency room without being seen, he died. Healthcare workers from the hospital reported that they had assumed Sinclair was drunk and "sleeping it off" (Geary 2017). These and many other repeated events (e.g., forced tubal ligation; see Boyer and Bartlett [2017]), have clearly shown that there is an underlying social dimension to the racism against Indigenous people in healthcare and other public services in Canada (Boyer 2017) as is the case in many other countries too (Gee and Ford 2011). They show without any doubts that social consensus about certain stereotypes represents a barrier to moral awareness. In these cases, problematic attitudes are so normalized that many are blinded to them, like the orderly who misrecognized the racist remarks as well-intended and thus not morally problematic in Echaquan's case.

Implicit Biases (Again)

Earlier in this chapter, implicit assumptions were treated as a cognitive barrier, but such implicit associations can be reinforced by broader social forces such as organizational culture (Reynolds, Leavitt, and DeCelles 2010). In the context of business ethics, a study by Reynolds and colleagues found that one's implicitly held view of the business context in general (as moral or as immoral) interacted with both subtle and explicit contextual cues (e.g., reading different versions of a memo from a fictive president to yield moral and immoral behavior; completing an insurance claim for the appropriate or an overestimated amount) (Reynolds, Leavitt, and DeCelles 2010). This study and its accompanying literature review support an "interactionist view of automatic moral behavior" because "an implicit assumption about the moral nature of business can interact with contextual cues to shape moral behavior" (Reynolds, Leavitt, and DeCelles 2010, p. 752). The authors concluded that employees are "reflexive interactionists, moral agents whose moral behavior is shaped by the interplay between their implicit assumptions about morality and the contextual cues, both explicit and subtle, that pertain to moral behavior they receive" (Reynolds, Leavitt, and DeCelles 2010, p. 754). This is telling experimental evidence showing how cognitive patterns can be associated with broader social context. There is a sense in which this comes as no surprise given how, for example, a social phenomenon like racism has pervasive interpersonal, cognitive, and affective consequences and implications. However, this kind of experimental evidence illustrates the transactional nature of our worldviews and the deep implications of our relationships with each other and to our physical and social environment.

Gateways to Social Aspects of Ethical Awareness

Social barriers to moral awareness can be powerful and deeply embedded in habits of thought and practices because they are reinforced systemically in policies, educational materials (e.g., in healthcare), and business practices. They may even appear to be insurmountable given that they, by definition as social barriers, are beyond the purview of any single individual. Moreover, social barriers can be embodied in entrenched forms of discrimination, mistreatment, and neglect because a dominant culture does not necessarily pay attention to the interests of those who are marginalized. Sometimes they are part of efforts to actively and systemically discriminate against or make invisible those who are not part of the dominant culture (e.g., Canadian assimilation policies and practices). But when we consider the progress accomplished in many societies' awareness of the interests of marginalized populations, including people with different sexual orientations, ethnic and linguistic minorities, and so on, there are positive signs in the various real-world changes we are seeing too. But clearly progress has not been gratuitous and it is the result of arduous social struggles for recognition. Major social changes require major efforts because ingrained patterns of thinking and behaving need considerable resources and energy to be changed; but should the endeavor of social changes fall on the shoulders of those who are marginalized? This would be a rather unfair position. Rather, it falls on everyone to change their individual habits and patterns of thought—as individuals, groups, communities, countries—and to change things that are social in nature.

So, how can social barriers to moral awareness be addressed? In discussions about human morality and ethics, the focus is often on individual capacities and behaviors, but there is an equal share of burden on context and the development of "morally aware contexts" that afford the agent (Gibson 1977) the ability to engage in morally justified behavior and to support their moral awareness. Here, strategies that commonly foster social and institutional change in areas such as public health (e.g., health awareness) and organizational science are compelling.

The magnitude of the effects of social barriers to moral awareness means that those who are responsible for these broader issues (e.g., individuals in authority and in decisional positions such as managers and administrators) bear some responsibility in tackling organizational cultures that perpetuate problematic implicit associations (Reynolds, Leavitt, and DeCelles 2010). Within organizations, interventions to consider include dedicated training and socialization that—assuming that emphasizing the moral nature of business diminishes perceptions of immoral behavior—explicitly inform employees about the existence of immoral behaviors in the context of business to correct this misperception (Reynolds, Leavitt, and DeCelles 2010). In healthcare settings, organizations can engage directly with people concerned to allow them to tell their stories and raise awareness about the moral matter they have encountered as users of healthcare services. For example, in ongoing research in my research group, we are working with people who have rare diseases in order to sensitize the healthcare community to the numerous ethical problems they encounter throughout their life trajectories, starting with a common "diagnostic odyssey," which entails five, ten, even 15 or more years to receive a diagnosis, after which they can better seek medical help (Quintal et al. 2023). Organizations can also use measures of a phenomenon like moral attentiveness to monitor their organizational culture and determine who needs training the most and where to place those who have strengths therein (Reynolds 2008). Being cognizant of how factors can change the awareness of moral situations, organizations could calibrate interventions based on their members' levels of moral awareness (Reynolds 2008).

Mechanisms by which moral awareness can challenge moral unawareness and allow progression in sensibilities to moral situations are crucial. Democratic institutions that allow open debate that challenges existing ways of defining situations (e.g., about sexism, racism, ageism) and open dialogue on the values at stake are crucial to ensure a progression of learning and growth of moral awareness. However, even in democracies, the most marginalized people may still not be fully heard; the concerns and needs of the majority are likely to be prioritized, thus reinforcing their marginalization. Accordingly, freedom of speech and communication (direct or mediated) that allow for the existence of a public sphere are essential to the processes by which social barriers to moral awareness are put into question.[12] In particular, the protection of journalistic sources (e.g., people who are disempowered, marginalized) and the protection of dissidents and whistleblowers is part of a healthy democratic culture in which moral matters can be articulated and discussed publicly despite how they may affect the interests of powerful economic, institutional, or corporative interests. Transparency and accountability of organizations and groups is also a means by which questions about the values of corporations and groups can be raised. Having democratic judicial and regulatory systems in place to hold accountable those who are responsible for fueling problematic attitudes and practices is another crucial factor. Without sounding alarmist, these basic features of democracies have come to be taken for granted and are under great pressure.

In democracies, social change is an inherently political process because changes need to have legitimacy, and be argued for, in order to gain public and economic support. At the same time, there are some outstandingly courageous people who put the best of their energies into making such changes happen and become catalysts for change, which is challenging when that change is a very long and difficult process. A good contemporary example to illustrate this phenomenon is whistleblowing and how highly mediatized cases can have an impact on institutions and cultures. In Canada, a major whistleblower, Ken Pereira, who denounced enormous abuses and corruption in the construction industry in provincial and municipal politics in Quebec, faced a tremendous number of barriers (many social and structural in nature) to denounce this situation publicly (Pereira and Bouchard 2015).

Eventually his efforts—supported by journalists like Marie-Maude Denis who courageously reported on this controversial story—led to the creation of a provincial governmental inquiry, but he basically left his career as a union leader and an industrial mechanic behind him and suffered numerous physical threats and severe psychological stress. Although the situation was not known to the broader public, it was also not seen as a problem at all by corrupted leaders who could operate with impunity thanks to the complacency of those around them. It took a lot of courage to break the *omertà* around this situation and problematize it so clearly and vociferously.

There is no doubt that a social climate of dishonesty and complacency creates huge barriers to the recognition of a situation as morally problematic. The costs associated with the situation denounced by Pereira were huge: millions and millions of dollars wasted, as well as the priceless toll on human lives in terms of broken careers because of how conflicts of interest and systemic abuses lead to unfair situations and biased decisions of all sorts. Aside from such moral heroes (and there are many examples around us and before us), there are also less visible and more discrete citizens who can in many ways participate in supporting positive social and cultural changes. This is, in a sense, where democracy takes all its meaning as a form of self-governance. Indeed, there is a deep connection between social aspects of moral awareness and the ability to sustain a democratic culture (Fouke 2009).

There are limitations to this chapter. I want to remark that I reviewed literature assembled under four clusters of barriers to moral awareness, somewhat consistent with some proposed typologies (Miller, Rodgers, and Bingham 2014). Furthermore, these barriers map rather well to some remediation strategies that focus more precisely on the different aspects (e.g., affective, cognitive). However, other sets of barriers could exist, and the proposed typology could evolve. For example, "moral recognition" (Reynolds and Miller 2015) is possibly influenced by biological factors (e.g., gender[13] and age). For example, women appear to exhibit on average greater moral sensibility (You, Maeda, and Bebeau 2011), and there are relationships between age and moral sensitivity that have been studied with unclear conclusions (Reynolds and Miller 2015, p. 114). What such findings mean or should mean (e.g., are they biological caused or consequences of socialization based on prevailing ageist and sexist attitudes?) is unclear and requires more investigation. Cognitive fatigue and depletion can also reduce moral awareness (Reynolds and Miller 2015). For example, one study has shown that moral awareness is greater in the morning than in the afternoon (Kouchaki and Smith 2013). These kinds of factors may point to distinctively biological and physiological aspects (and related barriers) of moral awareness, although like any complex human behavior, it is best understood in multidimensional ways.

Conclusion

Morally problematic situations can be ignored deliberately or nonconsciously, yet moral awareness is a fundamental precondition of any further ethical analysis or deliberation. It is a mediator between morally problematic situations and the engagement in ethical deliberation. The initial awareness and understanding (or misunderstanding) of a situation as morally problematic does not necessarily represent a single episode, nor is it necessarily only contained in the initial recognition of a given problem or situation. Importantly, it can be corrected or adjusted through processes of ethical inquiry, which involve reconstructing an understanding of the situation in a more inclusive way. Furthermore, moral awareness can be cultivated as ethical awareness to direct the ability to recognize morally problematic situations that impact human flourishing. This entails admitting that our common moral sense is not perfect; it is informed by upbringing, education, social norms, and practices for the very reason that our starting point, our moralities, are based on traditions handed down to us. This chapter reviewed four sets of barriers to moral awareness, as well as potential remediation strategies. Affective barriers to moral awareness are obstacles created by the inability to feel the existence of moral tensions and key aspects of moral situations. Cognitive barriers to moral awareness are obstacles created by the inability to apprehend moral situations or certain aspects of moral situations because of cognitive processes such as implicit associations. Relational (interpersonal) barriers to moral awareness designate the inability to see the other person as a person, take into account their interests, and relate to others as human beings in interpersonal relationships. Finally, social barriers to moral awareness are obstacles created by systemic and contextual factors that impede the ability to recognize the interest of others and lead to the perception of the situation as a genuine moral situation. Barriers are not fixed or unchanging. For example, triggers of moral awareness have changed dramatically as societies and cultures have changed their interpretation of a wide range of situations. Change or preservation of moral awareness are not desirable in and of themselves; it is the ability of moral awareness to detect situations involving hindrance to human flourishing, in the form of ethical unacceptability or ethical blameworthiness (see chapter 5), and bring about change based on that process that is important. In this sense moral awareness yields a reservoir of moral insights and moral sensibility, but these are also bound to evolve and be reflected upon as we learn to become more ethical human beings. In sum, cultivating moral awareness and developing an ethical awareness is an exercise that helps set the context for the search of solutions to moral matters.[14]

10 Scenario-Based Ethical Deliberation and Ethical Imagination

Ethics has long been viewed as a search for a rational system of morality that could rely on bedrock foundations. This account of ethics remains influential in academia but also among the general public, as well as journalists and policy-makers. Indeed, ethicists are often consulted with the expectation that they can define what is wrong and what is right. The model of ethical analysis that corresponds to this account of ethics is typically a deductive model of ethical reasoning. This common deductive approach has merits, notably those of simplicity and of exposing arguments underlying the justifications of behaviors. However, it falls short of offering a compelling account of why ethical deliberation is a creative, energizing process. Indeed, the deductive account of ethics seems to bestow the ethicist with the authority to state what is ethical based on their knowledge of ethics theories, principles, and arguments. But in the end, the response to a morally problematic situation is not an argument, a deduction, or an exercise in intellectual virtuosity. The response to such situations is a much more entrenched kind of action. It is an enactment (see chapter 12), an acting out of a vision of what is the preferable course of action in response to that situation. Following this line of thought, ethical deliberation is embedded in situational analyses and relies on the ability to mobilize creative resources to surmount moral challenges of everyday life. I argue that scenario-based deliberation, notably based on the insights of my mentor Hubert Doucet, taps into more holistic ways of understanding situations and of contextualizing the analysis of desirable futures. In the end, ethical deliberation is an exercise in generating possible response-scenarios, and of evaluating their merits in light of their potential ability to foster important aspects of human flourishing and well-being. Therein, traditional tools of ethical analysis have a place, but within a more holistic and creative conceptual and methodological ecosystem.

> The only thing worse than being blind is having sight but no vision.
> —Helen Keller

Chapter 3 on situated agency and chapters 4–7 on morally problematic situations introduced the notion that morally problematic situations represent the fundamental experiential unit through which moral agents encounter moral matters. In other words, moral matters, in their most concrete and vivid forms, are situations, and situations are lived experiential units of agent-context transactions that make up the existence of moral agents. The awareness of the existence of such problems is contingent on the ability of moral agents to recognize situations as such and may be impacted by barriers (chapter 9). Once this recognition occurs, the problem becomes, so to speak, an ethical one that is recognized as grounded in morality but requires more than the resources of morality as a system 1 implicit cognitive system (Kahneman 2011).

Ethics initiates a structured moral problem resolution process (as a system 2; see Kahneman [2011]) once there is a sense that a morally problematic situation does exist and represents something to grapple with. Obviously, one's final opinion on the better response to the situation could be corrected or invalidated by the reality of a situation. But the recognition of this fallibility—even after engaging in a structured deliberative process—is of paramount importance for deliberative wisdom and for any form of ethics premised on the positive contribution that experience can make to learning moral agents.[1] From a pragmatist perspective, the whole moral problem resolution process is best captured as a process of deliberation and imagination and not merely an implementation of application process (Fesmire 2003; Lekan 2006).

In the pragmatist tradition, ethical deliberation is grounded in the advocacy for the greater role of dialogue or the ethical role of communication, as seen in some accounts of deliberative democracy or discourse ethics (Habermas 1991, 1999). It is a creative, research-like, problem-oriented, and future-oriented process that is only partly guided toward consensus and the respect of formal rules of discussion (Ralston 2010). Ethical analysis is one of the components of ethical deliberation, but ethical deliberation describes a more inclusive and holistic process wherein different points of view are taken into consideration in the understanding of a moral matter. Ethical analysis is undertaken not solely to identify which principles are preferable but rather, more globally, how different action-scenarios are embedded in certain principles or values, as well as the many other aspects of scenarios that would make them feasible or not. The outcome of a successful deliberation is not only a decisive judgment, but the choice of a preferred action-scenario, that is, a course of action that will be enacted and is promising in terms of both its ethical acceptability and ethical praiseworthiness.

In this chapter, I explain why an ethical approach based on deliberation is subversive of the influential deductive and applied model of ethics. The transactional ontology on which the theory of deliberative wisdom relies calls for an instrumentalist epistemology and a dialogical approach to ethical and moral problem solving (see chapter 1).[2] Reflecting on the action-oriented and instrumental nature of human thinking, I introduce deliberation as an ability to rehearse potential scenarios of action (action-scenarios) wherein imagination plays a key role to support the critical analysis of possibilities. When applied to the context of ethics, ethical deliberation reflects an open-minded stance of mutual learning and companionship in envisioning and developing the flourishing life. Because it is action-oriented, human thinking is impregnated with an imagination of the future. Accordingly, ethical deliberation reflects this positionality and captures the ability to identify what is valued in human situations, to see what is possible and to talk openly about the merits of different courses of action. Socrates warned against the unexamined life but, without using such a conclusive and condemning statement, it is to be wondered if the unimagined life is a fulfilling life, because a flourishing life is a life of thinking about the best possibilities and of pursuing them.

Deliberation

The term "deliberation" needs clarification here. To deliberate means to consider a matter carefully and thoroughly, often via exchanges with others following a dialogical way of being and interacting (Widdershoven, Abma, and Molewijk 2009). Deliberation differs from discussion, at least when discussion can mean exchanging without being engaged with a given problem to solve. It is certainly different from arguing since arguing in favor of or against a moral standpoint can easily become an exercise in advocating for that standpoint. Ancient philosophy rightfully distinguished *dialectics* from mere *eristics*. *Eristics* designates the activity of engaging in arguments but with no commitment to finding the truth, most often to defeat one's opponent. *Dialectics* expresses how dialogue is undertaken within a joint search for the truth and therefore the possibility of admitting one's ignorance. Here, deliberation is thus aligned with the ancient idea of dialectics or dialogue.

Deliberation—as a process that encompasses the critical review of a problematic situation, the sorting out of possible options, and the critical analysis of action-scenarios in response to the situation—ideally takes place in an interpersonal context and involves those directly concerned by the situation or their representatives who have an internal understanding of the nature of the situation, as well as those less directly involved who have broader situational, scientific, or other relevant knowledge to bring to the table. However, there are times when actual deliberation—in this social sense—is impossible (e.g., an individual clinician facing a difficult and urgent dilemmatic situation; a businessperson having to quickly make the decision for their company during a brief meeting). In those cases, a shorthand, by-default version of deliberation can take place, and the actual physical meeting of a group may be replaced by some effort to imagine the points of view of others in a particular situation and analyze broadly the context one is in. As the title of the chapter suggests, imagination is a key ingredient in the ability to support a forward-looking exercise where different action-scenarios need to be generated and analyzed.

Different forms of deliberation exist or, at least, deliberation can allow contemplation of different objects, and it is the goal of this chapter to describe some of the more specific traits of and guideposts for ethical deliberation in contrast to deliberations about other kinds of choices or problematic situations (e.g., deliberations about political change, deliberations about economic policies). Ethical deliberation is also an equivocal expression: it designates a deliberation that takes moral problems as its objects, but the actual approach embedded in this process can be assessed from an ethical standpoint (for a proposal regarding the latter point, see Senghor and Racine [2022]). This being said, it would be important for the future of ethics theory and practice to investigate more deeply the mechanisms of deliberation in general, and of ethical deliberation in particular, and the obstacles that can frustrate it. Lastly, it will be clear that engaging in a process of deliberative wisdom does not always guarantee good will, open-mindedness, willingness to discuss with others, or a context of discussion that is ideal, but it at least grants the possibility of expressing different points of view without clear coercion or undue pressure. Just like any process, success depends on individual and group skills and abilities, as well as a host of empowering or disempowering factors (Gross, Gagnayre, and Lombrail 2020). My description and understanding of deliberation neighbors that of what others—more inspired by hermeneutics than pragmatism—call dialogue (Inguaggiato et al. 2019a; Widdershoven, Abma, and Molewijk 2009).

The emphasis placed on deliberation on ethics in this book is no accident. Deliberation is a hallmark of ethics as a tradition initially described and exercised in a dialogical form in ancient Greek settings (e.g., Athens) where a plurality of views about much everything existed. This pluralism is an aspect of ancient civilizations often set aside given the emergence and predominance of the rather centralizing and unifying European concept of the nation-state (Apfel 2011; Veyne 2016). Ancient pluralism reflected a situation of active commerce between people and the goods they wanted to trade, as well as more egalitarian conditions for Athenian citizens, at least for the exclusive group that free men represented. Indeed, the material conditions in which human groups and civilizations are placed triggers the development of questions about the right and good life to live as an individual and as a political community, notably when pluralism about the good life and just institutions is provoked. In other words, deliberation embodies a kind of open-ended and inquisitive relationship with the world and its central role in democracy explains why this political regime is more profoundly understood as a way of life with deep connections to ethics (Pappas 2008). Thus, beyond the specific choice or judgment generated by deliberation, deliberation more comprehensively designates the quintessence of ethics as an open-ended, reflective, and dialogical social exercise based on the consideration of facts (observations, experience) and critical thinking. The iterative nature of ethical deliberation does not imply that the outcomes of deliberations do not constitute a kind of progress and that progress is entirely relative. Rather its iterative nature signals that outcomes of deliberation can be the steppingstone for greater ethical wisdom in the search for human flourishing. Fallibilism and the recognition of the procedural nature of the search for knowledge is not an abdication to relativism.

Ethical Deliberation

Ethical deliberation designates both discrete events (e.g., a given deliberation), as well as the broader dialogical process by which an initial problematic moral situation triggers the need to consider and carefully examine alternate courses of action for their respective merits. As a process or an approach, ethical deliberation—as developed within the theory of deliberative wisdom—includes the broadening and deepening of the understanding of the situation, the envisioning of action-scenarios (notably based on the processes of imagination and dramatic rehearsal), and ultimately the reaching of a judgment or decision based on the comparative evaluation and analysis of potential action-scenarios for their ethical merits. The result of a successful ethical deliberation is the identification of a preferred or preferable course of action (action-scenario). The preferable nature of a given action-scenario is based on its ability to fulfill conditions of ethical acceptability and ethical praiseworthiness (see explanations in chapter 2). Accordingly, a preferable action-scenario does not induce undue harm (ethical acceptability) but ideally also fosters human growth and flourishing (ethical praiseworthiness). The specific moments associated with the process of ethical deliberation are further described in the next chapter.

Accordingly, an ethical deliberation is distinct—but not necessarily disconnected—from other topics, such as maximizing the economic wealth yielded by a certain course of action or increasing pluralism and diversity, although these goals can overlap with those embodied by the criteria of ethical acceptability and ethical praiseworthiness. However, an ethical deliberation can lead to questioning the contribution of economic growth, of cultural diversity, and so on, to the flourishing of individuals in a more open-ended and broader way. As readers will see, better-known values and principles (e.g., justice, autonomy) are built into considerations of acceptability (not causing harm) and praiseworthiness (promoting good and well-being). Thus, the theory of deliberative wisdom is structured such that, ideally, all ethical wisdom allowing more profound ethical analysis can be mobilized to assess the ethical merit of a given action-scenario. However, action-scenarios generated within deliberations are considered hypothetical insofar as their value needs ultimately to be reflected upon and tested in human experience. Indeed, an action—in spite of good intents and sound deliberation—that does not lead to the desired outcome reflects a suboptimal outcome and should be viewed as a learning opportunity. This line of thinking is sometimes described as "instrumentalism" because ethical concepts and principles are repositioned as being in service of enriched moral experience (e.g., successful resolution of moral matters) (Racine et al. 2019a). In the context of deliberation or dialogue, one's own ideas are thus considered to be open to change and enrichment (Inguaggiato et al. 2019b; Widdershoven, Abma, and Molewijk 2009).

The open-endedness offered by ethical deliberation may appear vague and unsatisfying, even frustrating, or disappointing. Here, one must keep in mind how ethics has often been viewed as the provider of ultimate moral foundations, or has been seen as the field that determines who is right and who is wrong, who are the "heroes" and the "villains." Consistent with this understanding, the task of ethics has traditionally been to identify and most importantly to argue for which vision of ethics is superior and can overrule that of others. This sometimes amounts to an intellectual cockfight of whose account of ethics can better guide other people's lives, other societies, and so on, setting the stage for authoritarian uses of ethics (Fesmire 2003). It may also lead to moral imperialism (Moreno 2004) and could support broader colonial thinking (Mills 1997). The deductive model of ethics reasoning, and the emphasis on ethical justifications and the search for indubitable foundations, has been a mainstream approach defining how ethics could contribute to resolving moral matters (Hoffmaster 2018; Hoffmaster and Hooker 2009; Lekan 2006). Find the right principles, and the problem will be fixed, or almost (for early critiques see Caplan [1980] and Toulmin [1982]).[3] This model was rather quickly abandoned in more applied circles of ethics, such as bioethics and business ethics, but as a paradigm to envision ethics it has not been replaced by a strong positive and constructive vision about how an empirically informed ethics can be a tool of human development (Martela 2017; Racine 2022). Accordingly, a deliberation-oriented ethics encourages a displacement of focus, a sort of Copernican revolution in what counts as being of greatest importance in ethics. Ethical principles and justifications matter, but their true importance is how they are instrumentalized for human flourishing. Their "warranted assertability," their truth conditions so to speak, is to be found in their instrumentality insofar as they contribute positively to human experience and human existence (Dewey 1941).

Thus, one should not fear that ethical deliberation opens the door to relativism since deeply cherished values—if they are worthy at all of being cherished—are reflective of conditions and experiences that allow humans to flourish (Czeżowski 1953).[4] For example, justice is a primordial value to allow each a chance to human flourish and ensure conditions of fair process. Principles of justice are therefore not threatened by an ultimate focus of deliberation on flourishing. Rather, this focus gives it a more profound meaning and brings support to, for example, democratic institutions that promote greater participation of individuals in an active and fulfilling life. Likewise for autonomy, which is valued for the very good reason that intrinsically motivated actions are highly rewarding to individuals (Ryan and Deci 2000) and have been found to be associated with flourishing and well-being (Chirkov et al. 2003; Ryff 1989; Ryff and Singer 2008). The validation of the evaluations of good courses of actions can be most appropriately examined via open discussion and what is often called intersubjectivity. This point is well articulated by the Polish philosopher and phenomenologist Czeżowski:

> So it is not an assumption of the objectivity or of the absoluteness of evaluations that is a necessary condition for the constructing of an ethical theory: it is something else—and that it is, that evaluations should be intersubjectively communicable and verifiable or, to speak, more plainly, that different researchers should be in a position to communicate to one another their respective evaluations and to ascertain among themselves whether those evaluations are in agreement. . . . Generally, even men of strongly differing cultures, when they forget about the struggles and diverging interests that separate the ones from the others, and when they deal with one another in a simple human fashion, can find a common idiom when it comes to conceptions of good and evil. (Czeżowski 1953, p. 169)

Importantly, ethical deliberation is not the be-all or end-all of ethics. Ethics without deliberation is incomplete, but ethics it not exhausted by its deliberative moments. Deliberation designates a process in which solutions to the problematic situation are envisioned, debated, compared, and scrutinized in light of their capacity to promote human flourishing in keeping with the elaborations of the previous chapters on moral awareness and situational understandings of moral matters. However, the resulting preferred scenario falls short of resolving the situation itself, since the envisioned course of action will need to be enacted and eventually the effects of the well-intended scenario assessed and appreciated. The fact that a course of action was generated and vetted via deliberation does not ensure that it will be more ethical or make a positive eudaimonic contribution. It could fail. However, it is true that deliberation does play an important role in ensuring that some ethical pitfalls are avoided, such as conceiving narrowly of morally problematic situations and opting for biased and personally/morally preferred actions.

In sum, the process of ethical deliberation—when properly conducted—reflects a genuine commitment to human flourishing and improving our hypotheses of how we can bring it about. Because of its stated commitment to the evaluation of action-scenarios based on their ethical acceptability and praiseworthiness, it is a process that helps assess whether given actions are conducive to human flourishing beyond ensuring that they are acceptable, which is not necessarily to be dismissed, as we will see below. Because of this open-endedness, ethical deliberation fosters processes of mutual learning and sharing about experiences and why certain experiences are valued and valuable. It therefore embodies the important idea that ethics is fallible, intersubjective, and is thus an iterative, empirical, and experience-driven process. Indeed, there is no point in deliberating about options and courses of action if one's own thinking is presumed to be infallible or the mere reflection of one's individual arbitrary preferences. However, there is always a risk that deliberation be tricked or implicitly biased toward a preferred moral stance. This is why I articulate rough guideposts to establish the conditions and the process of a sound ethical deliberation in chapter 11 (see also Senghor and Racine [2022]).

Imagination

Imagination is a term that evokes contrasting understandings. It is readily associated with creative arts and aesthetics, perhaps with an undue connotation of frivolity. It is a capacity that is understood as being detached, sometimes isolated from real-world experiences and realities. Implicitly, this account of imagination presumes a subject-object epistemology, where knowing about things is contrasted to simply imagining things. It also upholds a rather descriptive and objectifying stance toward the nature of human thinking granted that imagination is envisioned also as a form of thinking or a part of thinking.

A contrasting account of imagination (and of knowing) is offered by the instrumentalist epistemology developed by pragmatists like Dewey and Fesmire (Fesmire 2003). If our existence is a set of experiences that are themselves transactions between organism and environment, then even what we see as our more purely agential capabilities like knowing, thinking, deciding, and imagining are also—in their nature and the result of—transactional processes (Dewey 1941). Likewise, if our mode of being is transactional, our ideas are projective or purposive in nature because they point to ends-in-views. They express what we seek and desire, that is, what we value in our existences and continuous transactions. They embody visions of a future, explicitly or more implicitly. In this sense, our thinking is fundamentally imaginative or projective in the sense that it embodies forward-looking visions. Dewey writes, for example, that "ideas are essentially intentions (plans and methods), and . . . what they, as ideas, ultimately intend is prospective" (Dewey [1908] as cited in Martela [2015, p. 196]).

The presence of imagination—following such a transactional ontology—is a matter of emphasis in all that we do. All our ideas and claims are projective at least in the minimal sense of upholding meaning that can, in principle, be validated in experience and used to guide our actions, even though this meaning may be so well asserted and validated that it becomes trivially true and commonly accepted as such (e.g., the earth is round). In the latter case, there is no genuine shared doubt, there is only ignorance (e.g., someone claims the earth to be flat) since no serious scientific observation points to the contrary. However, more clearly representative of imagination are the processes by which our ideas are somewhat freed from their habitual capturing of previous or current transactions. They are allowed to roam and question the ends we seek and whether the means we seek are fit to desirable ends. The emphasis on the projective and purposive nature of human thinking amounts to the understanding that imagination is a process of bracketing common habit and the ordinary intelligence of things. It consists of essentially taking distance from one's (or a collective's) own lived situation, examining what it means and what it purports, and asking whether this is valuable or not, and what would be desirable instead. Exercised in the context of moral problem solving, imagination can be subversive because it can imply that there are problems with the status quo, with reality as it is, because imagination allows us to envision possibilities that may be discrepant with current practices. I leave aside how creative arts engage with imaginative capabilities, how they can nurture them, what they mean to our lives and the future of our societies, and how they can enrich our worldviews and existences; although I must acknowledge that Dewey discussed their importance in depth (Dewey 1934/1958). Within such a pragmatist perspective, imagining possibilities (or the "possible") is a quintessential element of a free existence, a life that can detach itself from the habitual, and importantly, enact possibilities. As Kierkegaard wrote, "Life can only be understood backwards; but it must be lived forwards" (Bouchier-Hayes 2005). While imagination nourishes a forward-looking ethical existence, imagination severed from all connections with experience can become delusion, hallucination, and even alienation as one's imagined possibilities become incompatible with their reality. Unbridled imagination becomes schizophrenic-like insofar as it, etymologically speaking, represents a splitting (*schizo*) of the mind (*phrene*) between the experienced and the imagined. There can also be tremendous frustration and existential damage caused by the inability to enact an imagined possibility. The inner tension caused by this inability can nurture regrets, humiliation, and frustration. Broken dreams for one's existence and one's community are some of the most difficult situations to live, and are well described in the literature on biographical disruptions or ruptures in the context of chronic illness (Bury 1982), though this has much broader application.

It is of utmost importance that imaginative processes be undertaken not only as an individual task (at least in the context of ethics) but as a task of foreseeing futures in ways that tap into the possibilities desired by different human beings and different communities. Otherwise, the imposition of a dominant imagination of the future risks perpetuating the forces (e.g., colonialism, capitalism, racism) that constrain not only the past and present but also the future of human beings and therefore the ability of many human beings to live a life of their own. This is the powerful and touching message of John Lennon's 1972 subversive song "Imagine," which invites listeners to a joint, peaceful future as a humankind. In other words, we can moralize and condemn or glorify the past, but regarding the future, we have the ability to project and construct. Without these imaginative capacities, without any ability to look beyond the past and present, ethics would be powerless.

Ethical Imagination

Ethical imagination is the use of imaginative abilities of human thought and human agents to understand, discuss, and envision paths to human flourishing. Imagination, in this context, is not so much about the development of a grandiose vision of the good life, a fantasy. It is rather an ability, a resourcefulness, used to question the situations of everyday life: to ask what their meanings are; to interrogate oneself, others, and one's society regarding their practices and how they can support the well-being and flourishing of human beings. Eleanor Roosevelt, former First Lady of the United States and leading human rights activist, is credited for stating that, "The future belongs to those who believe in the beauty of their dreams." Ethical imagination, in this light, is the beautification and enrichment of our dreams through dialogue and exchange. It is very much possible for the creative arts to be ethically imaginative in this sense or at least foster imaginative skills. This is the case of the various art forms used by minorities across the world to voice their plight in music and in visual arts, notably to denounce various forms of discrimination and marginalization. For example, Quebec's Quiet Revolution in the 1960s saw companion political forms of arts (e.g., literature, music, poetry) being deployed to not only reclaim a future for a given linguistic and cultural legacy of French Canadians, but to exercise and renew this culture in new forms (Pâquet and Savard 2021). A similar revival occurred when Catalans claimed (in the early twentieth century) and reclaimed their political and cultural future after the end of Franco's long fascist dictatorship (1939–1975), which marginalized their language and their culture. Their partial recovery of political autonomy fueled cultural renewal in literature and other forms of art (Miller and Miller 1996).

Despite the importance of the formal and creative arts, the kind of imagination I have in mind is perhaps more basic and broadly shared among human beings and does not require special artistic talents. Ethical imagination is a general imaginative capability applied to the question of the good life, of flourishing and well-being. Open-mindedness is certainly a prerequisite for ethical imagination since allowing oneself to be questioned in our beliefs and identity by the events we live, discussing freely (or as freely as possible) difficult situations, and discussing various courses of actions—some of which may not be squarely within one's interests—requires the ability to open up and consider new ideas and new possibilities. But imagination—at least within a deliberative process—is more than simple open-mindedness because it also entails a participation in the elaboration of the future and the ability to generate different options for action. It also requires knowledge and a practical sense about what is realistic, feasible, and within one's ability to act in a given context. Indeed, the creative component of moral problem solving calls for the intellectual and practical capacity to look beyond current situations and foresee those that can help us get beyond current crises.

I have insisted numerous times on the *collaboration* required to identify, discuss, and resolve moral matters, but the imaginative aspect of moral problem-solving taps into processes of *co-elaboration*. Sketching out what could be a preferable situation or a good course of action requires reciprocity and the willingness to listen to other points of views and an approach that considers these differences less as obstacles and more as possible enrichments and learning opportunities (Inguaggiato et al. 2019a, 2019b; Widdershoven, Abma, and Molewijk 2009). A person's reluctance speaks to what matters to them, what they value, and what they see threatened, imperiled, or compromised. Achieving a better understanding of the situation requires attending to experiential aspects of human situations. Otherwise, it becomes very difficult to move forward because participants in an ethical deliberation cannot imagine what they could gain from the solution or how it could be beneficial to them and concerned parties. Mutual concessions are likely unavoidable in a process that is built on the unstiffening of moral perspectives, but this can be very difficult if a party feels like they are unheard, not listened to, marginalized, or deeply compromised such that they stand to lose something. When historical contexts shape individual and collective narratives in ways that justify resentment, progress can be very difficult because the sense of justice and fairness required to proceed is heavily challenged. How can one co-imagine and co-elaborate a future when different experiences and interpretations of the past and present divide people?

Preexisting challenges to the development of joint imagination can be difficult to handle and undermine the climate of trust and respect required for a dialogical process. An illustrative example of difficult contexts is provided by the way Canadian healthcare systems treat members of Indigenous communities. I cannot even skim the surface of the profound impact of successive French, English, and finally Canadian colonialism on the cultures, languages, and sociopolitical structures of Indigenous peoples, but it is nothing short of a cultural genocide (Truth and Reconciliation Commission of Canada 2015). This mass destruction of human possibilities prolonged itself through the tentacles of the Canadian state. This most notably took the form of the residential schools initiated in the 1880s and operated until very recently[5] (Hanson, Gamez, and Manuel 2020). This was done in the ignorance of most Canadians who only recently learned about them; there are still many who deny and willingly ignore their impact. Although the last Canadian residential school has been closed, the destruction of Indigenous possibilities certainly continues. It will not surprise readers that systemic discrimination against members of Indigenous communities actively pervades public services such as Canadian healthcare systems—as shown by the cruel deaths of certain persons (e.g., Brian Sinclair, Joyce Echaquan as discussed in the previous chapter) and by everyday dealings with these healthcare systems—showing the extent of harmful experiences within these systems (Davis and Tait 2003; Boyer 2017). Imagine a difficult context in which a moral matter surfaces between a patient who is a member of a Canadian Indigenous community and a White physician, a member of the dominant community. How will that patient develop trust and feel respected, especially with the intersectional and compounding effect of the preexisting physician-patient power dynamic that exists and furthers historical-systemic injustice and violence (Barned, Lajoie, and Racine 2019)? To overcome mistrust and suspicion, which are historically and still warranted, forms of reconciliation need to happen, and that reconciliation requires an understanding of the past in order to be able to see a global joint future and all the specific situations where visions about what is the good and preferable thing to do give rise to important differences about the future and the ways of living together (Truth and Reconciliation Commission of Canada 2015). This is described by Métis scholar, Caroline Tait, and her colleagues as forms of microreconciliations in everyday relationships and service use (Tait, Mussell, and Henry 2019).[6] There is thus a very deep connection (see also chapter 1) between the narrative understanding of existence (past and present) and the ability to imagine a future. Narrative understandings of the meaning of situations help unleash the potential to learn from each other, foresee and imagine a future together in specific situations, and elaborate a general orientation. The construction of such narratives is supported by episodic memory, which captures the process by which our situated experiences constitute who we are while the imagination (and eventual enactment) of scenarios is the way we create a future that is ours.

Scenarios and Scenario-Based Deliberation

The first part of this chapter explained the concepts of deliberation and imagination and how I proposed—following others—to see the important connections between them, especially when problems to be pondered involve deep valuations, such as in ethics. Now I turn my attention to scenarios and scenario-based deliberation as a way to build from insights about situational agency and the workings of episodic memory and imagination.

Given the transactional nature of human experience, including human thinking, I have already explained how human moral thinking is action-oriented and that the outcomes of ethical deliberations are less about achieving *certainty* (e.g., about ethical conclusions and ethical judgments) than about opening up *possibilities* of existence and action. In this sense, ethical thinking and, *a fortiori,* ethical deliberation where human thinking is mobilized toward seeking solutions is fundamentally an imaginative task. Solutions are firmly rooted in situational understandings of problematic experiences, but solutions are also an envisioned promise of new equilibria with respect to the tensions at stake in a given situation; solutions remain a promise until they are enacted and evaluated for their merits and limitations. In this sense, solutions are forward-looking exercises. They are hypotheses to test.

To my knowledge, the most compelling language to designate such hypotheses, following others such as Dewey and my mentor bioethicist Doucet, is the concept of scenario. Dewey first explains scenario development as being constitutive of ethical deliberation (Caspary 1991):

> Deliberation is a dramatic rehearsal (in imagination) of various competing possible lines of action. It starts from the blocking of efficient overt action, due to that conflict of prior habit and newly released impulse to which reference has been made. Then each habit, each impulse, involved in the temporary suspense of overt action takes its turn in being tried out. Deliberation is an experiment in finding out what the various lines of possible actions are really like. (Dewey 1922, p. 190)

In other words, deliberation is a creative effort to envision different possibilities; it is a simulation exercise. In many ways, Dewey was preceded by the writings of Hume and Smith. For example, Smith exposes in *The Theory of Moral Sentiments* (1759) that imagination is essential to both understanding other peoples' moral sentiments and for ethical judgment, notably in the ability to project what a situation would be like for another person (Smith 1759). He also uses the idea of a theater to explain moral judgment and the interplay between the actor and spectator perspectives, noting their eventual ability to "sympathize" (Müller 2016).[7] (For more about the theatrical aspects, see Griswold Jr [1999] and Marshall [1986, pp. 63–70]).

A few contemporary scholars—somewhat independently from Dewey, based on their own observations and experiences—have come to the conclusion that ethical deliberation is structured around the development of action-scenarios. This is explicit in Doucet's scenario-based ethics method, which has been used in circles of French-speaking health ethicists in Quebec and Switzerland (Baertschi 1998; Dion-Labrie 2009; Doucet 2004). Within such an approach, possible courses of actions are to be generated, considered, and analyzed, then enacted and evaluated.[8]

Doucet's method is one of few that proposes to structure deliberation around the generation and analysis of action-scenarios with the explicit goal of avoiding Manicheism and dualism (e.g., for or against a certain action or option) frequently encountered in concrete case discussions and moral deliberations where there are often tensions and difficulties to discuss openly. Similarly minded proposals can be found in the work of Dutch scholars such as Guy Widdershoven, Suzanne Metselaar, and Bert Molewijk rooted in hermeneutics and pragmatism (Hartman et al. 2019; Inguaggiato et al. 2019b; van der Scheer and Widdershoven 2004; Widdershoven, Abma, and Molewijk 2009). Others have also alluded to the scenario-based nature of human ethical thinking in ethics. For example, Hoffmaster explains, based on Abby Lippman's and Mark Fraser's original fieldwork, that parents who project living with a child with certain disabilities imagine a future in the context of antenatal genetic counseling; they do not appeal explicitly to ethical principles but rather envision scenarios and assess their worth based on the projection of who they (the parents) are and what they will and want to become (Hoffmaster 2018; Lippman-Hand and Fraser 1979a, 1979b, 1979c).

The next pages are dedicated to further arguing that action-scenarios are fundamental units supporting deliberation and that there are definite advantages and virtues to espousing such an orientation to ethical deliberation.

Action-Scenarios as Fundamental Units Supporting Deliberation

Moral situations represent the most meaningful unit of moral experience (see chapters 3 and 7) because this conceptual unit allows us to usefully identify experiential building blocks of human existence in ways that reflect how they are situated and embodied. Correspondingly, morally problematic situations need to be resolved with a family of concepts that preserve the situatedness of moral experience in concrete contexts and the experience of agents. The concept of scenario best represents the forward-looking and projective nature of the deliberative process since scenarios are contemplated and deliberated as potential courses of action that would resolve the morally problematic situation.[9] The emphasis on scenario development within ethical deliberation does not mean that traditional ethical analysis is useless or should be dismissed. Rather, traditional ethical analysis plays a more circumscribed and less determinative role within a broader understanding of human behavior and of moral problem resolution via deliberation.[10]

The current theory proposes that progress toward resolution of moral matters be seen as a creative exercise to mobilize the interests and values of those involved to first generate different action-scenarios in response to the morally problematic situation. Even if it entails an appeal to the imaginative abilities of moral agents, this process needs to be transparent and trackable, at least in the sense of being explicit and relying on evidence, experiences, and argumentation. Ideally, action-scenarios are generated collaboratively with those who will need to carry out an action, or at least taking into account their views and opinions about the nature of the moral situation, and then the appropriateness of the response in mind. This engagement with those involved is worthwhile not only because it will correspond to the situational understanding of moral situations (see chapters 4–7) and thus increase the efficacy of the process, but also because it will allow moral agents to develop their abilities to understand and resolve moral matters. It also reflects that the attempts to understand, make sense of, resolve, and deliberate on situations is an interested inquiry, that is, an inquiry based on vested interests. It is a transaction, an existence-affirmation or enactment effort to overcome a difficult situation and build the strength to do so in a way that projects and enacts one's values and identity. Thus, scenarios—taken here as projections of futures analyzed for their desirability in ethical deliberations—are consistent with the embodied, contextualized, and existential nature of moral matters. Indeed, genuine ethical deliberation—because it is concerned with valuations—involves a process of reassessing the position of moral agents, including their understanding of themselves, their context, and their experience within situations. Deliberation thus starkly differs from common moral behavior where one can rely on custom or tradition that provides some foundation for a behavior or attitude. It engages and mobilizes those concerned as ethical agents, that is, as moral agents interested in doing better and pursuing a flourishing life.[11]

The concepts of scenario and scenario-based deliberation help avoid some of the pitfalls of traditional ethics of relying on principles by preserving a contextualized understanding of the situation and proposing solutions that are connected to that understanding. The common deductive approach of moral judgment has been heavily criticized for being very hard to conjugate with practice such that practitioners have proposed more bottom-up strategies (e.g., the renewal of casuistry[12] [see Jonsen 1991, 1995; Jonsen, Siegler, and Winslade 2010; Jonsen and Toulmin 1988]) and methodologies that bring attention to the rich feature of cases (e.g., such as the narrative ethics or feminist ethics focused on context and relationships [Charon and Montello 1999; DeRenzo and Strauss 1997]). Furthermore, empirical research in cognitive science, notably, has challenged some of the assumptions of the traditional principle-based model, for example, by suggesting that noncognitive aspects of morality significantly outweigh the more explicit, conscious, and deliberate aspects of morality (Bargh and Chartrand 1999; Haidt 2001). Ethics theory needs to recognize the importance of this body of knowledge about morality, yet the task of ethics is to provide, were we to use Heideggerian terminology, the capacity of *Dasein* to move from a state of thrownness and realize itself based on the understanding of possibilities is a process of projection; the difference is that Heidegger specifies that projection "has nothing to do with comporting oneself toward a plan that has been thought out" (Heidegger 1927/1996 p. 185). This ability for self-projection allows morality to be recreated and reinvigorated once a difficult problematic situation occurs. This existential process is part of the existence of the moral agent as explained in chapter 1. Accordingly, ethics must work with some appreciation of how morality functions, but it must guide this force to a new destination using what Dewey called "intelligence," that is, system 2–based knowledge and deliberation. In this sense, the resources of implicit biases and other forms of subtle psychological influences remain an unwieldy and controversial tool to use because agents are acted upon without their conscious awareness (e.g., as proposed in behavioral ethics [Bell, Dubljević, and Racine 2013]) in contrast to explicit deliberation and reasoning, which advances ethics itself as a way to preserve a connection between thought and action (Maxwell and Racine 2016). This is why I stress the role of deliberation as a method and describe deliberation as a grounded effort that needs to be cognizant of situations (agent, contexts, and their interactions) and remain concrete and actionable, much like what is recommended by case-based approaches, relational approaches, narrative approaches, and contextualized empirical ethics approaches.

The Value of Action-Scenarios

Envisioning moral problem solving as a scenario-based exercise has several merits to foster efficacious moral problem solving and important contributions to stakeholders' flourishing. In this section, I explain these advantages, which allow me to further unpack the concepts of scenario and scenario-based deliberation. To do so, I contrast this deliberative approach with the more standard deductive model.

Actionability Principle-based approaches tend to displace the gaze from moral matters to the analysis of principles, progressively moving to more abstract matters. While the consideration and analysis of ethical principles has a place within deliberation, it must be situated within a broader context to increase its usefulness and connection to actual contexts and actions (Baertschi 1998; Caspary 1991; Gouinlock 1978; Brown, Kitson, and McKnight 1992; Stolper, Molewijk, and Widdershoven 2016).

Scenario-based deliberation and its terminology are instructive because they clearly position action as the focal point of deliberation. In other words, the need for deliberation is triggered by difficult experiences, and a return to situations and the actions to be taken in response to the situation are called for. Imagined or projected scenarios have some endpoint in sight. This does not mean that deliberation could not contemplate the status quo as a potential scenario since the status quo may be the right course of action given the nature of the problematic situation and the other feasible courses of action. In contrast, other ways to structure ethical deliberation around the analysis of principles or cases may lack this integrative function provided by a focus on the response.[13] Indeed, one of the drawbacks of methods based on the analysis of principles is that the principles themselves become the central point of attention—and sometimes not even the experiences to which these principles refer—and thus do not provide a clear sense of the operation upon the problem situation and a direction for resolution of the situation itself. Focusing on principles and other components of moral epistemology as abstract constructs can also suggest that they are ends-in-themselves and that the inquiry is about "fixing" moral thinking rather than approaching moral thinking as though it is part of a situation to be solved. This remark is rooted in the observations and experiences of a great number of ethicists who have struggled with the use of abstract ethics theory (Altman 1983; Callahan 1973; Fesmire 2003; Hoffmaster 2018; Miller, Fins, and Bacchetta 1996; Widdershoven, Abma, and Molewijk 2009). To be fair, some contemporary scholars who have advocated for principle-based methods do explain how principles should be "specified" and interpreted in context (Beauchamp and Childress 2001). However, in practice, these steps are not often followed (Racine 2013; Racine 2016). Furthermore, some of the more advanced principle-based theories do not explain how knowledge about contexts or theories and empirical knowledge about agents or contexts could be helpful in any meaningful way. There remains a fundamental orientation toward application that is only somewhat sensitive to the moral competence of moral agents and their understanding of moral situations (Stolper, Molewijk, and Widdershoven 2016; Widdershoven, Abma, and Molewijk 2009). Engagement in more open-ended deliberation, for example, in generating action-scenarios, opens up ethical reflection and emphasizes the important role that moral agents play, as well as their growth and learnings as moral agents as they participate in the elaboration and evaluation of different action-scenarios (Aiguier and Cobbaut 2016; Aiguier et al. 2012; Aiguier and Loute 2016; Inguaggiato et al. 2019a). Scenario-based deliberation is not a panacea but is helpful in bringing attention to the creative and actionable nature of ethical deliberation and a focus on actual responses. It also explains why this actionability is so important given how human moral existence is shaped by values that are themselves inherently tied to our action-bound existences as moral agents.

Situatedness and experience Standard accounts of ethical judgment and reasoning tend to stress the more cognitive aspects of ethics, sometimes even reducing them to such considerations. Fesmire describes this as the "central dogma of ethical theory, namely that it rests on revealing or constructing a moral bedrock that tells us the right way to think about moral problems" (Fesmire 2003, p. 55). Likewise, Susan Sherwin explains that traditional ethics theory is focused on offering foundations (Sherwin 1999), and Hoffmaster describes this as the deductive model of ethics (Hoffmaster 2018).

In contrast, experience-oriented theories like the theory of deliberative wisdom call for a much larger place of the experiential, and for a positioning of ethical thinking in relation to actual lived experience of moral agents. For example, understanding how emotions can serve as triggers for the existence of a moral matter or a morally problematic situation in a family or a set of specific or similar cases is helpful. Understanding that an emotional trigger prompted a request for ethics consultation can mean that the deliberative process will pay due attention to that emotion and why a person felt that a situation was shocking, saddening, or surprising, notably because psychological theory and neuroscience indicate that emotions serve as initial interpretations of situations (Damasio 1994; Molewijk, Kleinlugtenbelt, and Widdershoven 2011; Racine 2004). Likewise, if similar situations (e.g., about end-of-life decisions) commonly trigger certain emotions and feature certain common factual aspects, deliberative processes will be enriched by a proper understanding of this aspect of these morally problematic situations. The ethicist will ideally be able to tap into this empirical knowledge to guide others in their understanding of the situation based on the different stakeholder perspectives (Pyeritz 1990). In contrast, principle-based methods do not typically start with such an existential and empirical grounding process and therefore have to accommodate for, after the fact, the existence of these emotions and experiences as being constitutive of the situation. Moreover, there is no integrative perspective to which greater empirical (experiential and scientific or experimental) understandings of situations can actually contribute to moral problem resolution strategies. In contrast, the current theory, and pragmatism more generally, make such situated experiences their very starting point as embodied in the concepts of morally problematic situation and scenario.

Collaboration and engagement Standard accounts of ethics and of moral reasoning and moral judgment stress the expertise of ethicists as moral experts (Altman 1983; Fesmire 2003; Hoffmaster 2018). In this respect, the knowledge of ethical principles, ethics literature, and its general theories position the ethicist as someone who can rather easily be considered a typical expert who can point to the moral wrongs and rights, and exercise authority in these matters (Parizeau 1995). In contrast to this simplified account of expert-based ethical judgment, granting a greater role for ethical deliberation as promoted by the theory of deliberative wisdom (and other like-minded theories) also calls for a greater role for co-learning. It situates the role of the ethicist as a facilitator of moral growth (Aiguier and Cobbaut 2016; Aiguier et al. 2012; Aiguier and Loute 2016; Inguaggiato et al. 2019a, 2019b; Stolper, Molewijk, and Widdershoven 2016; Widdershoven, Abma, and Molewijk 2009). Given that outcomes are sought, an emphasis on the ethicist's expertise would become a problem and an actual impediment to the growth of deliberation stakeholders as moral agents (Aiguier and Loute 2016; Doucet 1994; Inguaggiato et al. 2019a; Stolper, Molewijk, and Widdershoven 2016). In comparison, scenario generation and scenario building can provide the opportunity to mobilize the personal and specialized (technical) knowledge of different individuals toward an ideally deliberated and shared goal in which they are the protagonists. As a collaborative and engaging process, it also provides an opportunity to understand different scenarios and why those involved and participating see things differently than others and to develop orientations to that effect. As I described in chapter 2, situations, *a fortiori* morally problematic situations, are experienced by persons. This means that, ideally, deliberation and resolution methods make room for the accounts of the situation from different perspectives to help reconstruct the actual nature of the situation and the meaning of its different aspects. In order to develop such a shared perspective, collaboration and engagement are essential ingredients and strengths of ethical deliberation.

The exercises of scenario building and scenario analysis inherent to deliberation help broaden the discussion to different views and engage stakeholders as agents. In contrast, methods of ethical analysis focused on principles may come to quickly privilege interpretations based on academic ethics scholarship (Stolper, Molewijk, and Widdershoven 2016) and the values of the people who have developed it with their important positional biases related to, for example, gender, class, race, and education, throughout the tradition of ethics. Pushed to the extreme, such principle-based interpretations can be, in the words of the bioethicist Autumn Fiester, "weaponized" against those without expertise who are involved, since most ordinary people would not necessarily interpret their own situation as a conflict of such and such ethical principle in a way that the expert in ethics could (Fiester 2015). As I discussed in chapter 6, this language can become a source of alienation (Callahan 1973). Likewise, the terminology of "cases," although very useful within the theory and methodology of casuistry deployed in the context of medical ethics, which has many merits, can give the impression that only the clinician's account—and in particular the physician's account—of the situation is authoritative and trustworthy (Racine 2016). Few people actually view themselves or their situations as cases, and when this is true, it can reflect how they are derogatorily perceived by representatives of public services (i.e., being just another case or another number). Seemingly innocent and quasi-neutral concepts like "cases" and "ethical principles" may not be generous enough to seize the more fluid and unformed understandings that most stakeholders will have of what they are facing. Accordingly, the terminology of "scenario" may be useful to foster collaboration and engagement in the deliberation process and help move from experienced problem situations to scenarios for future action.[14] Ultimately, what is needed in ethics theory is an ability to speak to moral experience in such a way that theory helps articulate what the problem is about and helps to solve it in a meaningful way. When theory is premised on strawman arguments (caricatured descriptions of the experienced situation), the deliberation is on very bad footing. Unfortunately, much of the emphasis on very abstract virtuoso-like development of abstract ethics theory has been sidetracked from these more mundane but more meaningful goals (Fesmire 2003). The collaboration and engagement embodied in scenario-based deliberation can be a counterforce to this dominant and rather detrimental orientation, which tends to make ethics exclusive and disembodied rather than inclusive and engaging.

Creativity and ethical imagination There is a sense in which most everyone will agree on the value of rational foundations and transparent argumentation for ethics; unfounded and irrational reasoning is problematic, as are gratuitous claims and conclusions. However, there is a deeper issue with this emphasis on rational argumentation. Mainstream ethics theory has often reduced ethics to ethical reasoning and argumentation. As Fesmire notes, debate about sound reasoning and foundational principles "is synonymous with 'doing ethical theory.' Normative ethicists hope these principles and procedures will tell us what we should do, to at least specific precise limits on what we can and cannot do" (Fesmire 2003, p. 65). The result is a rather unimaginative view of ethical reasoning where valuable ends pursued are somewhat predefined in the principles that contain them; the world of supreme goods and values is a closed world, and ethicists have knowledge of it. For example, the gaze of ethics can be directed to how we can implement justice, autonomy, and dignity but this can sometimes be to the detriment of the actual possibilities related to the actual situation. A gap between theory and practice can easily manifest.

Contrary to terminologies and views of ethics as being concerned with conceptual soundness, intellectual rigor, clear and indubitable principles, pillars of moral reasoning, and so on, the theory of deliberative wisdom sides with humbler but much more realistic views of ethics concepts. Following combined insights of pragmatism (Racine 2017) and feminism, ethics concepts can be used as lenses (Sherwin 1999) and tools. They help grasp what is morally at stake and help evaluate the merits of different action-scenarios, but they do not monopolize the entire imaginative process of scenario building. The concept of scenarios borrows from theatrical terminology to convey the creative—and sometimes dramatic or tragic—dimension of moral problem resolution (Müller 2016). Like in other fields of activity such as economics or human resources, resolving a genuine (morally) problematic situation is not only a matter of understanding the situation, although this understanding may help greatly and be required. It is also not only about correct reasoning or being right about a preferred principle or theory. Resolution of such problems, rooted in differences of valuations and experiences, requires those involved to work on establishing some shared and agreed on direction that may not be obvious from the start. This shared horizon—in Gadamerian terminology[15]—may involve compromises, mutual learning, and a fair amount of related frustration and misunderstandings. Such an effort needs to mobilize flexible thinking and tap into processes (e.g., imagination) that allow human beings to consider things differently and genuinely incorporate the interests of others (Molewijk, Kleinlugtenbelt, and Widdershoven 2011; Stolper, Molewijk, and Widdershoven 2016).

Most ethical situations can be "resolved," although perhaps not always ideally or in a manner satisfactory to all parties. Readers should note that resolution here does not entail definitive closure and the resorption of all moral worries found in a situation—this point will be made clear in chapter 12 on ethical enactments and resolutions. For example, in the case of abortion, the proposed resolution opted for by many Western societies does not do away with all dimensions of the morally problematic situation experienced by all citizens.[16] There can be a moral residue from favored solutions where not everyone will be equally re-situated in a satisfactory manner. Some still believe that abortion is immoral for different reasons, some of which may be rather metaphysical, while others are more directly related to the welfare of women. For example, while there have been concerns that women having an abortion may suffer from psychological and moral distress, the evidence suggests that this may not always be the case (Biggs et al. 2017). Having an abortion can actually relieve psychological and moral distress from an unwanted pregnancy. It is still often not an easy choice for women to make and may lead to a certain amount of distress as well, though. Moreover, some of the stigma associated with abortion is related to its surrounding acceptability, thus pointing to the social aspects of morality, which actually shape the experience of abortion (Biggs, Brown, and Foster 2020). The proposed resolution adopted by many Western societies has found that legislation and institutional resources granting the right of women to abortion services is the most defensible option, although it does not mean that it is perfect, especially if women still suffer from social stigma when seeking an abortion and have difficult access to the service and support.

Stressing the creative process of deliberation may suggest that deliberation "makes up" whatever reality one wishes to see. This is far from my intent. In a similar debate about the nature of responses offered by the then nascent field of neuroethics, I clarified, alongside my colleague Judy Illes, that innovation (from the Latin *innovatio*) etymologically does not mean creating *de novo* (or *ex nihilo*) but designates something like the renewal of tradition, which is also often misunderstood as designating something that is stifled and unchanging. However, etymologically, *traditio* refers to what is given or handed to us and what we pass on to others (Illes and Racine 2005). Accordingly, describing ethical deliberation as a creative process stresses its human nature, how it is dependent on our always-limited capabilities to envision different options, to give the chance to all parties to express what they view as preferable. Importantly, it also avoids the misunderstanding that ethical reasoning can be done in ivory towers and that ethics experts can offer their ready-made opinions single-handedly without any dedicated effort to understand the situations at hand and to engage with others in their resolution. In the ethics literature, the customary focus on moral judgment makes it sound like that judgment could be exported to different situations, to different persons, and even to different communities without any effort to validate its relevance and desirability. Rather, I stress that solutions need to be envisioned and imagined. When this imaginative component of future-oriented action is recognized, it becomes more apparent what is sought, and the process of deliberation can thus enrich the recognition of the role of human agency in the fabrication of futures.

Empirical grounding Common views on ethics and its emphasis on the rationality and justification of ethical reasoning seem to offer an objective and rigorous foundation for ethical judgments. However, one of its crucial flaws is that its emphasis on the critical analysis of moral language has severed it from actual action, situations, and existence. This explains partly why empirical ethics has come as such a refreshing development in the last decades to help ethics reconnect with human experience (e.g., Borry, Schotsmans, and Dierickx 2005; Wangmo et al. 2018).

From the perspective of the current theory, the frequent emphasis on intellectual virtuosity and the general lack of applicability of ethics theory is a major problem because it yields a rather weak, abstract, disconnected, static, contestable, fragile, and, in the end, more or less useful approach. Even professional (and honest) ethicists report having a hard time applying highly abstract ethics theories to their own lives (Callahan 1973) and when they do, it can lead to arguments that are overly narrow (e.g., Singer 2011) because of its limited attention to what human beings consider to be important in their lives. Scenario-based deliberation, without being an empirical data-gathering method in and of itself, calls for the empirical grounding of understandings of human situations. In this respect, scenarios are grounded, factually rooted, embedded, and embodied productions.

As a dialogical practice, deliberation promotes a way of approaching moral matters and ethical thinking therein as being situated and connected to experience and likewise for eventual solutions, which need to be connected to situations and rooted in people's existence (Widdershoven, Abma, and Molewijk 2009). In contrast to more conventional approaches, the theory of deliberative wisdom makes it clear that empirical knowledge about agents and about contexts can be of great relevance in understanding the nature of moral situations and engaging in deliberation toward meaningful resolution via the mobilization of the knowledge toward a situation. This also increases the feasibility of enacting ethical scenarios because a productive use of imagination is focused on realizable possibilities, not disembodied fantasies. When engaging in a deliberation, participants are constrained by the contextual barriers that make up a situation. In other words, as the realist Flanagan has argued, moral "oughts" imply moral "cans" (Flanagan 1991; Flanagan 2009). Importantly some "cannots" are not genuine cannots in the sense that they are barriers that we perceive as such but are not true barriers, or are barriers that need not be overcome. Also, grounding deliberation in experience and factuality is by no means a call for any kind of conservatism relegating ethics to simply replicating situations that currently exist. Ethics' relationship to the understanding of the world is different; it is less about understanding the world in and of itself than understanding the world in order to realize one's potential within it and envision realizable possibilities. In other words, "necessity is the mother of invention," as attributed to certain translations of Plato's *Republic*. Likewise, ethics wants to understand constraints, but navigate around or through them by supporting the imaginative capabilities required for innovation. Accordingly, the ongoing discussion about the is-ought fallacy, about dichotomies between facts and values, about normative and empirical ethics can become subterfuges to protect ethics theory from real-world tests and relevance by allowing the separation of ethics theory from actual existence (Racine and Cascio 2020). The vocabularies of facts and values are indicators of our different apprehensions of reality; they are tools to understand realities, but they are not reality itself. Valuations are part of human experience, and ethics theories that sharply separate valuations and their different intellectualized forms (e.g., values, principles) subsequently have the problem of returning to human realities with an unhelpful tool that initially moved them away from the nature of human experience. Possibilities are amenable to critical discussion and where more traditional ethical analysis can play a role—although not necessarily a sole role—in the evaluations of the merits of scenarios of action.

Holistic evaluation Finally, another advantage of the epistemology and terminology of scenario-based deliberation over narrower understandings of ethical deliberation based on moral judgment and moral reasoning is that it helps integrate different facets of situations within the transformation of a problem situation to a resolution via a preferred action-scenario. Moral judgments about a situation are easy and transportable shorthand for resolutions, but they typically do not carry forth any understanding of the situation, of the compromises made to get there, the comparative effort between options that were undertaken, and so on. All these pieces are important components of the puzzle to understand the situation and the process by which it can be resolved. In contrast to a strategy focused on ethical principles, the terminology of scenarios helps to encapsulate options and their evaluation—including arguments for or against them—within concrete contexts. Sometimes this terminology is not adopted explicitly in deliberations, but the process followed allows for the generation of embodied and embedded thinking.

This holistic contribution is well illustrated in many real-world deliberative efforts, ranging from more mundane conversations to highly structured public deliberations (Abelson et al. 2003; Doucet 2001; Fishkin 1991). For example, the remarkable resolution of debates about assisted dying in Quebec, the Canadian province in which I live, represents one of the better moments of 225-year history of democratic parliamentary tradition, and exemplifies its strengths and limitations. This process involved public consultations and hearings, although perhaps not enough and perhaps not to the extent wished by some notable experts on the question (Doucet 2015).

While far from perfect, it was a nevertheless a transparent process wherein different legislative options were considered, and one that was guided by the desire to avoid undue suffering and give patients agency in deciding whether they should be granted the possibility of being assisted in putting an end to their suffering. In the end, the scenario of allowing medical assistance in dying was granted to adult patients who could provide consent. A significant amount of legal ("Loi concernant les soins de fin de vie" 2014) and clinical guidance (Collège des médecins du Québec 2017; Medical Assistance in Dying Research Group 2017) and mentoring is in place to avoid pitfalls such as hasty and unprofessional medical aid in dying. For some, it did not go far enough (e.g., not covering assisted dying in the context of incapacitated patients). For others, it clearly went too far by granting patients the possibility of receiving help in dying, but for most citizens and clinicians, it appeared to be the reasonable and most ethical scenario.

With the growing experience of this process, there is now a current desire to grant this possibility, in the form of advance directives, to those who will eventually suffer from cognitive impairments (e.g., Alzheimer's disease), allowing them to make this decision before being considered incapacitated with regard to consent (Government of Canada 2021). Discussions are also ongoing for allowing medical aid in dying to mentally ill patients suffering from conditions such as severe depression, schizophrenia, or anorexia (Bryden 2021). The enforcement of the law was postponed in March 2024. The proposed initial resolution, which is being monitored by a provincial oversight committee (in the case of Quebec) (Assemblée nationale du Québec 2021) can now cover more ground since the experience has been positive, although always difficult for families and clinical teams from a personal and human standpoint. Such processes of deliberation allow us to consider not only what principles should be guiding us—this task could likely be settled in a few hours—but to imagine as a society what it would be like to accept such a proposal, what it takes to offset some of its drawbacks, and what we need to monitor the situations. For pure theoretical ethics, this may seem like a trivial or unchallenging intellectual exercise. However, the task of envisioning what we desire is not in itself an intellectual task; intellectual resources need to be mobilized in such a process oriented toward action, but it is richer than a simple academic exercise. The failure of carrying out such a process can jeopardize the rich and holistic thinking needed to envision, as concretely as possible, options for our futures.

The concept of scenarios helps reflect this embodied, situational, existential aspect of moral problem solving, which is much more than mere rational ethical reasoning. Moral matters are as broad as human problems can be, as meaningful and entrenched as they can be in our existences, in our biases and customs, in our values and in our ways of seeing the world. Overrationalizing the entire process is bound to induce reductionisms of different sorts. Adopting language that tempers the temptation of pure rationality and favors a more humane standard of reasonableness may reduce confrontations and encourage collaborations, which can become difficult when a party is judged to be irrational.[17]

Conclusion

This chapter described how ethical deliberation—as a general orientation—stands astray from the more deductive and top-down modes of reasoning where the search for foundational principles has served for some as the equivalent of the quest for the Holy Grail. In this paradigm, deliberation is a thinned-down process of identifying the principles to apply to a given case. Therein, ethical analysis in a narrow sense (e.g., understand what principles mean, which one is most important) is the strategy favored. In contrast, pragmatism suggests a much more open-ended deliberative process wherein situations as units of moral experience are first understood and then discussed openly to gather stakeholder perspectives on them. Deliberation here represents the ability to generate response-scenarios to this situation, and imagination plays a fundamental role in the move from a past and present orientation to a future and forward-looking orientation. The dialogical processes implied in such deliberation reflects a transactional account of human existence while the imaginative component captures the action-oriented nature of human thinking. Fruitful ethical deliberation is the effective exercise of these capacities in the search for paths of human flourishing. I explained how thinking of ethical deliberation as scenario-based deliberation has several merits, many of which correspond to the situational understanding of moral existence (chapter 1) and moral matters (chapters 4–7) sketched previously. The next chapters further describe important moments in ethical deliberation and criteria for evaluating response-scenarios.

11 Key Moments in Ethical Deliberation

The previous chapter explained key features of ethical deliberation and ethical imagination. It also emphasized how scenario-based deliberation is a compelling way of understanding ethical deliberation as a process that gets us from the experience of a morally problematic situation to the envisioning of promising courses of action. In the current chapter, I describe three important "moments" in ethical deliberations. I refer to these as moments to avoid the more linear idea that there are steps or an algorithm rigidly guiding ethical deliberation. The first moment consists of broadening and deepening the understanding of the situation. There needs to be a time to construct or reconstruct the understanding of a situation in order to pin down what it is and who is involved. Doing so calls for a recognition that so-called facts are not mere facts because moral matters are premised on interpretations of a situation, which is itself reliant on the narratives of people involved. Taking into account narratives and lived experience supports the identification of what is at stake and what is not in a given situation. The second moment is the envisioning of action-scenarios. To imagine and elaborate an action-scenario, time must be given to allow participants in a deliberation to depart from experience such that they actually have time to think about what they experience. Then, progress can be made to envision different action-scenarios. Once action-scenarios are named and identified, the deliberation can proceed to the collaborative enrichment and refinement of the scenarios. The third moment is the act of coming to a judgment or a decision based on the comparative evaluation of scenarios. I integrate a basic "feasibility analysis" of scenarios such that any unrealistic and implausible scenarios can be removed from further analysis. Then, ethical analysis proper can be carried out. This analysis is structured by two tests related to two criteria to evaluate the ethical nature of action-scenarios: ethical acceptability and ethical praiseworthiness. These criteria are described as complementary such that neither alone exhausts the nature and meaning of ethics or human flourishing. Both are needed and balance each other. The use of these criteria is illustrated through the use of examples, and the limitations of their proposed use are discussed. Without being a recipe, these moments of scenario-based deliberation connect to the previously described situated account of agency, the situational nature of moral problems, and the proposal to see the acting out of scenarios as enactments (as described in subsequent chapters).

> As for the future, your task is not to foresee it, but to enable it
> —Antoine de Saint Exupéry

In the previous chapter, I explained how deliberation is essential to a form of dialogical ethics like the theory of deliberative wisdom that relies on a transactional account of human existence and human morality (see chapter 1). In this chapter, I want to explain how deliberation is not an end in itself but foremost a process by which those concerned pay attention to the problematic aspects of the moral situation and recognize that the problematic nature of moral situations is inherently perspectival or positional. In response, the proposed process by which solutions can be envisioned is participatory and moves away from the influential application and deductive paradigm encountered in normative and applied ethics (Altman 1983; Hoffmaster 2018; Winkler 1998). As a participatory process of engagement and self-reflection, ethical deliberation cannot be broken up into completely separate steps because it is a rather organic process. The common idea of an algorithm-like linear flow of tasks—sometimes encountered in more rationalistic ethical analysis methods—would be contrary to the ideas I have already explained with respect to situational insight of moral problems and of moral and ethical awareness. They are also contrary to the spirit of a deliberative ethics that acknowledges the fallible nature of all aspects of our ethical understandings, discussions, imaginations, enactments, and evaluations. This commitment to fallibilism and to a corollary epistemic humility invites a more iterative framework (Racine 2016). Miller and colleagues judiciously describe it as a corkscrew: it contains both a linear and a circular/iterative aspect (Miller, Fins, and Bacchetta 1996).

Despite this caveat about the artificiality of separating ethical deliberation into steps, there are important moments that can be identified in most ethical deliberations, and it is the intent of this chapter to describe these moments, which are introduced by general questions in figure 11.1. The key moments in an ethical deliberation are (1) broadening and deepening the understanding of the situation via stakeholder consultation and face-to-face discussion; (2) envisioning action-scenarios, notably based on the processes of imagination and dramatic rehearsal; and (3) reaching a judgment or decision based on the comparative evaluation and analysis of scenarios for their ethical merits. These are not necessarily sequential steps to follow like a recipe, but rather ingredients that are likely to make up good ethical deliberation. They are recognizable under different vocabularies in various ethical deliberation methods (Baertschi 1998; Dion-Labrie 2009; Doucet, Larouche, and Melchin 2001; Hartman et al. 2019; Inguaggiato et al. 2019b; Widdershoven, Abma, and Molewijk 2009; Gracia 2001, 2003). (They are also premised on the initial and prior recognition of the existence of a morally problematic situation, which acts as trigger for ethical deliberation.) There is likely an advantage to thinking of them as building from each other as long as this is not interpreted as a straightforward or linear process. Deliberation is not like a train leaving its station with tracks leading to a prespecified destination. It is a journey, and the destination and the collateral learnings and observations can be as important as the destination itself. This chapter does not offer methodological and practical guidelines for deliberation, properly speaking, but it provides an orientation for how to develop such guidelines in the form of methodological guideposts (see also on this matter and like-minded approaches in Stolper, Molewijk, and Widdershoven [2016] and Widdershoven, Abma, and Molewijk [2009], as well as Hartman et al. [2019] and Svantesson et al. [2014]).

Moment 1: Broadening and Deepening the Understanding of the Situation

Once a situation is considered to be morally problematic for an individual, a group of individuals, or a collective such as a political community, the different aspects of the situation need to be reconsidered. This entails moving from the initial moral anchoring of the situation as a first-person experience to a phenomenon to be considered intersubjectively to critically assess if the problem is indeed morally problematic. This is a bit like a psychological decentering or distancing process (Bernstein et al. 2015), which the theory of deliberative wisdom requires, much like other case analysis methods in different areas of ethics (reviewed in Racine [2007]) or other theories of deliberation (Brown, Kitson, and McKnight 1992). This process is sometimes referred to as an "appreciation of the situation" (Brown, Kitson, and McKnight 1992) or as a process of case presentation and formulation of the problem (e.g.,

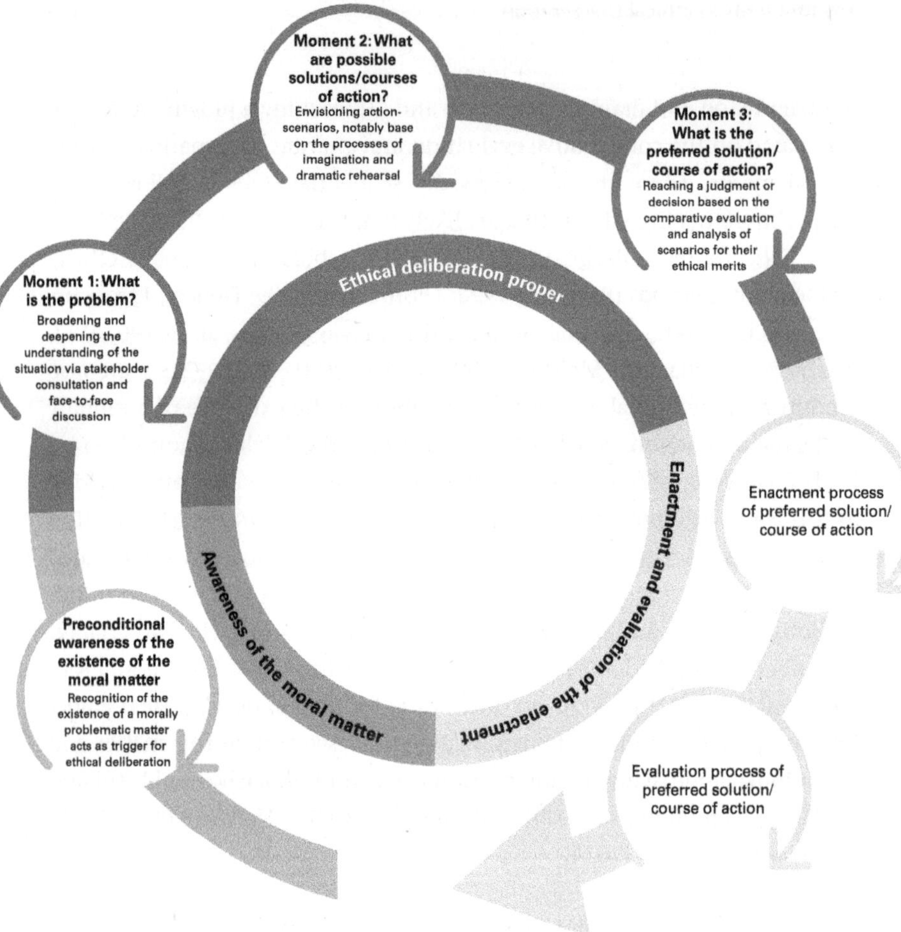

Figure 11.1
Overview of important moments of ethical deliberation. Inspired by Brown, Kitson, and McKnight (1992).

Although for simplicity's sake the figure portrays a linear process, I recognize that the tasks involved in each step are fallible, can be revised, and can interact with other moments and tasks. The first arrow is a gray arrow moving from the recognition of the morally problematic nature of a situation (as discussed in chapters 3–9) to a more open-ended engagement process of broadening understandings of the situation because deliberation typically presupposes that the existence of a moral matter is acknowledged. It is possible that more matters will be discovered or that matters will be discounted in the end, but the initial triggering is rooted in experience; it is a malaise that may be compounded, diminished, or enriched in the process of deliberation. Likewise, the remaining gray arrow pointing to ethical enactment as the last moment indicates an "end" to the process of deliberation and the beginning of the process of enactment described in subsequent chapters (chapters 12 and 13).

moral dilemma) within that situation or case, especially in the clinical or business context (Stolper, Molewijk, and Widdershoven 2016).

This process of broadening and deepening represents the distancing move from the moral experience to an ethical process where initial intuitions and experiences are put into dialogue with other perspectives and scrutinized with the help of facts and scientific evidence, if available. Obviously, this exercise will take different forms depending on the nature and scope of that situation. For example, if we take the healthcare setting to give ourselves some concrete parameters, it can entail the following nonmutually exclusive actions:

- An individual clinician confronted with a morally problematic situation involving a single patient will further examine the situation through consultation with colleagues and the involvement of advisers who can bring forth the experiential aspects of a situation (e.g., social worker, chaplain, bioethicist) to get a better sense of the actual problem and its nature.
- A group of clinicians (e.g., a clinical team) who are wrestling with a difficult situation may want to have a more structured process within a few meetings where the actual situation is first articulated through an open-ended process in which different members of a team or those touched by a situation are heard and have the chance to explain why they think the situation is problematic or not.
- A system of care where one or several potentially problematic situations have been reported may call for relevant authorities to examine, investigate, and offer some resolution to this problem. The level of the issue at stake and the existence of established processes may imply that the mechanisms for hearing stakeholders may become more structured with formal interviews, town hall meetings, committee meetings, note-taking, research into the relevant issues, and so on.

For each of these response-scenarios, there are better practices that may be distinct from each other. Yet, despite the different levels involved (individual, team, community, or system), there is a similar need to instill a process of openly defining the nature of the problem, as well as initiating a process of creative problem solving where different options or action-scenarios will be critically examined. For example, even the comprehensive Canadian Truth and Reconciliation Commission adopted a dialogical style where narratives were gathered across Canada to understand the experiential aspects of the life of Indigenous peoples (Truth and Reconciliation Commission of Canada 2015).

The attempt to broaden and deepen the understanding of the situation can be further divided in different "submoments," namely (1) reconstructing the understanding of situations, (2) interpreting the facts and understanding the narratives within situations, and (3) defining the issues at stake.

Reconstructing the Understanding of Situations

Understanding a morally problematic situation can be complex and limited by positional biases or misapprehensions. Acknowledging the perspectival aspect of this understanding of situations is crucial because it triggers a set of related epistemological reflexes to ensure that those limitations are always acknowledged and that measures are adopted to mitigate them. For example, the suggestion that one can simply get the "facts straight" is bound to be a large oversimplification since the facts will be tied to a person's experience of the situation and what the events mean for that person.[1] This is why, within an ethical process, the understanding of the situation needs to be reconstructed from initial understandings into one that is more critical and takes a distanced stance away from one's or a group of people's initial moral instincts and reflexes. It is also why I warned against the reduction of situations (see chapters 4 and 5) to a preferred standpoint (e.g., clinicians seeing the situations they face as "cases," or public servants seeing citizens as dossiers).

Moral awareness is the mechanism supporting the initial triggering of our moral sense as described in chapter 8, but this does not mean that all triggers are justified; they may even reflect different forms of bias and prejudice toward those of another group (e.g., ethnic, linguistic, racial, gender). Accordingly, understanding the situation does not mean to simply rely on the initial mechanisms of moral awareness to come to the assessment that a situation is morally problematic. That understanding needs to be scrutinized. For example, under the guise of alleged moral problems can be bad faith or vengeance at work to embarrass people. In this sense, the initial sorting out of the situation is not only a sorting out of facts, it is also a way of sorting out personal sentiments.

A good illustration here is the interpretation of certain discursive situations as offensive, which can be an important signal for problem situations (e.g., problematic language to describe people with disabilities or with nondominant gender and sexual orientations). The wielding of such offenses, notably in the English-speaking world, plays an important role in public spheres where the role of the individual and their values may be greater than in value-based republics (e.g., France) where a commitment to a set of civic values (e.g., *liberté, égalité, fraternité*) is seen as constitutive of the public sphere. There is a growing tendency to problematize and scrutinize the feeling of being offended—and why a person or group feels offended—such that experience-based and emotion-based understandings of a situation can be propelled to definitive understandings of problem situations. The threat of the power of unscrutinized claims of those offended is now seen by some as a threat to freedom of speech and the ability to discuss ideas in the academic and public spheres (Fish 2015; Ibbitson 2017; Rose 2017; Williams 2021).

Some ideas and words are associated with terrible harms and injustice, such that respectful, nonharmful, and constructive settings need to be created to discuss them and sometimes perhaps avoid them in the presence of some for whom these ideas and words are traumatizing. But the evolution of academic environments toward the creation of spaces free from offense raises important questions about what cannot be said and discussed because there is a risk that—while accommodating for sensitivities and experiences—we actually euphemize history and its legacy by "canceling" ideas disagreed with. I am not claiming that all those offended are wrong or that they are right, for that matter, because it depends what they are offended about and the reasons why. I am simply making the point that feelings of being offended need to be articulated and examined through reflection and dialogue to learn what they are truly signaling with regard to human flourishing. Problem situations cannot only be defined based on someone's impressions but call for a more critical interpretation if we are to make meaningful ethical progress. However, sometimes the open discussion needed to get there is simply too difficult for some to participate in.

I recognize that very difficult life experiences (e.g., racism, sexism, physical and sexual abuse) call for great discursive sensibility such that some need to be warned and spared from exposure to certain topics but, in the end, at the societal level, sensibility needs to be mobilized toward thinking about the situation and developing ways of being that surmount the harms at the root of offenses. We need to develop a positive vision of what we want in order to move beyond problematic categories, and this needs to be reasoned out and discussed with others. Policing, challenging, and revising language can be a step forward, but it needs to be built on understandings of situations and the context in which discourse arises (Cohan and Howlett 2017). Otherwise, changing or censoring discourse may give the impression of tackling an issue without addressing the broader social and economic aspects of some situations. In other words, changing the discourse is likely part of a solution, but our understandings of situations should delve deeper and claims of offense may limit the ability to do so. Imposing a new orthodoxy risks perpetuating systems of coercion in new ways instead of creating opportunities for joint futures and self-reflection. At the same time, dismissing the hurtful and offensive character of words and actions as being overly sentimental or as being all in people's heads, is a clear wrong.

Interpreting the Facts and Understanding the Narratives within the Situation

This brings us to the next methodological guidepost, which is that facts do not exist independently from narratives and interpretations. Interpretation is a part of a fact, as facts are experienced and the interpretive meaning given to lived experience is connected to personal narratives that are shaped by a person's history, which is itself shaped by the context in which life takes place. Here, cultural, linguistic, and other forms of diversity in identities can shape what different aspects of a situation mean to another person. Attempts to generate intersubjective understandings of a situation have been investigated by theoreticians inspired by Gadamer's hermeneutics (e.g., dialogical ethics) (Hartman et al. 2019; Ricoeur 1992; Widdershoven, Abma, and Molewijk 2009). This tradition has the benefit of moving away from the more Cartesian rationalistic understanding of deliberation (see discussion in Ralston [2010]) fostered by proponents of discourse ethics (Habermas 1991, 1999).

It will suffice to say for our purposes here that this initial apprehension of the situation carries forth implications, namely that the facts and the interpretations of facts need to be explored through the perspectives of different stakeholders involved in the situation, as well as those who may need to have a say. This perspective is narrative in structure in the sense that it is not only about explaining facts but also about understanding the meaning of experiences and events within the worldview and life trajectory of others. This narrative orientation sets the stage for envisioning how an eventual re-situation proposed through the elaboration of response-scenarios will fit in the projected existence of people's narratives. In a sense, the resolution of the moral difficulty is a form of healing and reconciliation in response to a break in narrative, granted that the deeper meaning and significance of certain events or situations within the existence of those concerned is acknowledged. This general methodological prescription does not guarantee that this form of narrative intake in the deliberation process will be done fairly and appropriately. Narratives can be collected and heard sometimes to then be better manipulated to harm the very people expressing them. Accordingly, the focus on human flourishing is crucial for more hermeneutic and pragmatist approaches, without which no cardinal orientation can guide the reconstructive process. This is what Ricoeur captures as the "aiming at a good life lived with and for others in just institutions" (Ricoeur 1992, p. 172). Within the theory of deliberative wisdom, this orientation—although open-ended—is further substantiated in terms of developing one's capacities, that is, developing oneself as a human being capable of authentic and autonomous actions (Ryff 1989). Aiming at the good life with others and in fair and just conditions is a main ingredient in developing conditions for human flourishing (Sen 1989).

Defining the Issues at Stake

Once a situation has been reconstructed, or at least enriched, through dialogue and team discussions, formal hearings, or simply taking distance and reflecting on different aspects, there is gradual movement toward the definition of what is at stake, that is, what is most important in the situation (Brown, Kitson, and McKnight 1992; Stolper, Molewijk, and Widdershoven 2016; Widdershoven, Abma, and Molewijk 2009). It could be a patient's well-being, a healthcare professional's sense of commitment; it could be the honesty toward a consumer, or, for instance, the positive or negative outcomes subsequent to certain specific actions. At this time in deliberation, it is not important to necessarily single out a most important feature but to identify those that are the "game-changers" and "game-makers" for different stakeholders. This process may at the same time lead to heated discussions and tensions. There could easily be disagreement about what is most at stake, and this is important to take into account. For example, a patient may consider medical aid in dying a most respectful and humane gesture that best respects their autonomy while, in contrast, a physician may take this as disrespectful toward a person (or their life) and inconsistent with their deontological obligations. Pinning down what is at stake in those difficult situations necessitates insight and open-mindedness not to impose a single vision (i.e., of one single party, of the moderator of a deliberation) that defines the situation and the important stakes. When such an understanding is formed through conversation, group deliberation, or a public hearing, a summary of that understanding should be tested and validated by the moderator with the input of those involved to ensure that their views are well reflected and not depicted in a strawman form.

Importantly, there are likely no simple rules of thumb to critically assess whether a problematic situation is a true positive (truly a problematic situation) or a false positive (a situation initially considered problematic but found not to be one after further consideration), and such language is misleading in its simplicity. Indeed, the notions of purely false negatives and false positives runs against the complexity of human experience because if some stakeholders experience the situation as morally problematic, it is possible that it may be determined not to be, but their initial experience should not be invalidated and should be an important driving factor. In chapters 4–7, I reviewed and discussed features of morally problematic situations in order to better define the contours of such situations. In this process, it can be useful to use existing ethics theories (see chapter 4, in particular) to identify whether some issue or stake can be articulated as a moral problem. This is what feminist philosopher Susan Sherwin (1999) describes as the use of ethics theories as lenses and the feminist pragmatist philosopher Mahowald dubs "so many ways to think" (Kegley 2013; Mahowald 1994). This view is also consistent with ideas proposed by Dewey such as envisioning ethical principles as shorthand, as pointers, for significant values and meaningful experiences (Dewey 1922). Likewise, in a paper discussing everyday ethics, Zizzo, Bell, and I propose a framework where different ethics theories can be "instrumentalized" as tools for perceptual apprehension of moral problems (Zizzo, Bell, and Racine 2016). Likewise, the ADC model of moral judgment (Dubljević and Racine 2014), following Dewey (1966), posits a multidimensional (pluralistic) view of the moral domain in line with three major families of ethics theories. They are theories focused on agents (A), theories focused on deeds (D) or acts, and theories focused on consequences (C). Additionally, for those involved in a deliberation about a potentially problematic situation, as well as those leading the discussions, some appreciation of basic features of any situation could help guide the initial understanding (see table 3.1 in chapter 3). Questions related to the persons involved (e.g., their intents or goals, their relationships), the action(s) at cause or being considered (e.g., their nature, their implications), and the different circumstances that give texture to the situation (e.g., in such and such business or health environment, with such and such resource constraints at hand) can also help unpack what the issue at stake is.

In sum, there are many ways that guidance can be developed to help identify salient moral problems through the analysis of the structure of moral situations, through tripartite integrative models of moral intuitions and moral judgment (Dubljević and Racine 2014), or through the integrative models of the use of ethics theories in guiding the identification of morally problematic situations (Zizzo, Bell, and Racine 2016). With respect to the theory of deliberative wisdom, these tools are envisioned as overlapping ways of operationalizing what matters to persons and what touches on their well-being and flourishing. In the end, talking to people and understanding what matters to them is the starting point, although ethics is also an invitation to critically reflect on what matters to us; it is not a mere consolidation or validation of our interests. It is an invitation to ponder them in light of what they actually lead us to.

Moment 2: Envisioning Action-Scenarios

Once a situation is understood by a group of people as being problematic and requiring a response, the process of envisioning scenarios to overcome this problem can begin. Here I use the term "envisioning" consistent with the meaning of theory as *theoria* (as a vision), and likewise, imagination designates a capacity to perceive or develop situated possibilities. Accordingly, the process of envisioning action-scenarios is in an active process. In the previous chapter, I elaborated on important virtues of scenario-based deliberation and the concept of scenarios as a projected situation, notably as a holistic and grounded unit of deliberation that responds to the nature of situated experience and situated moral agency. The concept of scenario-based deliberation expounded in the previous chapter also supports a process of deliberation as a strategy to move from experienced moral situations to the genesis of action-scenarios.

The development of action-scenarios engages forward-looking thinking and taps into the world of possibilities based on processes of imagination, much like other kinds of (creative) problem solving in other sectors of human activity (e.g., technical, educational, clinical, business and management, economics) (Mumford et al. 2010). Now, I turn my attention to more concrete aspects of this important moment in moral problem solving: the generation of action-scenarios. Three more specific "submoments" are reviewed, namely

(1) departing from experience, (2) freeing the imaginary to consider various scenarios, and (3) enriching and refining scenarios collaboratively.

Departing from Experience

I emphasized the importance of moral experience in previous chapters to the point where it could be seen as an anchor from which we cannot be freed. Given the situational account of ethical agency presented (see chapter 1), one could deduce that our existence is rooted in lived situations such that we remain confined to them. However, the role of deliberation and the elaboration of action-scenarios in response to situations is to allow us to look within but also beyond the situation, so to speak. In this sense, after having paid much attention to understanding a problem and validating understandings of the situation, there needs to be a *Gestalt* shift to imagine solutions. To do this, participants need to understand that deliberation could be uncomfortable. Discomfort can arise because someone is tied to an understanding of the situation that commits them to a certain kind of response. They only see things one way. But single-minded views are bound to be challenged throughout deliberation. It may also be uncomfortable because moving forward, no one's understanding of the situation, no one's existential point of view is necessarily favored *a priori*. This gap, this feeling of uneasiness or perplexity, of letting deliberation go, is a condition of growth and learning about others and about oneself (Higgins 2010). However, it may generate anxieties, especially if deliberation about a given scenario creates discomfort. It must be explained to participants that the departure from the situation is intended to better come back to it. In the end, no rash, unthoughtful, or unsatisfactory scenario will be recommended, so there needs to be some confidence that the process will eventually get participants to a valuable destination. But in the meantime, it may engender feelings of loss, uncertainty, perplexity, uneasiness . . . all the emotions related to not really knowing what to do and discussing openly existential uncertainties. This uncomfortable perplexity (Socratic *aporia*) was famously described by Plato in the *Meno* when Socrates asks his interlocutor, Meno:

> Have we done any harm by making him perplexed and numb as the stingray does? . . . Indeed we have probably achieved something relevant to finding out how matters stand, for now, as he (realizes) he does not know, he would be glad to find out . . . Would he have tried to find anything out before he fell into perplexity and realized he did not know and longed to know? (Plato 1981)

The ability to learn from others, to open ourselves to others and to the future, is associated with well-being and flourishing (Lambie 2014). The willingness to be open to new experiences and learn is one of the distinctive features of a positive and constructive life orientation, but it is premised on the willingness to understand and to admit one's ignorance or need to know. In this sense, the process of letting go of certain certainties, dogmas, and interpretations is to an extent unavoidable within genuine ethical deliberation (Senghor and Racine 2022). Therein, adopting a learning perspective is favorable to the moral agent and to their growth. But growing can be painful. It can also be perilous. The task of ethics is to make growth possible in order to overcome and while overcoming moral problems. It must instill conditions for deliberation to foster critical self-refection on everyone's stances and help people learn from each other (Higgins 2010). This is obviously easier said than done and cannot be boiled down to formulas or recipes.

Freeing the Imaginary to Consider Various Scenarios

It is one thing to shift the gaze from the actual problem to eventual solutions. However, to actively generate possible solutions, flexibility, open-mindedness, and the ability to consider various options is needed. This entails freeing the imagination to let our thinking move away from actual experience, which is sometimes difficult and emotionally charged, to then envision scenarios of what we would like to pursue as worthwhile goals and activities in an open-ended manner. Although a critical analysis of scenarios in terms of their merits needs to take place, the initial focus should be on freeing imaginative capacities and tolerating some lack of structure and clear direction in the form of ethical brainstorming. Focusing on opening up perspectives and loosening up can help to unstiffen perspectives and provide room for discussion and creativity. There must be collaboration in co-elaboration.

There are many different ways to promote creative moral thinking, and each person, each group of persons, each society is bound to have its own preferred methods and tricks that help free imaginative capabilities. For example, in Canadian Indigenous cultures (Muirhead and de Leeuw 2012), the arts can be a valued way to reflect and imagine what is desired and valued. Their contribution may be less immediate and perhaps more impactful on long-term life goals than when dealing with immediate and critical situations (Herman 2021). However, some medium of situation-bound art forms and creative activities may help unleash creative potential even in moments of turmoil. Creative arts may also have the additional benefit of bringing to the forefront a more leisurely approach to problems that appear unsurmountable, and thus reduce the anxiety of thinking about the future and about change, which may not be welcomed equally by everyone. In this sense, the creative arts may help tame problems and help reconciliation with the future insofar as they help mobilize a forward-looking perspective, one of the features of the arts emphasized in art-based therapies and interventions (Dalley 1984; Kapitan 2018).

There are certainly ways in which arts can be used within discussions, but most deliberations will take the more conventional form of an exchange, a discussion, or a dialogue. In thinking about possible action-scenarios, the goal is to eventually go back to what is esteemed as valuable, worthwhile, and crucial. If the moral problem at hand is about how to fairly handle a hiring process, then one (or a group) must start imagining what a fair hiring process would look like. If the moral problem is a personal moral question about how to reconcile obligations toward different persons (e.g., child, elderly parent, spouse), one must imagine how engaging in certain activities and commitments will impact one's relationships and the welfare of others around them.[2] In all cases, the idea is to return to what is valued and to start envisioning what could be possible responses to the situation. In the course of doing so, Doucet encourages spontaneity such that participation in a deliberation does not feel restrained to certain scenarios and to promote an open and constructive process (Baertschi 1998). This recommendation is advisable to avoid prematurely stiffening the exchange. It is possible to encourage each participant to voice a scenario or to use a white board (or other visual support) to note proposals made by participants as recommended in certain deliberation methods (Stolper, Molewijk, and Widdershoven 2016). In a nutshell, this aspect of ethical deliberation resembles a brainstorming session where hypotheses are generated without too much consideration of their full merits and implications. At this moment, all ideas should be welcomed (no wrong answers, at least not yet) and no detailed explanations of new ideas are needed, reserving for later the judgment about the merits and implications of scenarios, thereby compiling a list of creative, original, and outside-the-box ideas. All suggestions are valuable (at minimum, they will help contrast and compare options) and a greater volume of ideas will likely be useful, including generation of spin-off scenarios or adjustments to preferred scenarios.

To encourage pluralistic thinking and to unstiffen perspectives and the entire deliberation process, Doucet suggests that at least three potential action-scenarios be identified. This strategy helps to operationalize the insight that modern-day contexts are plural and that many options could be reasonable and even good (Baertschi 1998). It is also a response to the fact that moral problems tend to be approached as dilemmas and ethical discussions as debates, where different sides (usually two) argue against each other for the right and the wrong answer as most saliently illustrated by the culture wars and their impact on ethics discussions in the United States (Callahan 2005a; Hunter 1992) and now increasingly internationally (Duffy and Skinner 2021). Furthermore, it helps participants realize that each of these options could be potentially a combination of interesting features and trade-offs presented by other options. Accordingly, the "rule of three" proposed by Doucet is not set in stone, but it should be viewed as an encouragement to think in terms of the plurality of potentially viable and defensible options. It thus helps to instill distance-taking and encourage discussion about the possible benefits of different options (e.g., when groups are divided into yes/for and no/against camps). It also fosters a sense of inquiry, a form of Socratic *elenchus*, a dialectic as a form of collaborative search for knowledge based on the love of wisdom. This contrasts to rather a mere debate between two opposite positions or other forms of dualism or an eristic, the love of argument for the sake of arguing. The existence of multiple possible action-scenarios and responses triggers a need for an inquiry, that is, a dedicated effort to scrutinize the situation and possible responses to it. Accordingly, it can be a useful rule to always envision at least three options to mobilize one's own or others' creative moral thinking skills. Importantly, the exercise is focused on the actual courses of action at this stage and not on different ethical principles or different aspects of ethics theory. It thus helps to encourage projection into the course of action, which is the experiential realm in which ethics exists in its fullest sense because action is the realization of human existence from a transactional standpoint.

Not only is the process of scenario envisioning the beginning of a certain catharsis for participants, it is also when participants can start envisioning themselves as partners in a shared venture where they—as family members, colleagues, or citizens, depending on the context—will be called on to learn from each other and continue their progression as human beings. This process may take time and may need to accommodate pauses and silences where people do not speak (or when there is a lack of clear direction). In complicated public dialogues, progress may be measured in years and even decades. Great care is needed to ensure that those who feel less comfortable speaking be provided time and invitations to speak, potentially offering alternate ways of communicating (e.g., writing instead of speaking aloud, if the latter is a barrier). Indeed, a whole range of personal or social factors may hinder participation in scenario generation. As it can be imagined, the role of moderators (e.g., professional ethicists, trained facilitators, and mediators or others skilled in this task) in such deliberations and at sensitive times where imagination is freed is highly valuable. The purpose of moderation is not to enforce preset theories and concepts, but rather to facilitate exchanges and guide the person, the group, the collective in the process of generating different scenarios. This has been well described by Widdershoven, Molewijk, and colleagues who have articulated a vast practical experience and reflection on ethics as a dialogical process, including about the role of the ethicist therein, which as one can imagine resembles Socrates' *maieutic* (midwifery), a process by which Socrates considered that he helped give birth to truth as truth was already in the person so to speak (Inguaggiato et al. 2019a, 2019b; Widdershoven, Abma, and Molewijk 2009).

Enriching and Refining Scenarios Collaboratively

Once scenarios have been proposed—and even while they are proposed—the enrichment of the different aspects of these scenarios can take place. While a more formal evaluation and judgment should be postponed to avoid stiffening the imagination and open discussion that frees imaginative capabilities, further fleshing out of scenarios in terms of, for example, their justifications and implications can take place. However, while thinking and talking about individual scenarios or when moderating group discussions about scenarios, there should be a more formal pause in the process to ensure that each scenario is given sufficient attention and its implications are spelled out. This process, given the brainstorming spirit of the initial steps of an ethical deliberation, should be fair such that each scenario is given some charitable attention. In this process, there could be some cleaning up and sorting out of scenarios such that some are amalgamated with others, while others may be discarded because they are deemed too impractical or unfeasible to merit further attention. Some may want to withdraw their proposed scenario from the discussion because they feel it is unjustified or not worthwhile to pursue given the other options on the table. However, in the spirit of openness and collegiality, and for the sake of conducting a rigorous deliberative exercise, the dismissal of scenarios should be postponed until their more formal assessment is carried out (see the third major moment in the next section).

The separation of scenario genesis (coming up with scenarios) from scenario enrichment (trying to make the case for each scenario) is somewhat artificial, but there are virtues of separating them, perhaps more so when the discussion is heated and deep tensions are at stake, or the direction is particularly unclear or controversial. This is the case because such tense contexts stiffen the imagination needed to generate scenarios freely and to really try to envision the future openly. Accordingly, the effort to separate out the proposal of scenarios from their more thorough consideration is even more important to ensure leeway in the deliberative process.

Practically speaking, moderators of ethical discussions can list identified scenarios as they surface and then revisit them individually. In this context, participants are asked to imagine a scenario, list its implications, and identify its justifications. Doucet recommends that each scenario be fleshed out in terms of its implications (both advantageous and disadvantageous) and that a key moral value or principle be identified as to why a given scenario would be deemed worthwhile. I am not sure whether pinning down a single motive is preferable to identifying a few key reasons. Identifying what is valued within a given scenario and why it would be deemed worthwhile seems sufficient without emphasizing too much the need to pin down a single motivation or to formulate it in standard academic ethics discourse (e.g., as a recognized ethical value). Simply put, the principle guiding this moment in the deliberation is that each envisioned action-scenario be seen in the most favorable light. This does not mean that disadvantages are neglected, but that the full merits of a given scenario be envisioned. Dewey also recommends seeing this task of deliberation as a dramatic rehearsal of what such and such scenario would look like and would do (Caspary 1991). Doing so requires humility and self-restraint from participants because many may entertain preferences for a given scenario, but the task is not to advocate for a given scenario at this point in time but to give it full consideration as a course of action. A certain generosity toward other points of view is also required to sustain this process. Ideally, this process is again a collaboration in co-elaboration where participants in a deliberation help each other in fleshing out implications of scenarios. By doing this together, participants in a deliberation learn from each other, think with each other, and therefore also prepare to act together with a better understanding of the concerns and thoughts of others. As I have stressed elsewhere in this book, the viability of ethical deliberation, and ethics more generally, is premised on the good will of participants and their hope that the process and result will be helpful to them and to the situation. Skilled and seasoned ethics moderators may be helpful when the climate is too tense for participants to engage collegially in such a process, especially when participants have little experience in doing so. Ethics as a field can nurture such confidence and reassurance by pointing to precedent successes in deliberation. It can explain the many virtues of deliberation and the value of learning from each other. However, the powers of ethics are limited when the search for agreement and mutual benefit are not in sight. In such cases, formal arbitration or coercive norms imposed by the law can become the only way of moving away from a deadlocked or entrenched conflict. In the case of relationships between communities and nations, it may become a last resort to claim rights and moral status. For example, in Canada, the courts have largely supported the claims of Indigenous peoples, to the detriment of the Canadian Government, thus ensuring a form of ethical acceptability while the Canadian Government could move more proactively toward an orientation of ethical praiseworthiness (Saul 2014). Living with an unresolved problematic situation is also a possible outcome, but it is bound to generate distress and tension. While moral aspects of human existence are quasi-inescapable (e.g., even bandits and thieves will have their own understandings of reciprocity and fairness), ethics is not mandatory and cannot be imposed. It is a decision and a journey to take.

Moment 3: Coming to a Judgment or Decision Based on the Comparative Evaluation of Scenarios

Once different action-scenarios have been generated and refined, they must be compared and contrasted for their respective ethical merits. Now, it is true that ethical deliberation is ultimately, according to the theory of deliberative wisdom, focused on the impact of different scenarios on eudaimonic flourishing, namely doing the best thing, the right thing, the good thing. This singular focus may initially come across as vague, unconnected to ethical theories (the specific principles and the values they promote), and difficult to operationalize. Accordingly, I set out to provide more details about these points in hopes of showing that ethical deliberation aims to focus on and seek insights into human flourishing. Ethical deliberations build from existing ethical wisdom of human beings instead of suggesting that human flourishing is disconnected from ethics theory and is a stand-alone entity. In so doing, I keep in mind that pragmatist ethics is less an ethics theory in the traditional sense of normative ethics theories than it is "a method for understanding better—or reconstructing—already existing theories, and more generally, a method that enables greater awareness of our actual moral life" (Serra 2010, p. 7).

Pursuing this task of reconstruction, the theory of deliberative wisdom proposes that insights from previous ethics theories be operationalized as two different standards or tests of action-scenarios: ethical acceptability and ethical praiseworthiness (Racine 2010).[3] As I will explain, this process helps eventually identify a preferred action-scenario and thus lead the path to the enactment of a preferred response-scenario. However, prior to this ethical analysis, I propose that a preliminary "feasibility analysis" be undertaken to help sort out some of the possible ethically worthwhile but realistically impractical scenarios. Such scenarios may be those that are impossible to realize at a certain time or place. In such cases where a scenario is ethically preferable but unfeasible, it is best to recognize it as such and justify explicitly why a certain preferable scenario is left aside to avoid deeming such scenarios ethically inappropriate. This will help situate the conversation in the realm of the possible such that it is clearly understood that it is material circumstances that make a scenario impossible, even though the ethical merits and justifications of these scenarios may be revisited or made possible in other circumstances. In the end, action is the existential integrator where material conditions and values are embedded and perpetuated.[4] Scenario-based deliberation and ethical analysis helps reflect this inseparability of values from human realities and is differently oriented than more theoretically driven strategies that tend to isolate values from actual experiences and courses of action.

Feasibility Analysis

In order for different scenarios to be discussed seriously and considered as legitimate options for ethical action, they must feasible, even if only minimally (i.e., not impossible). Now, this is a tricky suggestion to make. First, discussions about feasibility can be contaminated by biases and preferences for a given action-scenario such that one can easily start dismissing a scenario as unfeasible because of more normative, value-laden reasons and not due to more factual considerations. (I recognize that there is a spectrum of fact and value, but it is nevertheless helpful to make this distinction to pursue an intersubjective assessment.) Second, the assessment of feasibility can veer into conservatism if no change to actual practices and situations are made, including uses of resources and energy granted to pursue a given scenario. For example, some scenarios may require asking for additional resources from those administering an institution. Or they may require people to do things differently or to change tasks or roles among different individuals.[5] What I have in mind is a rather generous and charitable exercise that is perhaps more focused on identifying scenarios that are implausible or impossible rather than actually assessing and comparing the feasibility of different scenarios. This helps give full attention to the scenarios that are feasible, if only minimally. In a sense, the exercise is more an assessment of nonfeasibility of scenarios or the identification of unfeasible scenarios than those that are feasible.

Given all these obstacles, would it not be preferable to simply not propose a feasibility test at all? This option would certainly make it easier in a certain sense and avoid the hassles of entering a discussion about what is practically feasible and, perhaps more importantly, what is clearly unfeasible. However, in my eyes, in general—since there could be exceptions or situations where this is less obviously needed—it is better to stage at least an initial conversation about feasibility before engaging in a deeper discussion about the comparative ethical value of potential action-scenarios. There are several reasons for this. First, not engaging in a critical analysis of (non-) feasibility could undermine the credibility of the process. Discussing unrealistic scenarios may lead to disengagement or a decrease in interest and attention of anyone minimally realistic and practical in spirit. Second, if feasibility is not discussed explicitly, it will likely keep playing a role, perhaps an implicit but insidious role, where some scenarios will be critiqued and undermined because of their lack of realism and not really for the ethical value they hold. Also, having participants know that there will be a point in time in the deliberation where feasibility is addressed (and not too late in the discussion) will reassure them that practical aspects will be tackled explicitly and meaningfully. Third, discussion about feasibility can help address some remediable defects of action-scenarios such that an actual scenario could be adjusted to make it more feasible and therefore the discussion about its ethical value will be done with a credible and actionable version of the scenario. Thus, a given scenario is not lost prematurely and can be salvaged, especially if it is deemed to reflect certain important values. Furthermore, those who would incorrectly judge a scenario to be unfeasible could be corrected and thus could eventually be rallied to the cause; they would likely be harder pressed to contest the feasibility of a scenario at the very end, after its ethical analysis *sensu stricto*. This encourages buy-in from most individuals, a focus on feasibility when it is relevant, and a focus on human values when that topic is later addressed directly. Finally, tackling feasibility ensures that the unfortunate outcome of an ethically desirable but unfeasible scenario is preempted.

In principle, all scenarios vetted after this moment of deliberation should be deemed feasible enough, although some details may need to be attended to, formal agreements for resources sought, and so on. It is possible that a given scenario will require an actual process of enactment where agreements are brokered and negotiations established to ensure the feasibility of a scenario. It may also be the case that a fallback scenario will end up being preferred because resources or authorization will not be given to a preferred action-scenario. Scenario-based deliberation allows for these possibilities and ensures clarity in why certain preferred scenarios were initially adopted but then abandoned for practical reasons. Even though this may seem suboptimal or like a defeat, it is perhaps a learning opportunity for an institution or a society to see workers or citizens wanting a given change in the longer term. These preferred but abandoned scenarios may herald the future because of the commitment they reflect to certain values and changes unfeasible now, but in sight for the future.

It is difficult to propose specific conditions to assess feasibility generally. Clearly, each individual, each institution, each context, each society has different resources and capabilities that make certain changes more or less difficult. Some environments may have a good self-understanding of what is a possible or an impossible change (e.g., because they have experience dealing with similar problems); others may not. It is also the case that overcoming what initially seem to be insurmountable obstacles and situations is often a source of tremendous pride and learning. As the famous Black orator and adviser to American presidents Booker T. Washington said, "Success is to be measured not so much by the position that one has reached in life as by the obstacles which he has overcome" (Booker 1997 [1900] p. 39). Overcoming obstacles gives us strength and confidence and builds capacity; it is central to the understanding of resilience (Luthar 2006). At the same time, there is wisdom in thinking of another idiom originating in the same period—when nineteenth century men would offer to each other chewing tobacco—to be careful to not "bite off more than you can chew." Feasibility, in the context of ethics, involves taking risks inherent to changing habits and ways of thinking and acting, but it cannot be an exercise in recklessness.

Ethical Analysis

Once certain scenarios are deemed unfeasible (through the process of feasibility analysis), the remaining potential action-scenarios need to be considered for their ethical merits. While traditional ethical analysis has typically structured this exercise based on the consideration of ethical principles, the proposed scenario-based deliberation anchors considerations of ethical merits to actual action-scenarios, and it centers on human flourishing in those scenarios and less so on traditional ethical principles. Accordingly, although ethics theories can be used to analyze and discuss values outside of concrete action, this is impossible from the standpoint of the agent because values do not float above human realities but are actually embedded therein; they are part of transactions. Thus, scenario-based deliberation encourages a grounded consideration of values within actual courses of action, not outside of them (see previous chapter). There is merit to highly detached analyses of values, but again from the standpoint of the user, a reconciliation with experience needs to happen. Moreover, the focus on human flourishing—which is here understood in an open-ended, nonteleological[6] way of calling for mutual learning—implies that there are often deep uncertainties regarding the actual merits of various courses of action in effectively supporting the development of fundamental aspects of human flourishing (e.g., autonomy, positive relationships with others). The recognition of this uncertainty is here viewed as a strength because it means that humility is in order and that we should be open to the fallibility of our enacted scenarios, to learning about the process of thinking about the better life, and that we should see the outcomes of our actions for what they really are.

To structure the consideration of ethical merits, I first present two ethical criteria—ethical acceptability and ethical praiseworthiness—that allow us to structure and operationalize the evaluation of response-scenarios from an ethical standpoint; these have been articulated previously (Racine 2010). Then, I compare this approach to more traditional methods of ethical analysis to situate this proposal with respect to traditional ethical analysis.

The Distinct Criteria of Ethical Acceptability and Ethical Praiseworthiness

I have previously proposed two criteria of ethical assessment of action-scenarios based on two fundamental nonmutually exclusive orientations of ethics: ethical acceptability and ethical praiseworthiness. I provide some background on these concepts as first introduced in my monograph *Pragmatic Neuroethics* (Racine 2010), and also expanded upon in previous chapters (chapters 1, 6, and 12) of this book. As readers will see, these concepts are tools that allow us to incorporate insights that concern two distinct orientations of ethical thinking. Some content of the following section is inspired by previous work (Racine 2010).

Ethical acceptability The concept of ethical acceptability relies heavily on the insights provided by the principle of not doing harm to others, nonmaleficence (*primum non nocere*). It also finds its source in negative obligations that do not necessarily entail positive requirements of action from the moral agent but often require one to refrain from imposing their own views and aspirations upon others (see table 11.1). The search for ethical acceptability is reflected in Berlin's account of negative liberty[7] (Berlin 2002) and Mill's famous harm principle ("The only purpose for which power can be rightfully exercised over any member of a civilized community, against his will, is to prevent harm to others") (Mill 1859/1956, p. 13). When striving for ethical acceptability, we seek permission to do something within the parameters of an existing moral framework while respecting others around us. Hence, sources of ethical acceptability typically lie in extrinsic, or external, sources of moral insights, such as the law and social consensus. Ethical acceptability yields what some call a minimal or thin form of morality, notably because it is mostly guided by the interests of others (e.g., not inducing harm to others) while not really spelling out our positive obligations toward each other (e.g., what we should do to live as a flourishing person or what we should do to contribute to our communities). Accordingly, if I can do something without hurting others while respecting their freedom, I can usually pursue action; it is ethically acceptable or ethically permissible to do so.

Ethical acceptability is also consistent with a procedural view of the public good and commitments to liberal individual rights. In *Pragmatic Neuroethics*, I explain how this commitment of liberal moral-political thought usually constitutes the main argument advanced by proponents of individual freedoms and is likely, in combination with liberal economic forces and product marketing, a key ingredient of this position. Indeed, the importance of ethical acceptability has perhaps been best enunciated by liberal thinkers like Kant and Mill, who have stressed how the exercise of our autonomy and freedom needs also to take into account the interests of others. Their work has laid the ground for a modern rendition of societies based on the development of individual persons who have the freedom to develop themselves. They have argued for the importance of respecting the freedom of others at the same time as they advocate for individual freedom and autonomy. These thinkers—and many others—did not stop their thinking with ethical acceptability. Although they are renowned for the formulations of ethical acceptability in the form of the categorical imperative and the harm principle, respectively, they also formulated a positive view of what autonomy and liberty are important for, namely the development of our own potential as human beings (Kant 1991; Mill 1956/1859). It would thus be profoundly unfair to label these thinkers as thinkers of ethical acceptability (only).

Ethical praiseworthiness Ethical praiseworthiness reflects another important and complementary orientation of ethics and human morality. Beyond respecting the criterion of ethical acceptability, ethical praiseworthiness calls on us to question if we should ethically and ideally pursue a given action or scenario. It relies on some of the moral insights captured in the ethical principle of utility or of beneficence and other positive obligations that entail prosocial behavior and the engagement of actively pursuing our own welfare, as well as promoting that of others. Ethical praiseworthiness is what many call the "thicker" (Geertz 1973; Williams 1985) consideration, often found in some more conservative views (Racine 2010). When pursuing ethical praiseworthiness and evaluating our behaviors in that light, we seek to fulfill positive responsibilities and duties toward others beyond what could be called for by mere convention or by simply not interfering with other people's welfare. Hence, the sources of ethical praiseworthiness typically lie in intrinsic, or inner, sources of morality, such as empathy, intrinsic motivation, and self-reflection, and are beyond the requirements of most normative obligations found in legal documents and other forms of social agreement or social consensus. In comparison to ethical acceptability's "minimal morality," ethical praiseworthiness yields an "ethics of ideals" or an "aspirational ethics" because it entails the active projection of who we want to become and how we help others become who they are and want to be (see table 11.1,

Table 11.1
Two complementary concepts of morality

	Ethical acceptability	Ethical praiseworthiness
Reformulations	Can we (ethically) do this? "Minimal ethics"	Should we (ethically) do this? "Ethics of ideals"
Inspiring principles	Nonmaleficence	Beneficence
Obligations	Negative obligations	Positive obligations
Sources	Extrinsic sources of morality, regulations and obligations	Intrinsic sources of morality, ideals and responsibilities
Developments	Ethics of the last 40–50 years: ethics as a mode of regulation	Ancient ethics and positive psychology: ethics as a way of life

Table adapted from Racine (2010)

adapted from *Pragmatic Neuroethics,* which offers a comparative overview of these two concepts).

The complementary nature of ethical acceptability and ethical praiseworthiness In most contemporary ethics theories, ethical acceptability is rather well addressed while ethical praiseworthiness is more contested, although most ethics theories (e.g., Kant, Bentham, Mill) provide insights about both (see illustrative examples in table 2.1 in chapter 2 about insights provided by various theories and table 11.2). For example, Kant teaches us how it is ethically unacceptable to treat each other as means to an end and invites us also to cultivate a view of each other that nurtures our ability for autonomy and self-governance. Both insights help spell out what is the most ethically desirable course of action and a comprehensive ethical analysis includes both orientations.

Despite my attempt to insist on the complementariness of both orientations, they can coexist or generate tensions and conflicts. For example, as a citizen, I am obliged to duly pay my income taxes to support social programs and services for my fellow citizens (mostly reflecting an ethical acceptability orientation), but provided that I have some ability to do so, should I give some of my income to charities beyond established obligations (mostly reflecting an ethical praiseworthiness orientation)? Another example is that healthcare professionals often face situations where they are reasonably satisfied with how health information is understood but may disagree with the decisions of patients (or of their surrogate of patients)

Table 11.2
Simplified illustrations of different insights of common ethics theories about ethical acceptability and ethical praiseworthiness*

Ethical acceptability (negative obligations)—things to avoid	Ethical praiseworthiness (positive obligations)—things to pursue
Aristotle (Aristotle 1999)	
Live a life consistent with the avoidance of extremes.	Pursue actions and paths of life that are most consistent with virtue, which may require siding with one side of the extreme as required by the situation.
Kant (Kant 1785/1998)	
Don't treat others (only) as means.	Treat others as persons who have autonomy and the ability to self-govern and facilitate this ability in others.
Mill (Mill 18591956)	
Avoid pain and harm to others.	Seek happiness (well-being) for the greatest number of individuals.
Rawls (Rawls 1971)	
Avoid unjust and unfair treatment of individuals.	Pursue a vision of justice as fairness as a structuring ideal of collective life.
Gilligan (Gilligan 2001)	
Don't be indifferent or insensitive to others (lack empathy toward others).	Attend to the needs of others; be engaged and respond to others' needs; transform the needs of others into positive obligations.

* The purpose of this table is to illustrate simply how we can learn more about ethical acceptability and ethical praiseworthiness in ethics scholarship. My statements are simplified here for this purpose. A reanalysis of insightful ethics theories could be undertaken to further operationalize these two basic orientations and expose the limitations of this strategy.

because they have diverging views on their own welfare and well-being. How far should they go in challenging their patients in their beliefs about their health to instill reflection on the broader picture and deeper values consistent with an orientation toward ethical praiseworthiness? How can they avoid paternalism by inviting patients to reflect more holistically on a situation or avoid simply respecting their patient's wishes without asking any questions, an attitude more in line with an orientation toward ethical acceptability? Countless examples could be pulled from other contexts, such as our attitude toward the environment or the way that workplaces are organized and the goals pursued by what we refer to as the "private sector" (although there is much public interest in that sector). Much everywhere we can envision possible tensions between our habits and their relationship to ethical acceptability and to ethical praiseworthiness. However, this tension is a productive one. To truly do good, we must not harm others, and to not harm others, we must also help ourselves and others grow and develop as human beings. Accordingly, a simple equation between acceptability and the dont's (or negative obligations) and ethical praiseworthiness and the dos (or positive obligations) is too restrictive since ethical acceptability and ethical praiseworthiness are only roughly approximated with these categories of obligations.[8]

Despite the complementariness of ethical acceptability and ethical praiseworthiness, there are several good reasons why many contemporary ethical theories settle for the "ethically acceptable." Centering ethics theory on the avoidance of harms to oneself and others offers a compelling account of what we need to settle on as a vision for the good. Also, it may be easier to clearly identify and agree on harms than on goods. Also, focusing on what is acceptable and especially what is unacceptable leaves some leeway for individuals to choose what to do with their autonomy, that is, to operate within a set of conditions that allows them to pursue their interests and values. Some influential ethics theories also consider that ethical praiseworthiness is beyond human achievement or is (too) relative to individual preferences and thus not easily identified or defined (Habermas 1991, 1997, 1999). As a result, they claim that we need to settle for a criterion of agreement on the conditions that make ethical judgments acceptable (e.g., conditions of the universalizability of ethical norms) without being able to specify a concrete direction or content related to any actual situations. However, identifying harms and what should be avoided or what represents minimal conditions of ethical acceptability only gets to part of the answer of the ethical conduct of human behavior as, once this is said, many options can remain, calling for an engagement with what we esteem to be valuable goods to pursue.

Fortunately, evidence has shown that the pursuit of human flourishing and a praiseworthy life is definable and refers to concrete dimensions and abilities in human lives, and that these abilities are very generally associated with well-being, even across cultures. In a review of over 25 years of research in this area, Carol Ryff—one of the pioneers of modern empirical eudaimonic research—explains how well-being is associated with physical and mental health, community engagement, purpose in life, and autonomy (Ryff 2014a). It is my hope that I can show and explain that ethics does not need to settle only for ethical acceptability but can embrace the search for ethical praiseworthiness, although ethical praiseworthiness is clearly a process, and not as easily ascertained as the fulfillment of the conditions of ethical acceptability (see notably chapter 2 on this topic).

I propose here and elsewhere that the two orientations of ethics are not opposed but complementary and necessary for an account of the ethical life (see figure 11.2) (Racine 2010), akin to how others have made similar arguments based on similar distinctions (Callahan 1994, 2005a, 2005b; Parens 2006). Irrespective of the final ability of these two orientations to serve as the two basic defining orientations of ethics, they help integrate and operationalize insights from a range of ethics theories and their teachings about what is wrong and what is right. Pulling from the insights of various theories about what is ethically acceptable and what is deemed ethically praiseworthy helps transform these theories into tools for the analysis and the refinement of action-scenarios. It also helps differentiate what is driving a given action and its ability to respect ethical principles (ethical acceptability) as well as its ability to yield more and promote the greater well-being of those concerned (ethical praiseworthiness).

From an ethical standpoint that encompasses both orientations, ethical acceptability can be viewed as a necessary but minimum condition. Surely, before actively seeking to promote a particular vision of the good (ethical praiseworthiness), we must ask ourselves if we are causing any harm or hindering anyone's freedom (ethical acceptability). However, it is clear that ethical praiseworthiness implies additional considerations based on a substantial view of morality (in opposition to a "thin" liberal view) committed to the

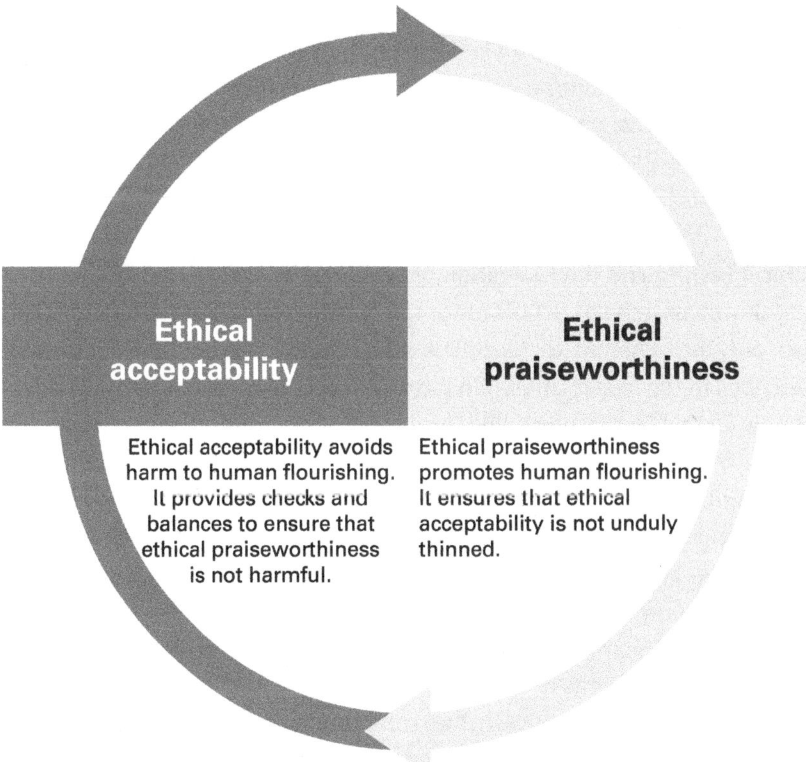

Figure 11.2
Mutually constraining (synergetic) model of relationships between ethical acceptability and ethical praiseworthiness.

analysis of the impact of the enactments of ethical scenarios on relationships and communities as well as on moral agents.

My proposal is that both orientations mutually inform and cross-check each other insofar as ethical acceptability provides a way to assess the acceptability of different views on ethical praiseworthiness, while ethical praiseworthiness invites us to reconsider what we judge to be acceptable and unacceptable. Accordingly, the theory of deliberative wisdom does not commit to a predefined view on human flourishing as entailed by conservative theories (Cohen 2006). Rather, it promotes a redefinition of human flourishing based on experience and deliberation (or dialogue) such that the pursuit of well-being and flourishing should be supported by proven and effective strategies that are agreed on via deliberation. Language about criteria could here be misleading since a criterion implies that there is a single entity to test for or to assess. The proposed two criteria are actually multifaceted and require the consideration of a set of issues that fall under them.[9] These more specific issues could be called "tests" of the criterion.[10]

Ethical Testing and the Evaluation of Scenarios

Within the exercise of deliberation, true testing cannot be carried out. This can only be achieved in its fullest sense during a process of enactment described in the next chapter. What can be undertaken in the course of deliberation resembles "face validity" in social science and psychology research insofar as a given scenario will be predicted to be conducive to certain valued or detrimental outcomes. Real testing takes place when a scenario is enacted. Deliberation—at least the third key moment in deliberation focused on evaluation—is an effort to screen scenarios for their predicted desirable and undesirable properties. It is an important step but of course no substitute for action itself. Moreover, I propose that, procedurally speaking, there is value in separating out merits of various scenarios in terms of ethical acceptability, then ethical praiseworthiness, before combining the results of these analyses. This procedure helps acknowledge and operationalize these two distinct ethical orientations. Since they are not necessarily opposites, they can reflect different sets of insights about the ethical thing to do and the preferable ethical scenario, while offering a path to the synergizing of the two orientations in order to improve scenarios.

There is no preset calculus or balancing between ethical acceptability and ethical praiseworthiness such that considerations of ethical acceptability would always supersede those of ethical praiseworthiness or vice versa. In some cases, breaching ethical acceptability may be warranted to promote a substantive and clearly defined praiseworthy good. For example, delaying the second injection for COVID-19 vaccination as was done in Canada—given that the first injection was deemed effective at 90–92 percent after two weeks—may have been justified given that the freed-up doses were used to immediately double the number of vulnerable people protected. It is true that there were violations of consent, agreements, and basic principles of contracting between parties at stake since a second dose was promised within a set period of time (CBC News 2021). This practice is ethically unacceptable, at least to those who were promised a second dose. However, this breach can be justified in terms of protecting additional lives, which appeals to ethical praiseworthiness and the saving of more lives. However, such violations of ethical acceptability are highly exceptional and need to be very well justified for concrete, achievable, substantive goods. In the case cited, due to the emergency and scale of the pandemic situation, in the management of the pandemic early on, not all precautions, evidence, and discussions were available or occurred before opting for this unacceptable but praiseworthy option, even though in the end it proved to be an effective and life-protecting scenario that expanded the coverage of those vaccinated with at least one dose. In other words, there was a prediction that this would prove to be a praiseworthy option and it did materialize as such, thus justifying encroachments to ethical acceptability. Otherwise, a swift reversal to the initial vaccination policy would have been necessary.

Procedurally speaking, and without premising the superior value of acceptability over praiseworthiness, I propose that there (1) be an examination of whether the options meet the basic conditions of ethical acceptability, then (2) whether some ethically acceptable options are also ethically praiseworthy or can be enhanced to carry out a vision of the good (a "supplement" of meaning to a scenario as a course of action), and (3) discuss and decide if an option is preferable based on the evaluation based on both criteria. This reflects the maxim of *primum non nocere*, that is, that in pursuing a vision of the good, one has to take into account the harms and other aspects that make a scenario ethically unacceptable. For each scenario, asking whether and how a given scenario is ethically acceptable entails an analysis guided by the harms prevented or fostered by a given scenario. Why is this scenario ethically acceptable or unacceptable? What are the harms at stake, notably those regarding important aspects of flourishing? What wrongs am I preventing or inducing? Similarly, asking whether and how a given scenario is ethically acceptable entails an analysis guided notably by the lack of significant harms, here understood as harms to human flourishing. How are basic harms avoided or mitigated by a given scenario? Likewise, a similar analysis needs to be carried out for ethical praiseworthiness. An example of such an analysis undertaken in the context of the phenomenon of cognitive enhancement can be found in the pages of *Pragmatic Neuroethics* (see chapter 4 in Racine [2010]).

Based on the results of testing of the criteria of ethical acceptability and praiseworthiness, four different options are possible (see table 11.3). Not all need to be represented in a given deliberation, and it can be the case that all scenarios considered are situated within a certain category (e.g., three or four scenarios that are minimally ethical; no highly ethical scenario being possible) (see table 11.3). I submit that an ethically unacceptable but ethically praiseworthy scenario is rather uncommon. There can be an exception in situations where legal or societal norms are profoundly unjust such that what could be deemed ethically acceptable based on conventions and moral consensus is clearly unpraiseworthy. However, determining whether unfair and discriminatory laws and social norms are actually ethically acceptable requires distinguishing between norms that are *morally* acceptable and those that are *ethically* acceptable, in keeping with the distinction made by the theory of deliberative wisdom.

Morally acceptable norms are those consistent with existing understandings of moral norms and behaviors, while *ethically* acceptable norms are those that can be defended by appeal to human flourishing and as a result of inquiry and critical analysis of acceptability. For example, discriminatory attitudes based on race, sex, and gender were at a given time part of the moral fabric of Western societies. Such attitudes still exist but are considered both morally and ethically unacceptable because of the wrongs they cause(d). As a general rule, the synergetic and interdependent model of the relationship between acceptability and praiseworthiness (see figure 11.2) implies that what is praiseworthy also needs to be acceptable. It would indeed be quite problematic and paradoxical if one considers that a flourishing form of human life would impede basic standards of ethical acceptability. This is why views about ethical praiseworthiness need to be cross-checked with standards

Table 11.3
Possible logical answers to ethical tests

		Ethical praiseworthiness test	
		Ethically praiseworthy	**Not ethically praiseworthy**
Ethical acceptability test	**Ethically acceptable**	*Highly ethical*	*Minimally ethical*
	Not ethically acceptable	**Unlikely ethical*	*Unethical*

Adapted from Racine (2010)
* See text explanation of "unlikey ethical."

of ethical acceptability; otherwise, there are risks that preferred views on the good induce significant harm to the flourishing of others. However, there are times where the ethical standards in place will seem to be impoverished, not well implemented such that what is considered praiseworthy will (appear to) violate conditions of ethical acceptability. Consider here Nelson Mandela's pacifistic resistance and protest against the Apartheid regime in South Africa (Lang 2004). The racist regime in place viewed itself as moral when it was left to its own internal interpretative moral resources because it enacted and perpetuated (very problematic) moral beliefs of racial superiority and segregation consistent with a certain entrenched and racist morality. But the views of Mandela were that a higher sense of justice called for reconciliation and a spirit of compromise toward a just and fair society even though more violent action could have been considered ethically acceptable based on a more basic *lex talionis* (the law of retaliation, "an eye for an eye, a tooth for tooth") understanding of justice. The Apartheid social and political regime was entrenched in specific moral beliefs, but it was profoundly unethical. Not all moralities are ethical and ethically defensible. In fact, few if any are in their entirety. This is why ethics is an unsettling force with respect to our moral habits.

Test 1: Ethical acceptability Comparing different scenarios and assessing which are ethically preferable involves reeling in the resources of ethics theories to help sustain this analysis and mobilize participants. The first question to ask is whether the course(s) of action involve any bad, any harm, any wrongs. As these should be avoided, the criterion of ethical acceptability usually supersedes the criterion of ethical praiseworthiness in practice, or at least tends to carry more weight and have a clearer impact. For example, the prohibition against killing is stronger (more legal repercussions, more moral expectations) than the obligation to save a life. Intended killing (murder as opposed to manslaughter) is broadly and severely condemned. Individuals who maliciously take someone's life can be incarcerated and severely punished (e.g., death sentences in some countries). The same holds true of institutions and corporations. If they are found guilty of causing death (e.g., by negligence) they can be held responsible for severe harm, as was the case of the former Montreal, Maine, and Atlantic Railway company that was fined for its criminal negligence that led to the horrific train accident in Lac-Mégantic (Québec, Canada). The accident led to a spill of over 7 million liters of light petroleum, causing a destructive fire and the deaths of forty-seven people (Marin 2018).

However, the obligation to commit one's resources to save the lives of others will usually be much more limited, and it is perhaps proportional to the consensus within a given society or culture on the criterion of ethical praiseworthiness.[11] For example, some legal systems such as civil codes (e.g., in Quebec) will explicitly prescribe some specific altruistic behavior in the case of someone whose life at stake. Citizens are expected to take reasonable (although not heroic) steps to help someone in such a life-threatening situation. For example, they could be charged (under civil law) for not calling for help for someone drowning, but they may not be required to offer that help by swimming to save that person because of the risks involved to themselves. A similar line of reasoning is often applied to the very rich inhabitants within their own countries (a.k.a., the top 1 percent) (Chomsky 2012). These citizens can grow impressive fortunes rivaling that of entire countries without having any legal obligation to others, although ethically, the situation is questionable. At what point does the economic and political power of someone and their ability to bring about important social changes favoring the flourishing of others become not only ethically (in the sense of ethical praiseworthiness) indefensible but also (in the sense of ethical acceptability) untenable? Likewise, the wealth and economic resources of wealthy states and their citizens could lead to dramatic improvement in the well-being and life conditions (often lifesaving) of poor populations if shared more fairly and equitably. However, the possibility of contributing to the welfare and well-being of others does not translate into a positive obligation. Several issues are at stake here, but one can see that in a rich state like the US, poverty continues to exist despite colossal fortunes accumulated by single individuals or families with huge impact on, for example, child poverty (much higher in the US than in many similarly resourced nations) (Paquin and Lévesque 2014). Despite the tremendous impact of the concentration of wealth, it remains that the obligation to share with others such immense resources holds no comparison to the prohibition to steal or to murder even when it concerns people who hold questionable attitudes.

As we will see, the pursuit of a eudaimonistic ethics can raise questions about the ability of ethical acceptability alone to fulfill the requirements of a meaningful ethics, inclusive of ethical praiseworthiness. Nonetheless, proposed action-scenarios need to meet the criterion of ethical acceptability. If they do not meet this criterion in some way, they would have to meet very clearly and with good arguments a criterion of ethical praiseworthiness to justify an overriding of ethical acceptability. As I explained above, this is likely only an apparent superseding of praiseworthiness over acceptability based on an understanding of ethical acceptability as a form of moral acceptability. Ethical praiseworthiness has the ability to expand the meaning of ethical acceptability once new standards of acceptability become recognized. Take for example, the *lex talionis*, probably of Babylonian origin (e.g., Hammurabi's Code, an advanced and influential set of laws of the ancient world) and also found in the *Tanakh*, the Hebrew Bible. This law is criticized by Jesus in the Sermon of the Mount in favor of what some interpret to be a more praiseworthy concept of justice, which frees itself of resentment and legal prosecution for vengeance and encourages forgiveness, peace, and compassion.

Take another example related to health ethics. The most convincing justifications for the participation of clinicians in medical aid in dying, and to considering the gesture ethically good, is found in the avoidance of harm and suffering of the person applying. The positive arguments for allowing someone to end their life as they desire would simply not hold up in comparison to the prohibition of killing. In this example, the harm of continued suffering is so great that it can be considered worse than death, thus justifying medical aid in dying. In this case, ethical praiseworthiness is not interpreted simply as propagating human life at all costs. It also takes into account the amount of suffering induced such that the painful life without meaning is considered by the patient worse than death. This understanding of what is ethically praiseworthy commands an amendment to the standard of ethical acceptability to not causing harm via the inducement of death. This is also why these interventions have been described as a form of end-of-life care in Canada because it is embedded in a caring purview that is focused on the welfare of the person.

However, if one takes away the caring perspective, we end up with merciful killing, which is punishable because it does not, by definition, fit the standards described by medical aid in dying, which requires the person's consent. All jurisdictions that have allowed medial aid in dying have also clearly stated the importance of suffering or pain (physical or psychological) in order to justify this act by clinicians. Switzerland is the exception because "the Swiss penal code considers assisting suicide a crime if and only if the motive is selfish. It condones assisting suicide for altruistic reasons" (Hurst and Mauron 2003, p. 271) but not for selfish motives. This account is rooted in the idea that suicide can be rational and can be an expression of honor and dignity. No special role of physicians is foreseen therein in Switzerland.

Test 2: Ethical praiseworthiness Action-scenarios can be considered not only for their respect of ethical acceptability but also for the kind of positive obligations and values that they allow participants to carry forward. As we saw, the criterion of ethical praiseworthiness captures such obligations or aspirational ideals, and therefore the remark above about the caveats to the terminology of criteria are even more relevant for ethical praiseworthiness since no clear-cut tests could provide a black-and-white answer with respect to human flourishing as discussed previously (chapter 2). However, there are clearly courses of action that, while fulfilling conditions of ethical acceptability, are also more amenable to promoting praiseworthiness and of being instruments of human flourishing.

To operationalize ethical praiseworthiness, one has to ask the complex question of whether the given course of action promotes human flourishing or not in terms of its consequences, involvement of agents, and actions pursued (Dubljević and Racine 2014). Here, we need to pause and acknowledge that the understanding of ethical praiseworthiness needs to be contextualized to take into account the nature of the problem situation and the possibilities at hand. However, dimensions of flourishing have been chartered by previous research in ethics and in modern psychology, which, despite worries that flourishing cannot be spelled out, have been shown to include concrete psychosocial dimensions of human life in recent psychological research (Ryff 1989; Ryff and Singer 2008) such as autonomy, purpose in life, positive relationships, personal growth, environmental mastery, and self-acceptance (see chapter 5 for a review of these dimensions and their relationships to ethical acceptability and ethical praiseworthiness).

The ability of human beings to direct their lives and make decisions that are aligned with progression on a path of flourishing is variable and may not only depend on individual psychological dispositions but also on social conditions, mental and physical health, and various physiological functions (e.g., wealth, sleep, diet) (Ryff 2014a). Importantly, the everyday valuations of concrete behaviors are building blocks in the progression of a person toward human flourishing and that of others. And to the question of whether specific actions and decisions are ethical (i.e., good, right, and defensible), human flourishing serves as a driving orientation to evaluate whether an action is conducive to growth or not. There is increasing research on interventions and behaviors that allow individuals to self-actualize and find fulfillment (Ryff 2014a). However, there is something of a disconnect between this research stemming from psychology (positive psychology) and contemporary ethics. It is true that ancient ethics and philosophy was guided by the cardinal goal of flourishing (Hadot 2002). However, the question of what it is to have a good life and a life worth living is often indirectly addressed in mainstream ethics, which tends to focus on the acceptable life or the life that everyone can live within the existing boundaries that societies and social institutions place on individuals and communities. In a sense, contemporary ethics' questioning of the good life is often premised on the need for a strong compatibility with existing modes of thinking and acting such that it does not overly disrupt them. The ancient ethics teachings and discussions were more confined (e.g., to certain groups of people) but delved much deeper. Discussions within the schools often took place among a limited number of individuals who formed schools of philosophy. They shared some basic assumptions and tried to live according to their principles. The peripatetics (following Aristotle), the stoics (following Zeno of Citium), and the epicureans (following Epicurus) are examples of followers of these schools, who embodied specific ways of thinking and ways of living following what they considered principles of a good life and teachings of their leading thinkers (Hadot 2002). From the Renaissance onward, ethics has become more of an abstract undertaking with less connection to one's real-life experiences. It has also become more encompassing and has tackled the entirety of the collectivity, notably in the form of political philosophy. Nowadays it is less common to envision ethics as a form of life philosophy or a living philosophy, although there are new developments in this direction (Racine et al. 2024). Ethics is now often conceived of narrowly, as a discipline involved in dealing with specific moral problems but with much less to say on the good life and the life worth living (Hadot 2002). This latter subject is judged to be too difficult and perilous because of the many lifestyles one can adopt and the many values one can uphold. However, international surveys about well-being and happiness point to rather identifiable explanations of variance, such as GDP per capita, social support, healthy life expectancy, freedom to make life choices, generosity, and freedom from corruption, as concrete and locatable factors at stake (Helliwell et al. 2020).[12] Systematically, countries that are able to effectively conjugate conditions of ethical acceptability with a vision for ethical praiseworthiness (e.g., being able to reach consensus and offer support for social support and healthy living conditions) top the list of happiness and well-being surveys (e.g., Finland, Norway, Denmark, Iceland, Switzerland, the Netherlands, Canada, New Zealand, Sweden, Australia). In some countries with high GDP and moral-political structures focused on ethical acceptability (e.g., individuals rights), there is less effective engagement in innovative ethical praiseworthiness (e.g., United States, United Kingdom) perhaps because of polarized public spheres (e.g., United States) (Callahan 2005a; Hunter 1992) or harsher neoliberal individualism (e.g., United Kingdom) (Paquin and Lévesque 2014).

The abandonment of the deeper role of ethics is understandable because of frailty of the valuations that are at the basis of theories that scaffolded them into more general theories. The quest for a good life remains a current concern, and research into well-being and human flourishing (see chapter 2) suggests that there are key components that form human flourishing for most people. Ultimately, the theory of deliberative wisdom advances that actions are ethical insofar as they help individuals and collectivities progress toward flourishing lives. This ultimate goal of ethics serves as its aspirational regulatory ideal.

Illustration of the use of the criteria of ethical acceptability and ethical praiseworthiness The synergy between ethical acceptability and ethical praiseworthiness can perhaps be best understood with further consideration of case examples. Here is an example of a clinical case shared by James Reese Jr. and Phillip Pearl in a coedited volume on ethics in child health:

> Carter, a previously healthy and developmentally normal 8-year-old boy, is evaluated following two separate and unprovoked generalized tonic-clonic seizures 6 months apart. Neurological examination, electroencephalogram (EEG) and brain magnetic resonance imaging (MRI) are normal. He and his family are being seen for their first visit with a child neurologist following referral by their pediatrician. The family has many questions about diagnosis and prognosis. They ask, "Do children die from epilepsy?" (Reese and Pearl 2016, p. 303)

What should the neurologist do when asked by the family if children die from epilepsy? It is true that there is a phenomenon of sudden unexpected death in epilepsy (SUDEP). SUDEP is rare in newly diagnosed epilepsy (1 in 10,000 person-years) but more frequent in cases of medically intractable epilepsy (1 in 100 person-years). There are many unknowns about SUDEP and neurologists in the US and Canada infrequently discuss the topic with their patients (Reese and Pearl 2016, p. 303). In this case, different response-scenarios can be spontaneously generated: (1) tell the truth about the real but very rare risks, (2) not disclose the risks, or (3) stage the truth (see table 11.4 for more details about these cursory scenarios). Ideally, this kind of deliberation could be supported by some consultation of colleagues and even some literature informing about the preferences of family members, the impact of the truth on families, and so on. For the time being, we will keep this morally problematic situation simple to illustrate the use of the concepts of ethical acceptability and ethical praiseworthiness in the evaluation of response-scenarios.

As the comparison featured in table 11.4 shows, scenario 3 has definite strengths in terms of its ability to respect basic conditions of ethical acceptability, as well as fulfilling requirements of ethical praiseworthiness. This is obviously a set of face-value considerations of ethical acceptability and praiseworthiness. Only the action itself and some assessment of the outcomes could tell whether the option was indeed preferable in the end. Based on the typology presented earlier (see table 11.3), scenario 1 could be considered at least minimally ethical; scenario 3, highly ethical; and scenario 2, unethical. Note that the comparatively (higher) ethical merit of scenario 3 is contingent on the existence of conditions that make the scenario possible. If it is not viable (e.g., no time to stage the truth), then scenario 1 would be preferable. Likewise, the comparative value of scenario 3 could increase if the parents are considering some immediate action based on the knowledge of the risks (e.g., having the child undergo an unproven and risky therapy based on stem cell transplantation or another controversial and invasive therapy). These remarks illustrate how finding the "preferable" scenario is a matter of dynamic equilibrium between sought values and the actual situation that can impact how these values can be realized. For example, scenario 3 may be ethically acceptable and praiseworthy but difficult to realize because of the time required to actually stage the conversation. If such time is not available, it may be that the clinician does not have the choice but to opt for scenario 1; but knowing that better could be done, they prepare the reorganization of services to help to deal with implications as they arise in the future with support information, support personnel in place, or referral to ethical consultations or other expertise. As an individual deals with problematic situations, the preferable action-scenarios can be progressively implemented and this direction pursued over time. However, taking into

Table 11.4

Example of a cursory scenario-based deliberation and analysis

	1. Tell the truth	2. Not disclose risks	3. Stage the truth
Shorthand description of the scenario	Share the actual large but highly unlikely risks (the honest, trustful approach) with a risk of inducing anxiety and needless worries (harm).	Not sharing the real risks (the beneficent, paternalistic approach) to avoid needless worries, with a risk of inducing distrust.	Take time (the approach of compromise) to eventually stage truth-telling (honest, trustful) but mitigate worries and feel the situation some more.
Feasibility test	Feasible	Feasible	Likely feasible (contingent on time available)
Evaluation of acceptability: Why is this scenario acceptable (A) or unacceptable (U)?	A: Respects the parents' or patient's right to the truth/need to have information. U: Possibility of inducing anxiety.	A: Avoids inducing anxiety. U: Borders on lying. Disrespects patient autonomy and informed consent and decision-making. Could lead to improper care.	A: Respects the parents' or patient's right to the truth/need to have information. A: Mitigates the impact of negative consequences (e.g., anxiety, psychosocial distress). U: Making the parents or patient wait for a response to an important question could induce anxiety and lead to lack of trust and the feeling that the doctor is being overly protective.
Evaluation of praiseworthiness: Why is this scenario praiseworthy (P) or not praiseworthy (NP)	P: Cultivates relationships based on values such as honesty and trust. NP: May not optimally consider the harms and risks induced by an abrupt and unprepared disclosure of the risks of sudden death.	P: Protects from anxiety, although paternalistically. NP: Does not cultivate relationships based on honesty and trust.	P: Cultivates relationships based on values such as honesty and trust. P: Involves compassionate and empathic consideration for the negative implications of abrupt truth. NP: None
Additional considerations related to enactment	Assumes that the risks induced by truth-telling are not too high (e.g., highly stressed and worried parents). Otherwise, the value of telling the truth could be diminished.		Requires ongoing relationships. May not be feasible if the practice setting does not allow for this or if communication is difficult and impossible to maintain or ensure in the future.

account relationships and contextual aspects of an action-scenario does not mean that anything goes (relativism). Rather, it means that foreseeing the realizable good is an exercise in situation-based thinking and deliberation.

Envisioning Enactment and Agents

Although the process of enactment is covered in steps subsequent to deliberation proper (see chapter 12), deciding who is in charge or who is responsible for leading the move from discussion to action is important. Even if an ethicist (or a skilled and trained moderator) may seem like a natural leader for a deliberation effort, it is unlikely that they will be in charge of enacting a response-scenario to a given situation. For example, a physician in charge of a healthcare team confronted with a difficult ethical question could make sure that the decision guides future plans of treatment. The vice president of a corporation may be tasked to see that the changes decided on are enacted in their section of activity concerned by the topic and outcomes of the deliberation. Or, someone may be tasked with talking to the institution's CEO if the scenario implies such a process. A small oversight committee from the initial group involved in the deliberation can be formed to monitor the evolution of the action-scenario. The change may actually only concern an individual, and the deliberation may take place within one person (e.g., the example of the lone clinician). In this case, the entire process of deliberation can be simulated within a person, perhaps with some discussion and/or consultation with others along the way. In certain circumstances, ethicists (e.g., a group of institutional ethicists, or an individual ethics consultant) may be called on to enact a response-scenario that involves teaching about ethics and raising awareness about a new policy. Many options are possible. Before the end of the deliberation, enactment should be discussed to ensure that a great deliberation exercise does not become futile. Enacting agents should be identified, as well as a plan of action and related steps and communications.

Closing Remarks on the Relationship between Traditional Ethical Analysis and Scenario-Based Ethical Analysis

At the outset of this chapter on key moments of ethical deliberation, it is important to first contrast a scenario-based method of ethical analysis, which situates the consideration of ethical dimensions of action within the actual context of an action, that is, in a way that captures the logic embodied in the motivation or intelligence of those undertaking the action(s), to a more traditional method of analysis, which could start outside of the consideration of the actual possibilities of action and the different courses of action. As described previously (chapters 6 and 10), a significant portion of the tradition of ethics scholarship has focused on the explicit and abstract consideration of values or principles, which in a sense has the very laudable virtue of simplifying ethical analysis to discussion about ethical principles and key values to promote (Baertschi 1998). It also provides clear normative benchmarks to evaluate whether a given action or decision (or scenario, ultimately) is ethical or not, something that empirical ethics approaches often lack or are unclear about. However, that strategy has significant downsides that I have already discussed in terms of applicability and meaningfulness.

The theory of deliberative wisdom proposes that ethical analysis be considered within a broader or more inclusive process of thinking about moral problem solving and the evaluation of the implications of proposed responses, notably in terms of learning and empowerment (Aiguier and Loute 2016).[13] Often enough, more bottom-up approaches seem to lack the sharpness that would make it easy to teach ethical content or to be able to articulate the outcomes of ethical analysis, that is, there are no clear ethics criteria, so anything could be thought to be acceptable, leading to a charge of relativism (Inguaggiato, Widdershoven, and Metselaar 2021). At the same time, ethical analysis can become so thin and abstract that connections to the actual situation are lost and the solution generated based on the results of this ethical analysis have a life of their own, disconnected with reality. This is clearly a dead end. Furthermore, a number of issues trouble the smooth application of deductive top-down models, such as conflicts between principles, doubts about the relevance of a principle, or conflicts generated because of fundamentally different perspectives about what is to be valued in a given situation (Doucet 2004). In contrast, a scenario-based method situates ethical analysis and deliberation in the context of potential courses of action (action-scenarios). This reinterpretation of the role of principles as expressive tools (rather than as argumentative foundations) sets the stage to use them to articulate what matters in a given scenario, as discussed in chapter 5.

Traditionally, ethics theories have claimed to offer the resources for guiding ethical decision-making. However, based on the experience of many users (Andre 2004; Anscombe 1958; Callahan 1973; Miller, Fins, and Bacchetta 1996), most of these theories are difficult to apply as proposed by their authors, leading to the suggestion to see them more appropriately as "lenses"

(Mahowald 1994; Sherwin 1999) or as shorthand for what matters in a situation, as Dewey recommended (Racine et al. 2019).[14] When one reflects on the nature of choices available to them, rarely does any choice or decision concern solely moral matters, although morality is likely always present in some implicit or (more rarely) explicit form. Situating ethics within action means that action is the driver and ethics is an instrument or a lens through which to look at action. Accordingly, the criteria or approaches proposed for ethical analysis and deliberation may seem thwarted by scenario-based methods such as deliberative wisdom because the range of ethical principles may not be a "free market"; it is constrained by the facts of the situation, the people involved, the resources available, and so on. In other words, ethics is contextualized; it is considered *within* the motivations of moral agents and with concrete contexts and actual possibilities, which are viewed as being part of and not entirely *external to* them. This means that ethical theory as it is commonly known is instrumentalized as explained previously: ethics theory needs to be a tool in service of guiding action toward good and right objectives and ideally good and right outcomes, too. Accordingly, deliberative wisdom posits that ethical deliberation, or ethical analysis, must take place with the actual courses of action considered and judged to be feasible. Within such a frame of mind, ethical analysis is a kind of analysis that focuses on the ethical merit of a given scenario in response to a problematic situation.

The ethical value or ethical worth of a given scenario is a broad question and will always need to be contextualized to understand what is possible in a given situation (i.e., what the options at hand are), as will the comparison of those action-scenarios based on what kind of value they respect and help promote in a specific situation. Within such an analysis, principles stand as shorthand for a valuation process and the perpetuation of the experiences that the principle encourages. Principles serve to identify behaviors that help promote a given value or are consistent or aligned with a given value (for such a pragmatist analysis of autonomy, see (Racine et al. 2021). As I explained, the current theory proposes that two distinct criteria can be used to help guide the analysis and comparison of action-scenarios: ethical acceptability and ethical praiseworthiness. I recognize that the tests of ethical acceptability and ethical praiseworthiness are tools and guides in the analysis of ethical merits of different response-scenarios. These conceptual tools do not replace individual judgment but help support a more structured and comprehensive analysis of scenarios. In the end, the right or good thing to do will hopefully align with considerations of acceptability and praiseworthiness and help motivate and facilitate defensible and praiseworthy behavior. And, even more importantly, the subsequent evaluation of the effects and implications of response-scenarios will be paramount in evaluating the ethical worth of any idea. The ability of scenario-based deliberation to more fully carry forth this latter suggestion is perhaps what most sets it apart from traditional ethical analysis. It is also one of its key advantages from the standpoint of deliberative wisdom, the others being its ability to ground understandings of moral situations in experience and to unleash ethical imagination. In sum, scenario-based ethical analysis attempts to operationalize a shift in gaze and perspective from the spectator perspective found in traditional ethical analysis to the standpoint of the actor or the user of ethics theory such that the meaning of theory itself shifts.

Conclusion

This chapter provided an account of the important steps involved in ethical deliberation: (1) broadening and deepening the understanding of the situation via stakeholder consultation and face-to-face discussion; (2) envisioning action-scenarios, notably based on the processes of imagination and dramatic rehearsal; and (3) reaching a judgment or decision based on the comparative evaluation and analysis of scenarios for their ethical merits. These guideposts remain rather general and in need of further specification for different contexts (e.g., healthcare, business, personal matters) and even for certain kinds of questions within these contexts. There are already some available deliberation methods highly consonant with the proposed approach, such as Doucet's method (Baertschi 1998) and the dilemma method and moral case deliberation pioneered by Molewijk and Widdershoven (Hartman et al. 2019; Stolper, Molewijk, and Widdershoven 2016; Svantesson et al. 2014). However, I hope that this chapter serves to connect specific guidance for ethical deliberations with an account of ethics and methodologies for improving the relationships between ethics theory and ethics practice. The next chapter sheds more light on the nature of ethical enactments in line with the account of ethical deliberation provided so far.

12 Ethical Enactments

Ethics—as described so far—is concerned with and involves thinking about experience and action. In order to do this, it must remain rooted in daily experience to make sense of it and to propel it forward. Ethics theory is therefore a kind of embedded thinking, inseparable from lived experience. In this chapter, I take issue with some of the traditional ways of understanding and talking about action in ethics, as well as the traditional way of thinking about the relationship between theory and action. Traditional ethics' focus on ideas needs to be combined with a view about the meaning of ideas as being part of actions and human experience. In presenting how this can be done, I rely on the concept of enactment, namely ethical enactments. Although some recent cognitive science theories have heralded the embodied, extended, embedded, and enacted nature of human cognition, these insights are not new to pragmatists. They are at the very core of pragmatism's understanding of human existence and human thought as explained by Peirce, James, Dewey, Addams, and Mead, most notably. This explains why this movement was so embroiled and productive in reforming public education, developing social work, and promoting democratic reforms of all sorts. In this chapter, I explain enactments in relation to a situational understanding of human behavior and psychology and the concept of deliberation-based scenarios (see chapter 10). In a nutshell, while deliberation constitutes dramatic rehearsals of scenarios, enactment is the process of acting them out. This theatrical terminology stemming from the work of Smith, and later Dewey, has definitive advantages in exhibiting the connections between theory and action from the standpoint of the user of the theory, who is simultaneously a learner. It helps explain the acting out of response-scenarios, the skills involved in doing so, and the contextual and temporal aspects of scenario enactment because enacting is a process and not only an outcome. This chapter also explains that ethical enactments and response-scenarios range greatly in complexity, while the enacted nature of these responses makes it clear that scenarios are fallible hypotheses. The important notion of re-situation is presented as the outcome of enacted scenarios, in line with a situational understanding of human psychology and behavior. This chapter articulates how the concept of ethical enactments supports the use of participatory methods of acting out, as well as participatory evaluation methods. In all cases, meaningfulness of enactments is key, and their success can be evaluated in this light, as explained in the next chapter.

> The road to hell is paved with good intentions.
> —English proverb

Academic ethics tends to entertain a bias that ethics reasoning is the end of ethics such that ethics could exist even without being effective (for a good review of this phenomenon, see Fesmire [2003]). It is in this sense that *doing ethics* is often boiled down to engaging in ethics scholarship. This bias is understandable because it is likely the component of ethics that can be most easily approached and is within the reach of scholarly work. More substantially, I believe this is the case because ethics is a most implosive field of scholarship as it directly concerns the meaning of life and asks questions about how our habits promote human flourishing. This activity is socially and politically disruptive, especially in oppressive and rigid societies where politics are far removed from human flourishing. If not addressed, this tendency or bias to reduce ethics to ethics scholarship, especially when it is promoted as a canonical truth or authoritative way of envisioning ethics, yields an incomplete picture and can only exist within a partial understanding of and generally ineffective approach to ethics. Not surprisingly, the mainstream tradition of ethics has dedicated limited attention to how different normative theories are conducive (or not conducive) to ethical behavior in practice and to the enactment of agreed on ethical scenarios, or more precisely ethically oriented actions-scenarios. More precisely, the development of normative models of what we ought to do has not been tightly coupled with studies about how moral agents actually behave and the different factors that can modulate the ability of agents to carry on with their plans, let alone the outcomes of these actions (Martela 2017). As a result, ethics has remained mostly confined to the realm of ideas since the actual outcomes of ethically oriented actions are often disregarded.

In this chapter, I first try to understand why ethics has been rather uninterested in action due to disciplinary biases, as well as narrow accounts of action (e.g., value-free accounts of human action). I then describe how ethical actions are responses to situations and constitute "enactments." I explain why "ethical enactment" is in a sense an *appelation d'origine contrôlée* in so far as it designates *sensu stricto* an action that has been assessed, appreciated, and evaluated for its ethical merits based on the outcomes, whereas, at the beginning of action, the response-scenario is a hypothesis or a hypothetically ethical action. Therefore, I sometimes use the expression "ethically oriented action-scenarios" to avoid attributing ethicality prematurely, that is, before the actual implications of an action are known. The remainder of the chapter describes the advantages of this meaning and conceptual displacement, notably by highlighting the connections between theory and action from the standpoint of the user of the theory who is also, according to the current theory, a learner by definition. The enactment terminology helps explain the process of acting out response-scenarios, the skills involved in doing so, and the contextual and temporal aspects of scenario enactment since enacting is not only an outcome but also a process from which we learn. Ethical enactments and corresponding action-scenarios range greatly in complexity, while the enacted nature of these responses makes it clear that scenarios are akin to hypotheses because of their ability to project an existential moment in the future and are thus fallible. The important notion of re-situation is presented as the outcome of enacted scenarios, in line with a situational understanding of human psychology and behavior. Although seemingly innocuous, changes of words (e.g., ethical actions to ethical enactments, solutions to re-situations, responses to response-scenarios) imply changes of meanings in the sense that these indicate changes in how certain ideas fit into our existence and into our worldly contexts. Richer accounts of action as it relates to human behavior, human situations, and human lives (e.g., in literature, social science, lived experience) can accompany this effort.

Ethics' Disregard for Action and Its Implications

It could be a book-long endeavor to understand why ethics, generally speaking, and even in some of its most applied contexts, seems to have been—and sometimes still appears to be—rather uninterested in human action to the point of favoring pure thinking, to the detriment of thinking related to experience and action. To understand this trend noted by numerous scholars, we would need to review why ethics has become heavily focused on ethical judgments, why it has privileged a focus on disembodied principles and values to the detriment of a more situated account of thinking about human morality, and why it is often an inward-looking academic discipline. This issue is of great importance, but instead of giving a full and definitive answer, I will pinpoint a few possible explanations, which are important to keep in mind when trying to think differently about ethics. For those skeptical of this rather sweeping characterization of ethics, I recommend a number of thorough reviews and discussions highlighting this trend as a dominant feature of ethics (Hoffmaster 1992, 2018; Hoffmaster and Hooker 2009; Lekan 2006), as well as how attempts to develop interdisciplinary and socially engaged forms of ethical thinking have been marginalized (Capps 2003; West 1989).

A first explanation could be found in the argument that ethics' lack of concern for action in a broader sense (e.g., the context of action, the outcomes of action) is due to an understanding that this is simply not what ethics is about. Accordingly, *doing ethics*, in the words of some academic circles, means by far and large thinking, theorizing, and writing about ethics theory and cases. It does not entail measuring the effects or real actions and policies or figuring out what to do within the existence of this or that person or getting involved in solving real-world situations. Ethics, for some, is academic by definition, and the work of ethicists can be completely removed from practice such that ethicists are not even positioned as users of the knowledge they generate. But what argument is there that this way of doing ethics is best? Leaving that assumption unchecked allows for the maintenance of a disciplinary bias wherein, from teacher to pupil, a way of envisioning ethics has gained traction, power, influence, and so on such that it perpetuates itself as the natural account of ethics.

A second answer is that traditional ethics has been rooted in a narrow and top-down account of human rationality modeled by modern science (Hoffmaster 2018). Rationality is synonymous with impartiality, universality, objectivity, and so on. What can correspond to this paradigm in the context of discussions about the worthiness of human actions in a moral context? The simple answer is that human beings operate on principles based on their commitment to proper reasoning and objectivity or impartiality in their search for the truth of the matter in ethics, as in modern science. How plausible is this account of human rationality, and what are its implications for ethics theory, methodology, and practice? Theoretically, ethics has put a significant amount of energy into identifying and discussing first principles, foundational principles that guide entire existences, entire societies, and even the entire world. I leave aside how this account of rationality meshes with imperialistic and colonialist orientations of the societies from which they originate, but it is no small detail that the authority of being "right" and of having found the ethical "truth" in principles could then be claimed by those who wield those principles as instruments of power and domination in micro or macro settings, now and for centuries (Azikiwe 1931; Fiester 2015). Methodologically, with this ideal of a rationalistic epistemology of modern science in mind, ethics has sometimes espoused in its most extreme forms the quest of finding axioms and algorithms that mimic mathematical knowledge (Rota 1991). The idea here is that truth could be found in the logical analysis of concepts and principles and that by far and large, empirical aspects of situations could be glossed over and considered unimportant in comparison to concepts judged to be quintessential to ethics (Toulmin 1982). Practically, this ideal model of ethics has supported an expert and authoritative model of ethics knowledge that views ethical action as the almost mechanical application of principles to situations, akin to the deductive model one could expect when finding and applying the correct mathematical formula to a given equation to solve it (Caplan, 1980; Hoffmaster 1992).

For pragmatists and others who envision ethics as a field whose value should lie in its fruits, this account of ethics is unsatisfactory, and is bound to not be able to deliver what ethics should deliver, that is, usable knowledge that guides human beings toward flourishing. It is not so much a failure of this or that specific theory, but a deeper meta-concern that the underlying paradigm on which it rests is deeply problematic because it relies on a dualistic object-subject epistemology for which ethical truth means universal and indubitable principles (Lekan 2006). While traditional ethics scholarship may appear weak, it has been effective enough to justify and maintain those in authority in academic departments, to publish scholarship, and to recruit those who think like-mindedly to create an academic ethics culture that self-perpetuates, notably in philosophy departments. However, what must be asked from a pragmatist standpoint is not if the knowledge wielded is useful for those who gain from its acceptance, but rather if it is an important tool for human flourishing. Asking this question means that ethicists cannot defend their work solely based on its conceptual purity but must also investigate its motivations and effects. It also means that academic ethicists must be invited to step down from their position as thinkers who do not need to be involved in and can even be disengaged from thinking about their own moral lives and the relationships between their theoretical work and their own experiences.

My critique of ethics may appear quite radical here and needs to be nuanced. First, there are many traditions within ethics and notable exceptions to the commitment to a highly rationalized and abstract approach. For example, MacIntyre (1984) has advanced the importance of intrinsic motivation found in practices, as well as of reflection on the goods pursued within concrete practices. However, his approach remains somewhat imprisoned by virtue theory and, following Carden's analysis, can be supplemented by the insights of pragmatist theory about habits and deliberation, to offer a much more open-ended account of ethics theory with respect to the diversity of flourishing lives (Carden 2006).

Second, some areas of applied ethics (for lack of a better word, or of sector-specific ethics) have engaged more directly with recent behavioral science. For example, scholarship in behavioral ethics at the intersection of moral psychology and most notably business ethics has focused on behavior for a few decades now. Perhaps this is the case because of the need for business ethics to ensure that enterprises that include large numbers of workers maintain a good reputation and ensure the proper and ethical working of the company. The tradition of management science is also deeply influenced by social science (e.g., social psychology, sociology), which encourages the use of models and strategies to understand organizational behavior and support common objectives to accomplish projects. In contrast, there has been a lesser impetus for capturing behavior in bioethics. This may be because healthcare professionals are regulated within their own professions, and individual professionals have considerable freedom. They have codes of ethics and a very long and strong tradition of ethical reflection (e.g., medical ethics and the Hippocratic oath, nursing ethics) (Duplantie et al. 2000; Fowler 2017). For example, physicians are often not in a typical employer-employee relationship in the institutions in which they work, thus creating greater complexity than in other settings. Accordingly, corporate visions about a certain healthcare institution's ethics (e.g., code of ethics for a hospital) are often secondary to the healthcare worker's own professional ethics. However, this means that there has perhaps been less attention attributed to the results and actions subsequent to ethical deliberation in bioethics (e.g., focused on tooling healthcare professionals to act ethically) in comparison to business ethics, where responsibility is often attributed to the most senior position (i.e., CEO, manager) of a large-scale human operation. Another line of explanation—not unrelated to the salience of professional autonomy in the health sector—could explore how health economics and outcomes research is typically poorly integrated in healthcare settings while outcomes tend to be crucial for most businesses (Mooney 2009).

Despite some perhaps increasing attention to behavioral aspects of morality in some headquarters of ethics (e.g., behavioral ethics), ethics can become siloed into ways of thinking that are not fully engaged with ideas of intrinsic motivation and human flourishing, such that what is sought is compliance with organizational values, rather than real engagement with critical thinking. This point has been well advanced by Gareth Craze, who argues that many contemporary business ethics and activities surrounding corporate social responsibility represent a thinning down of ethics and do not effectively tackle processes of dehumanization because of corporate insistence on strategic thinking and compliance (Craze 2020). Furthermore, the development of so-called behavioral ethics has emphasized behavior, but it has done so in terms of the automated aspects of human behaviors and their relationship to implicit cognition, typically underplaying the role of moral cognition (Maxwell and Racine 2016). Accordingly, despite some interest in more advanced insights about behavior, there are still different frameworks within which these insights are used. Some are more positivist inclined and favor instrumental uses of behavioral knowledge, while others resist this tendency and call for a more constructive and meaningful orientation geared toward human development and capabilities.

Thus, ethics can benefit from reconnecting with action, but it needs an account of human behavior and related theory that helps us understand the place of morality and human values therein. This requirement for accounting for morality does not imply that all values pursued in human behaviors are unproblematic; we need not gloss over problematic motivations of human beings, because we need to acknowledge and unpack these values and look squarely at all that is negative rather than remain oblivious to it. Hatred, jealousy, and similar emotions that fuel behaviors must be acknowledged and be taken into account. War and crimes of all sorts are actions undertaken by human beings and must be part of the picture. However, some movements in the disciplines most focused on human action (e.g., psychology, sociology, political science) have encouraged describing action without recourse to intents and values in their explanatory models. These movements have favored descriptions of behaviors rather than actions, at least in the more experimental branches of these fields. In other words, explanations of social behavior are often described from the third-person standpoint, thereby tending to diminish the subjective and personal role that one can play within situations and the role that motivation and resolution can offer (Taylor 1971). This objectivist (and objectifying) tradition may put off some who are deeply concerned with the rationale for actions and who operate within a paradigm where intents and values have profound importance, but sometimes to the neglect of the economic, social, and political factors that shape human situations. The neglect of action in ethics and the neglect of values in the scientific study of human behavior represent a Janus-faced phenomenon: the separation of knowledge from valuation and the separation of valuation from knowledge. In contrast, ethics needs an account of human action and human agency where valuations and values exist in order for ethics to build from these and help guide them in constructive and positive orientations. Hence, a richer and more comprehensive account of human action offers more to ethics than a mere reduction of action to behavior or to principles. It needs a description of action as part of and as a prolongation of human existence, including one's own existence as a user of ethics knowledge. Importantly, ethics knowledge must be knowledge aware of contextual factors such as social and political factors that may shape what we see and think of as knowledge. This is largely the rationale supporting the concept of ethical enactments.

Ethical Action as Enactment of Response-Scenarios

Ethics theory—although it should not be prevented from roaming freely just like other academic and creative pursuits—should be checked to see whether and how it improves human lives and human practices. Accordingly, ethics cannot afford to ignore insights from disciplines that study human thought and behavior. It needs to establish strong collaborations with psychology, sociology, management sciences, and so on. This does not mean that ethics is subservient to those fields and some of their preferred theories. Ethics may question some orientations encountered therein, regarding, for example, whether and how the knowledge generated helps human beings pursue more fulfilling existences. This existential and user standpoint in ethics theory and the moral life, as well as the role of agency in human flourishing, were briefly explained in chapter 1. The previous chapters on situated agency (chapter 1), human flourishing (chapter 2), and deliberation (chapter 10), set the stage for an account of human action as enactment when discussed in the context of ethics as I describe here.

Ethics Theory from the Standpoint of the User and the Need for Appropriate Vocabularies

Given the focus of ethics on human flourishing as described in this book and its reliance on a transactionalist account of agency, we must ask which views on human action and human behavior can support or conflict with ethics. Just like our views and words describing the individual human being (e.g., a person, a patient, a consumer, a client, or an agent) or those describing blocks of human existence (e.g., events, moments, or situations), the conceptual vocabulary describing what individuals do, matters. This vocabulary can foster the idea that we are static organisms to which life happens, or it can stress that we are agents in our existence such that our actions compose and perpetuate our existence. The latter idea is compatible with the project of ethics, while the former is not.

Ethics unequivocally commands an agential account of human existence because it is committed to the idea that we can orient our lives, that we can make choices, that we can open up to new possibilities. Ethics simply does not make any sense otherwise. This explains why seemingly neutral, value-free, perspective-free epistemologies promoted in the past (e.g., behaviorism) or ideologies that undermine the active role we can play in face of different aspects of our individual and collective lives (e.g., neoliberal accounts of globalization as a ineluctable process, deterministic accounts of historical materialism leading to communism, strong religious doctrines that require conformity with fatalistic beliefs leaving no place for human agency) have always created deep tensions in ethics and with ethics scholars. They undermine ethics itself.[1] If ethics is to support a life of flourishing and help in critically analyzing the impact of our habits in light of human flourishing, it needs to adopt an account of human action that is consistent with that intent and facilitates it.

Despite this requirement, ethics cannot fabricate evidence or understanding. For example, it cannot postulate a free will, a will entirely exempt from contextual influences. It cannot postulate a domain of human activity in and of itself or create its own world, but it is a project of making the world that of human beings, a project of humanizing the world.[2] Just like medicine upholds an ideal of health and sets in motion substantive research, and all kinds of other means toward that goal, ethics is also motivated by its own goals. No one will challenge the fact that medicine relies on biomedical and other kinds of knowledge mobilized toward health. In light of this goal, certain pieces of biomedical knowledge may prove unhelpful, even if they are true. But knowing what is false and incompatible with the goals of medicine is also helpful in the pursuit of health and mitigating disease. The same applies to ethics, which is motivated by a project, and to which knowledge about a very wide array of domains is relevant.

Embarking on the project of ethics implies that it instrumentalizes knowledge toward a given goal: human flourishing. Accordingly, there is a sense in which ethics is not necessary at all, as human populations existed with recourse to human moral resources for a very long time and progressively adjusted their traditions to new situations. Likewise, human populations evolved without contemporary scientific knowledge (albeit with different forms of knowledge) for a very long time. Ethics, science, and medicine are human-made and human-given projects. Hence, ethics reflects specific intents. Accordingly, at the very core of ethics is at least a minimal commitment to the existence and veracity of human agency and human reason such that we can choose our behaviors (perhaps not all of them, perhaps with difficulty) and that we can understand them and reason through them, albeit imperfectly. This fundamental commitment to enlightened human agency supports a certain stance toward the use of knowledge as grounded in the agency of human beings. This also commands a certain recognition that all human knowledge is the result of transactions with the world, and the understanding that what we designate as true beliefs, as knowledge, needs to be understood as being the result of such positioning and transactions such that the value of knowledge can be judged in this light (Dewey 1941). This by no means leads to relativism or subjectivism because what is considered to be objective is, following the pragmatist tradition, in fact, really intersubjective, and the tests we put on knowledge to ensure its objectivity are really tests of intersubjective observations and agreement (Misak 2013).

A pragmatist orientation to ethics theory is explicitly committed to the user perspective on that theory (Inguaggiato et al. 2019b). Accordingly, the ideas embedded in ethics and its key concepts reflect intents. They are not neutral; they are not value-free or perspective-free. As a matter of fact, the point of view of the user or agent is no less neutral than alternative accounts of the view of human existence as passive. All are reflecting commitments to certain values and making certain kinds of situations and lived experiences more or less probable based on these commitments. It is in this sense that ethics reflects intents and is projective in nature, just like other ideas which, through their conception and use, enable or disable certain possibilities. Given the projective goal-oriented nature of human thinking, imagination (see also chapter 11) plays a crucial role in understanding reality. As Derrick Bell, one of the pioneers of critical race theory wrote, "To see things as they really are, you must imagine them for what they might be" (Bell 1995, p. 898), which for pragmatists means their role and purpose. This practical orientation at the very core of ethics is misunderstood when scholars or others attempt to transform ethics into a basic science that would be perspective-free or could be discussed in the absence of an explicit statement of its goals. The user perspective on knowledge is the agential perspective. This perspective does not discount knowledge, it situates it in the context of its interests and how it represents helpful or unhelpful ways of thinking in light of its implications for human action.

The Concept of Ethical Enactments and Its Connections to Agency and Moral Existence

Following the above clarifications and discussion, the preferred formulation of action in the context of ethics theory, at least when action is guided by explicit deliberation, is that of enactment for different reasons. Enactments are transactions of a specific kind, transactions guided by a clear willingness to project a specific kind of experience of the world. I recognize here that my use of the term may be somewhat estranged from enactivist and extended mind theories, notably because of the influence of pragmatism on my fundamental transactionalist take on action and enaction (Gallagher 2014). I leave aside the more elaborate distinction between enaction and transaction, notably because actions are always transactions, but enactions or enactments represent the deliberate attempt to act out a given scenario or course of action as a projection of oneself.

First, enactment reflects the active role of the agent who plays out a scenario that has been envisioned through ethical deliberation and, ideally, thoroughly assessed for its contribution to human flourishing. This was well captured by Dewey who alluded to the "dramatic rehearsal" of scenarios in his account of deliberation in the sense that moral agents envision what scenarios could be, as well as their implications. The scenario terminology also brings to the forefront the insight that scenarios can be "studied," "rehearsed," or "projected" in imagination before they are "enacted" to ensure an execution in keeping with the original intents and the foreseen outcome. Accordingly, agents who enact are agents who act out a way of being that re-situates relationships between organism (themselves) and environment. Likewise, the construct of a response-scenario makes it explicit that some thinking effort is guiding the actual action to be enacted, that is, some social intelligence. The slightly awkward terminology of "response-scenario" also makes it clearer that we are alluding here to a specific theoretical construct; not all moral behavior qualifies as a response-scenario, since those scenarios are the outcomes of a deliberative process described in the previous chapter. In contrast to enactment, the vocabulary of "action" is less able to convey these aspects and does not reflect the theoretical underpinnings of deliberative wisdom and the distinctions described in previous chapters. Likewise, I have repeatedly used the term "agent," which has important implications in discussions about ethical action. In order to enable flourishing, ethics—as described in this book—commits to viewing as ideal the kinds of actions that allow human beings to express who they are and pursue an existence as they see fit. Enaction differs from action in the very sense of being a more elaborate, more deliberate, and more engaged way of acting that projects what matters to human existence. Enaction is a way to make this even clearer and reflect more explicitly that action is a projection of one's existence and constitutes the agent and vice versa, consistent with a situated account of ethical agency (chapter 1).

Second, the use of the concept of "enactment" avoids the simplified vocabularies that remove intent and volition (e.g., behavior, activity, gesture, movement), that (1) make action static by capturing only a portion of it, notably its results (e.g., a deed, an act); (2) depict action mechanistically (e.g., an implementation, an operation); and (3) stress particular connotations of success (e.g., an achievement, an accomplishment) or heroic agency (e.g., a venture, an enterprise, an endeavor), all of which could lead to misunderstandings. Also, as a response to a situation, an enactment, or any form of action as a matter of fact, is not a reaction, though both terms (enactment and reaction) designate behaviors developed in response to a given event. A reaction is, by definition, triggered by another action or event. Enactment, as a response, incorporates a higher level of integration and implies that the reaction is supplemented with some ability or some effort of calibration to answer the problematic nature of a situation meaningfully and not simply or more narrowly lead to another action. The vocabulary of "enactments," reflects that ethics-informed action captures intents and connections to the projected outcomes of scenario-based deliberation, as well as the need to think about ethics-oriented action—which is concerned with the welfare and flourishing of others. It is not merely movement or behavior that happens to someone or a reaction to circumstances, since ethics-guided action has purpose and intent. The concept of ethical enactments thus represents a holistic account of human action and is part of a long-standing pragmatist effort to avoid reductionism of human action to mere behavior (Dewey 1896).

Third, "enactments" designates the putting into action of a thought out idea, a plan, or a scenario. Following the present ethics theory, response-scenarios are generated through scenario-based deliberation. It is not merely an application of theory or an implementation of preconceived ideas. This latter formulation in particular removes authorship and the participation of agents in the development of ideas. It also does not do justice to the creative role of agents when they have to adjust along the way, when they are guided by a plan of action but need to take into account unforeseen or unthought-of aspects of their situation. Keeping in mind the concern of ethics for the welfare of others, ethics-oriented action must be envisioned in a collaborative way, where we not only look ahead, but look to where we come from and look sideways to cross-check with others if our actions make sense given their purpose, akin to developing not only a central but also a peripheral vision. This explains why straightforward, individualistic, and unilateral ways of thinking about human action, as well as the literal implementation of moral norms and preconceived moral ideas, do not represent action well from an ethics standpoint. Accordingly, the enactment terminology helps in stressing how meaning making is a social and engaged, ideally participatory, process (De Jaegher and Di Paolo 2007).

In sum, "ethical enactment" is a somewhat technical term embedded in the theory of deliberative wisdom connected to the ideas of rehearsal and carrying out action-scenarios. The concept of enactment belongs more to the domain of ethics and ethical theory than that of morality, since it invites side-stepping our common and often implicit habits of action in order to promote explicit thinking about the purposes of our actions. Viewing actions as enactments, or actions in light of them being or not being enactments, is part of the project that attempts to foster human flourishing where the ability of ethical agents to self-actualize is central.

Enactments and Re-Situations

Given the situational account of moral matters at the core of this book (see chapters 3–7), I have insisted on how ethical enactments need to be envisioned—within the process of ethical deliberation—as response-scenarios and then, in the course of action, as ethical enactments. Ultimately, the aim of response-scenarios when conceived, and then enacted, is to strike a new equilibrium such that the desire of the moral agent to grow and learn is enabled in the situation, fostered by a given course of action. In other words, the ideal enactments are those that foster human flourishing. These kinds of resolutions are not mere quick fixes to problem situations (although it is possible that some responses to complex problems be easy, sometimes almost obvious), but perhaps neglected or buried deep under layers of reality (e.g., implicit biases) that need to be identified and unlearned. Better defining what are successful, unsuccessful, and partly successful re-situations is the matter of the next chapter, but it is important to spell out here how the account of ethical enactment constitutes re-situations, at least from the standpoint of describing outcomes of enactments.

In a nutshell, re-situations are not simply solutions; they not only solve a problem but actually change the configuration of the situation, that is, either the agent, the context, or the transactions between agent and context. In explaining Dewey's concept of situation, Gallagher writes, correctly, that "the trick to solving a problematic situation is not simply to rearrange objects in the environment, but to rearrange oneself as well—to make adjustments to one's own behaviors. Indeed, any adjustment one makes to objects, artifacts, tools, practices, social relations or institutions, is equally an adjustment of oneself" (Gallagher 2014, p. 116). Successful deliberation has the power to offer keys to transform the situation in ways that resolve its problematic aspect(s), but this is not only a transformation of the agent but of the entire situation. In many ways, this outcome re-situates insofar as the agent, the context, and the relationships between them are changed to realign agential experience and aspirations and contextual resources toward a goal. In some cases, it is the understanding or meaning of the situation that changes such that the materiality of the context does not change, but the situation could be resolved because there is a new reconfiguration or alignment that is judged to now be the best scenario to pursue. This is the frequent experience that a fair number of moral matters (e.g., leading to ethics consultations) are caused by misunderstandings and misapprehensions between people

(Pescosolido et al. 2013). Regardless of the changes occurring primarily in contexts, agents, or both, from the onset of the problematic situation and the re-situation, there will have been changes such that the relationships between key components of the situation will be changed (see table 12.1). Accordingly, the agent will thus evolve in a changed situation, even if it is only their own attitude toward others that has changed.

From the standpoint of ethics, the hope is that a given re-situation will further the agency of the person such that it will further build or constitute the agent who will further flourish by enacting themself. As Nietzsche wrote, "What doesn't kill me makes me stronger" (Nietzsche 1889/1997, p. 6). Or, as formulated by Winston Churchill's resolutely agent-emphasizing philosophy, "A pessimist sees the difficulty in every opportunity. An optimist sees the opportunity in every difficulty." Although these are simplifications and generalizations, they capture the idea that successful re-situations are the result of ethical enactments that foster human flourishing and see the opportunity for growth through overcoming obstacles. This means that ethical enactments have a distinctive orientation. They have a clear purpose and goal, which sets them apart from other actions that are not purportedly driven by an explicit desire to foster human flourishing. Also, re-situations and ethical enactments are fundamentally creative moments of existence because they involve prior reframing of the understanding of a problem situation and the development of resources to overcome it. When successful, or mostly successful, they engender a sense of peace, of recovered calm, an ability to prosper, a sense of acquired strength gained in face of surmounting adversity or challenges. Re-situations foster a sense of overcoming, although paradoxically, there may be no competition at hand and nothing to win. The victory can be within oneself.

In sum, re-situations reflect the importance of viewing agency as being embodied and embedded in situations, that our existence is marked by these situations and re-situations, and that we grow as persons through our ability to overcome and thrive in changing and challenging circumstances. In practice, the identification of what counts as a "final" or "satisfactory" or "partial" re-situation (or whether a re-situation amounts to a resolution) is a difficult matter, often unclear for the agents involved, and is contingent on time, which is crucial to appreciate how robust a given re-situation is before it becomes problematized again.[3]

Table 12.1
Types of re-situations

Type of re-situation	Agent(s)	Context	Explanation
Agent–context re-situation (or global re-situation)	Change	Change	Agent(s) come(s) with a new understanding of the context, or of themselves in the context, and features of the context also change. These features of the context may fuel changes in the agent and vice versa.
Agential re-situation	Change	No change	Agent(s) come(s) with a new understanding of the context or of themselves such that the views on the context change but the features of the context do not change.
Contextual re-situation	No change	Change	Features of the context change but the agent's perspective does not.
Status quo	No change	No change	Agent perspective does not change and neither do the features of the context. In some cases, it could be simply that the status quo is maintained as a viable response-scenario.*

* Status quo as a genuine form of re-situation is sometimes a form of appropriate and satisfactory response to a morally problematic situation; I am here excluding the pure neglect or ignorance of problems as a strategy to response to such situations. The latter case would not be a re-situation but a denial of a problem.

Enactments as the Move Away from Prereflective Moralities

Ethics has mostly followed an intellectualist tradition wherein, as noted before, it attempts to provide "answers" to "ethical questions" (Racine 2016). I noted that, from a pragmatist standpoint, a response does not mean an "answer," which would be the solution to a "question." Indeed, it is the situation that is morally problematic; therefore, the moral matter is located within a situation, and therefore, the problem does not exist, so to speak, outside of situations as a floating question with one true correct answer. Accordingly, the response to a morally problematic situation should never be simply cognitive (e.g., an answer in the scholarly sense) but, in the end, should be thought of as a more integrative action or a response. It is true that an answer, in the narrow sense, is potentially the beginning of a movement toward action and will be unavoidably part of the global response, but it is certainly incomplete. If responding (in a behavioral and noncognitive sense) would remain the only possibility to situations, it would amount to a form of alienation because thought would be prevented from having any kind of effect. It would imply to generating a system of ethical thinking disconnected from existence and actual accomplishment of human potential and meaningfulness such that it becomes a system of false consciousness and resentment (Nietzsche 1887/1971).[4] Thus, an ethical response should be envisioned not solely as a common behavior in reaction to a problematic aspect of moral life but as a response to a morally problematic situation where ethics is solicited to prepare, through deliberation, a thoughtful and rather thorough plan of action.

The enactment of ethical responses is not synonymous with moral behavior *sensu stricto*. Ethical enactment is a term that should be used to designate ethical actions (which stem from system 2, deliberated and effortful processes) and not moral behavior (system 1, which stem from more automated and effortless processes). Common behaviors belong to morality, and enactment belongs to ethics, that is, the science that projects and re-situates in the face of morally problematic situations. By this I do not mean to say that some moral behaviors are not conducive to human flourishing and/or cannot be ethically defended. Some moral behaviors can have prereflective connections to human flourishing because most human beings implicitly know what is in their interests as flourishing individuals. That knowledge is imperfect and can be cultivated, but it is also important to recognize the connections between ethics as a science and common moral experience. Furthermore, some ethical responses can become engrained in morality and yield a revised morality. One can think here of some of the values guiding the evolution to vegetarianism and veganism (e.g., animal welfare, environmental impact, human health) from much prereflective and habitual omnivorism, sometimes based on no clear argument except habits in the Western world. It is increasingly probable that these alternate dietary approaches (for the Western world) will become a new norm and yield a new form of morality that will become commonly accepted and invisible just like current habits are. Indeed, the process of problematization of current Western dietary regimens is in motion (Dombrowski 1984; Racine and Prud'homme 2019). However, vegetarian and vegan diets are not fitting for everyone; they will need to be analyzed critically in terms of their actual implications for real people and their real implications for the environment. The same could be said of rights of linguistic minorities, ethnic minorities, sexual minorities, and so on; the respect and just treatment of these groups is now becoming morally habituated and discrimination against them is becoming increasingly condemnable. We are moving from the imposition (sometimes implicit but sometimes deliberate) of the prereflective morality of a majority, to more plural moralities informed by the science of ethics as I describe it. The reevaluations and reconstructions of human practices in the direction of more effective pursuit of human flourishing is the inherent task of ethics. In this light, the barriers to human flourishing take the form of habits that cause harm, such as social practices that discriminate based on biological differences (e.g., skin color, sex, age, physical abilities) and socially constructed categories (e.g., gender, ethnicity, disability).

Responses Are Enacted Scenarios with Temporal and Contextual Aspects

Response-scenarios are explicitly and deliberately elaborated ethical constructions of the future; they are projections to be enacted. Explicitly integrating these aspects clarifies that a response-scenario will "unfold" and will have an intrinsic temporal and contextual dimension.

Enactments capture that a previous deliberation took place, that an orientation directed the act, and that those acting are actually guided by this orientation and are thus enacting—they are trying to achieve a certain goal while reality unfolds before them. Enactments thus involve a temporal narrative such that a response is not simply an action but an unfolding process that can be redescribed as an unfolding story. This temporal dimension implies that an ethical action is really an enacted scenario; even if the action in question is completed in a short period of time, it will find itself within a sequence of events prior to and after the enactment, as exemplified even by basic motor programs, a concept that describes how basic motor behavior is schematized and enacted in quasi-automatic ways (Schmidt 1975). Appreciating how an enactment involves changing moral sequences of existence and attempting to reorient them toward trajectories of flourishing gives credence to the common understanding that although retrospectively history may be recited as a narrative from the standpoint of the actual agent, it is a much more troubled and foggy reality with unclear directions and many intersections to ride through. Indeed, the unfolding of scenarios means that there are some predictable and many unpredictable spin-offs and developments to enactments. Even if one has rehearsed a scenario and tries to envision how things will turn out, it is impossible to predict how the story will unfold and how it will be recounted and made sense of in the future.

In addition, enactment means that the playing out of a course of action takes place in a certain scene, a certain context. This context can be constituted by facilitating or impeding factors for action of all sorts; specialized knowledge about these factors can be gained, cultivated, and shared. When enacting a given scenario, there are bound to be adjustments to the intended act as the agents take into account other participants in the situation, as well as their surrounding physical and social context(s). Some amount of improvisation may always be needed to adjust a scenario to the vagaries of life and the unpredictable circumstances of human situations, much like a chef needs to adjust a planned recipe to the availability of ingredients. In contrast, the nowadays frequently used notion of implementation (e.g., as in some of implementation science [Bauer et al. 2015]) carries a form of unilaterality of execution that lacks such flexibility. It gives the impression that the move from intention to action is an uninterrupted and simple process and that whatever difficulties encountered are "barriers" to this process. Such mechanistic overtones of implementation theory are to be avoided in ethics, notably because they lack the openness to learn from the process of enactment. The paradigm of implementation also fails to explicitly convey that ethical enactment is partly an artistic feat—in the etymological sense of art as an *artem* or *techne*, a skill and *poeisis*—requiring skills to enable certain actions to become realities in a way that is sensitive to the context in which an action takes place to create something and generate meaning. To pretend otherwise and to insist that scenarios can be conceived of and enacted in acontextual ways (i.e., as mechanistic implementations) is to overemphasize certain aspects of agency (especially the individualistic and rationalistic aspects of agency) and environmental control, while in reality agents rarely fully control the conditions in which they enact a scenario. They may try to understand these conditions, they may be fully aware that a number of social, human-made conditions are at the origins of the situation, but they cannot as single agents be held responsible for everything. In fact, moving away from such an implementationalist model may be salutatory to envisioning a participatory and socially embedded form of ethics as a genuine meaning-making activity where enactment is a process of learning (De Jaegher and Di Paolo 2007).

Enactment Involves Skills

Modern philosophy and contemporary ethics have sometimes entertained the idea that actions can be discussed in isolation of context and of who carries out the actions. Many theories are formulated in an impersonal way, as laws and norms, which could be applied by anyone, anywhere. It is even clearer that systems of regulation based on generalized conventions (e.g., laws, regulation, morality, and some forms of ethics) take this path. We do need such simplified accounts of what matters in moral agency such that we can have reference points for the guidance of human activities. In this sense, it is widely held that certain acts are reprehensible in themselves. Murder, theft, and other similar deeds are usually, by default, seen as being reprehensible because they cause unjustified harm to others, and agents are held responsible for such wrongful actions. This insight has supported the idea that actions can be assessed in and of themselves because the context in which they occur and the entirety of the situation is deemed controlled for, *ceteris paribus*. But there are obviously exceptions to this rule of thumb, and it is much more realistic to entertain a view in which multiple factors such as actions (deeds), agents, and consequences of actions are considered to assess the ethical nature of a given situation of an agent (Dewey 1966; Dubljević, Sattler, and Racine 2018a; Dubljević and Racine 2014). Determining whether an agent's act induces flourishing or withering cannot always be solely assessed based on the consideration of a single factor like the nature of the act; a more holistic strategy must be employed in complex situations. Of importance is that an act does not stand disconnected from an agent and, moreover, that the act does not stand apart from the actual performance of the act by the agent. A seasoned actor will more readily be able to enact a given play with more punch and conviction than will an amateur. Skills of performance count, and the ability to transform situations of withering into situations of flourishing is also an ability that can be cultivated based on experience, knowledge, self-reflection, and so on.

Some rationalistic strands of modern ethics theory have tackled acts irrespective of who undertakes them, of who thinks about a given problem situation, and so on. This impersonal take on such a personal and existential matter has had the virtue of opening up a world of knowledge and possibility to those who would otherwise not be deemed meritorious or capable of such discussions in the past. Indeed because of its original elitism and selectiveness, ancient ethics theory would not have cared much for the life of underprivileged and marginalized people (e.g., women and slaves were not part of democratic political processes). Accordingly, modern rationalistic ethics theory has reinforced the point that we are equal because of our meaning-making abilities as individuals and the equal worth of human lives, despite our differences. This is an outstanding gain. However, rationalistic ethics theory has tended to overlook the connection between the kinds of persons we are and the kinds of things we do, including the ability to get something done because of its greater insistence on general principles and rules of conduct. But envisioning ethics theory as a tool of flourishing implies that the ability to understand that tool, to wield it, can be refined and cultivated, as well as lost or abandoned. In this sense, all aspects of the enactment process can be cultivated and nurtured as skills regardless of one's more natural initial talent in those tasks.

Ethical Enactments and Response-Scenarios Range in Complexity

One clear drawback of the concept of ethical enactments and its corollary concept of response-scenarios is that they may inadvertently suggest that scenarios are complex and elaborate stories, and that enactement requires special and obscure skills reserved for an elite. Within the parameters of the current theory, response-scenarios can describe a range of actions and strategies from the very simple to the more complex. In other words, complex deliberations do not necessarily command complex solutions; response-scenarios can be simple even though the discussion leading to their identification could be extensive and thorough. Likewise, a complex response-scenario could stem from a rather easy and simple deliberative exercise, especially if there is a time pressure to offer a response.

These points are important to stress since ethics theory and scholarship has a natural and explainable tendency—as an exercise of intelligent inquiry—to overanalyze and complexify most everything, sometimes to the detriment of the search for an effective and workable response to the situation. This point is particularly vexing when there are pressing time constraints, as encountered by first respondent workers who need to act in the middle of emergencies (Racine, Cioaba, Boehlen, 2023). Furthermore, a response-scenario favored at the outcome of an ethical deliberation could even be the existing status quo. Imagine an initially problematic situation where a businessperson is confronted with a difficult situation with high financial and moral stakes in which they need to decide very quickly (e.g., during a business dinner with new partners) whether to invest in new environmental technologies (e.g., a novel electric battery) to diminish environmental impact (e.g., carbon footprint) of a technology and replace carbon-based energies. In principle, the outcome of the short deliberation may be that at this time there is no strong argument to change the production processes and energy supplies because of the unknowns about the true environmental impact of the proposed batteries and their environmental impact in replacing current technologies. In this case, the favored response-scenario could be a form of "active or vigilant status quo,"[5] that is, to use as efficiently and as responsibly as possible current sources of energies and monitor the situation about the environmental impact of new replacement technologies such that the terms of the analysis conducted within the deliberation may change. This is an oversimplified example, but it suffices to show that response-scenarios can be simple or complex. Although the example presented requires quick deliberation, there is no logical necessity that extensive deliberation leads to complex response-scenarios. Indeed, it should not be viewed as a disappointment that thoughtful and substantive engagement in ethical deliberation leads to minor changes or even maintenance of the status quo (see table 12.1). Participants in ethical deliberations should be warned upfront that this is a possibility but that this would be because this response-scenario has been judged to be the most satisfactory and best justified at this time. A similar remark is needed to explain the possibility of an agential re-situation that leads to a change in thinking, but no concrete change in practice.

Response-Scenarios Are Fallible Hypotheses

Following tenets of pragmatist philosophy and the need to assess actions based on their outcomes and effects on moral agents and on situations, responses are always considered hypothetically ethical such that the assessment of their ethical nature is never solely dependent on intents alone (although as

Kant suggested, this could be the place where ethics is supremely located)[6] or consequences alone (defined narrowly) but on a more comprehensive or holistic assessment of the resolution of the situation that includes both the agent(s) and the context in which they evolve.

One definite advantage of the response-scenario terminology is that it implies very clearly (in contrast to the concept of action) that the response is contingent and fallible, that is, that the response could be misdirected, lead to suboptimal outcomes, or be dependent on the execution of a course of action that could go wrong. In contrast, the epistemology of action does not explicitly entail the built-in recognition of the provisional and deliberate attempt at answering a situation. Since a scenario is a projection, and is hypothetical, its enactment is always tentative and could go sour. This recognition of fallibility carries with it the requirement to be humble and recognize the limits of the intelligence provided by our best attempts at moral problem solving. It also means that outcomes matter tremendously, such that processes (see chapter 13) need to be in place to capture the effects of one's actions, as well the ability of a given action to resolve the problematic situation. These measures can range from the less to the more formal, depending on the scope of the situation at hand. For example, it could be for an individual moral agent (e.g., a nurse) to follow up with a patient or a family member to see how things went for the patient, following discussions about a sensitive or emotional aspect of care. For a trialed intervention within a research project, it could be a full-blown evaluation with specifically developed tools, allowing the assessment of outcomes from an ethical standpoint using elaborate open-ended and participatory approaches. Importantly, the recognition of the fallible nature of ethical responses and ethical deliberations supporting them is implied by the concept of ethical wisdom and growth described in chapter 2 because the idea that we can learn from experience implies that current approaches, and foreseeably those to come, will eventually be challenged and perhaps ruled out and we must thus be open to learning. Ethics as a scientific and growing area of inquiry is premised on this very recognition of the fallible nature of much everything about ethics. This includes its abilities to pinpoint specific moral matters, to be sensitive to and aware of new and upcoming moral matters, to foster open and productive deliberation on these problems, to make sense of how engaging in discussions about such problems contributes to the growth and development of people, and so on. Recognizing all these fallibilities is not defeatism, it is a constructive acknowledgment of the knowledge-informed nature of these important tasks. To claim otherwise and offer reassurance where it should not be provided could lead to arrogance and dogmatism.

The Impact of Terminological Clarifications about Ethical Enactments

It may seem unwieldy and inconvenient to describe the outcomes of ethical deliberation as "action-scenarios" and "response-scenarios," to talk about actions resulting from "scenario-based deliberation" as ethical enactments, and to refer to of deliberation as "rehearsal" leading to "re-situations." There is no doubt that simpler terminology could and actually must often be used in practice, especially when these concepts are used to guide the practices of those outside the field of ethics. Despite the seemingly endless detours of theory and reflection on our vocabularies, the repurposing of these words has a number of theoretical and methodological advantages.

The agential terminology of deliberative wisdom positions enaction as the focus, the integrator of the problematization and deliberation efforts, because action-scenarios are individuated on the basis of the action to be undertaken, including the motivations underlying it (the Deed), its consequences (C), and the people involved (Agents) explained by the ADC model of moral judgment (Dubljević and Racine 2014). Second, this terminology also makes clear that philosophical analysis and other aspects of ethics as a deliberate moral problem-solving effort need to produce response-scenarios as well as an assessment of whether these responses actually resolve or do not resolve the situation and to whose detriment and benefit. The evaluation of scenario-based deliberation is not only a conceptual affair. It is an engagement with experience. An integrative vision offering benchmarks to evaluate the value of ethics theory as a tool to resolve effectively and with honesty morally problematic situation is thus required. (These benchmarks include evaluating whether and how a given proposed action-scenario is lived by different people concerned and how effective it was in actually responding to the original problematic situation.) Using a vocabulary that makes all of this more explicit supports strategies that use the empirical and experimental sciences (notably psychology, anthropology, and sociology) to challenge some of the epistemological backbones of deliberative wisdom, since the understanding of "situations," "moral awareness," or "wisdom" can also be enhanced significantly by dedicated scientific inquiry as well as allegedly more theoretical and conceptual refinements.

At this time, whether categories and classifications of action-scenarios and ethical enactments are helpful to structure ethically oriented thinking and action is open to investigation and experimentation. In other parts of this book, I have elaborated on the need for ethics to be tailored to the field it is connected with (e.g., business, healthcare, politics) because assessing the values promoted by a given action implies a broader understanding of how this action fits within actual contexts and the possibilities afforded to the agents within those settings. Moreover, one could imagine that ethical enactments could be understood in terms of the scope of the situation involved (e.g., micro/interpersonal, meso/social, or macro/societal [Durand 1999]). In this way, actions that rely on interpersonal interactions may differ from those that involve much broader societal phenomena that more directly impact individually lived situations. Another strategy of categorizing ethical actions could be based on their desired effects and transformations (e.g., personal changes, interpersonal changes, work environment changes, societal changes) in areas such as moral awareness (empathy), skills for deliberation, and so on. Furthermore, they could also be classified according to the dimensions of flourishing that they promote (e.g., autonomy, environmental mastery, positive relationships). This orientation is already in motion in a wide-ranging body of work, not structured by a specific account of human flourishing. For example, there are interventions and actions that promote autonomy (Nagel and Reiner 2013), which could be pooled and examined together for their effectiveness as well as dignity-promoting interventions (Johnston et al. 2016), empathy-promoting interventions (Kiosses, Karathanos, and Tatsioni 2016), and interventions that promote relationships and engagement (Mackenzie and Stoljar 2000). This literature is spread across various domains of activity and is currently scattered. However, there are emerging efforts to pool and compare experiences in terms of their effectiveness across domains of activity—already undertaken to some extent as shown in the above-cited references. This categorization—consisting of assembling interventions under dimensions of human flourishing—is perhaps the most promising and telling, ethically speaking, since it is consistent with the instrumentalization of concepts like response-scenarios and ethical enactments toward what is ethically meaningful in terms of human flourishing. However, whether and how the theoretical apparatus of the theory of deliberative wisdom can actually deliver its promises stands to be determined by its community of users and those who should be benefiting from it.

Conclusion

In this chapter, I explained how ethical actions—in their fullest sense of being transactions with the world and having an experiential nature—have strangely been granted a limited role in ethics theories, which betrays their actual importance in practice. Different forms of disciplinary biases can explain why this is the case, although in my eyes, none can justify it. Moreover, actions, in common parlance and sometimes within ethics theory, are often considered to be ethical or unethical in themselves, irrespective of the context in which they occur or the outcomes they have. This simplification is sometimes useful but can also be limiting or too narrow to tackle complex situations. To move away from these problems and reflect an agential perspective in ethics theory, I advanced the concept of ethical enactment to describe how human action, guided by ethics, is a process of putting to the test a response-scenario generated in a process of ethical deliberation. I also explained how the user perspective of ethics theory calls for a displacement of ethics language to properly situate the terminology of ethics in the context of action and avoid losing sight of the user perspective and the actual problems. I also explained that successful ethical enactments are re-situations that propose a new equilibrium between agents and contexts. They have inherent temporal and contextual aspects, require skills for successful enactment, and can range significantly in complexity. Response-scenarios, and ethical enactments are fallible and their fallibility is assessed in practice. I hope to have shown how the reformulation of ethical action as enactment promises to bridge a specific account of ethical deliberation and ethical imagination (see chapter 10) with the transactional account of moral agency described in chapter 1. The next chapter explains how ethical enactments can be ethically evaluated in keeping with these and other components of the proposed theory.

13 Evaluating Ethical Enactments

Morally problematic situations are, by definition, in need of resolution. Not just any kind of enactment or any kind of response-scenario can count as a meaningful ethical resolution, and not all resolutions are fully satisfactory resolutions. Although this problem of assessing whether a moral matter is resolved is heavily dependent on how one envisions ethics, different ethical theories remain rather abstract and even silent on not only what would count as a meaningful resolution, but also on the methods of assessing this meaningfulness. In some more casual circumstances, nothing more than informal assessment will be possible or even be desired, such as relying on one's understanding of the resolution of the situation or on the feedback of those involved. Still, it would be useful to guide this evaluation—especially when formal assessment is desired or required—in ways that ensure greater intersubjectivity and greater holism and thoroughness to different dimensions of the resolution. In this chapter, I propose that the evaluation of ethical enactments be carried out in light of human flourishing. Following this, I revisit the six dimensions of human flourishing presented earlier as starting points for the evaluation of enactments and propose the six additional methodological or procedural guideposts mentioned in chapter 1 (e.g., holistic accounts of outcomes, nonreductionist inquiry, and participation and deliberation). I conclude with a discussion about formal and informal strategies to evaluate ethical enactments. The topic of the evaluation of ethical action is perhaps one of the most obscured in ethics theory and practice, and I hope this more tentative chapter brings further attention to this crucial topic.

> If someone is able to show me that what I think or do is not right, I will happily change, for I seek the truth, by which no one was ever truly harmed. It is the person who continues in his self-deception and ignorance who is harmed.
>
> —Marcus Aurelius, *Meditations*

It may appear quite strange that a book on ethics seems to need to go to great lengths to justify why evaluating the outcomes of actions matters from the standpoint of ethics theory, methodology, and practice. For most everyone in society, work and actions are aimed at achieving certain kinds of outcomes. This is very often the narrowed-down meaning of work itself—to accomplish certain tasks and attain certain results. Should it not be the case that ethics be concerned with action and how ethics deliberation impacts practices? Even in academic circles, for most other kinds of action-oriented sciences and professions, the evaluation of outcomes matter. Think here of medicine, a combination of science and practice. Think also of legal scholarship and its interest in the applicability of legal advice and legal analysis to guide, clarify, and understand actual judgments and policies. Similarly, think of engineering and areas of applied science where outcomes are central to the enterprise. Why then has ethics—which, as I described in previous chapters, emerged as a discipline specifically concerned with and connected to how people live, and most concerned with the profound and existential aspects of human lives—become unconcerned with actions and the evaluation of their outcomes? A pragmatist orientation certainly calls into question this oversight and invites a response.

In this chapter, I first revisit why ethics has been rather unconcerned by outcomes and their evaluation. This neglect is instructive for a form of ethics like the current theory that synergizes theory and experience. Then, I describe how human flourishing represents the focal point of the evaluation of ethical enactments and delineate initial substantive and procedural (methodological) guideposts that can help guide the development of evaluations, notably by avoiding certain pitfalls, such as neglecting contextual understandings of flourishing or using an authoritative view of science that reduces understandings of flourishing. I close the chapter by discussing how evaluations can take more or less formal forms using published examples, and discuss promising orientations for the evaluation of ethical enactments in light of the proposal made in this chapter. Recognizing the complexity of ethics outcome evaluation, due to the sheer complexity of human morality, means that no simple solution would conveniently dispose of the effort to think anew about what metric should be used to measure "success" and "satisfactory resolution" in a given situation. However, attempting to spell out what can count as meaningful enactment as done in this chapter could serve as a springboard for future developments.

The Challenge for Ethics to Be Concerned with the Evaluation of Outcomes

Although this chapter is focused on offering a proposal for the evaluation of enactments, it is worth stepping back to understand why ethics has been rather unconcerned with the evaluation of the outcomes of enactments, especially in the context of the kind of theoretical offer proposed in this book. There are many answers to this broad question, many of which are hard to evidence convincingly. One line of thinking would point to more technical and methodological issues. Is it perhaps that the field of ethics is not advanced enough such that this contribution will come if we simply keep going down the same path? Is it perhaps that ethics has not made sufficient connections with disciplines like sociology and psychology, which could provide resources to develop tools to evaluate outcomes? Or is it perhaps that there is something unique about the complexity of ethics such that there are in-principle arguments preventing evaluation of outcomes? Although likely plausible, there have been advances in so many areas of knowledge that it is hard to believe that no progress can be made if there is willingness to do so. So, a second line of thinking could be to ask the following: Why has ethics not cultivated a strong interest in outcomes and instead entertained a view of ethics that tends to disregard action, experience, and the evaluation of ethics' impact on them? (For authors making this point, see notably Fesmire [2003], Hoffmaster and Hooker [2009], and Hoffmaster [2018].) An unfair proposal would be to depict ethics as embedded in a theoretical academic culture because what is valued is the more abstract, purer forms of thinking instead of contextualized thinking connected to experience, contexts, and observations.[1] But there is perhaps a more concrete way of spelling out this explanation.

One hypothesis to explain the situation, which does not necessarily exclude the others cited above, is the disruptive nature of ethics. It is one thing to realize the existence of a morally problematic situation, and it is yet another to start engaging in ethical deliberation in order to surmount it. But it is also quite another to act upon that situation, respond to the initial concern, and then to have the humility to ask whether all that was done is useful and meaningful with respect to addressing the initial concern. This process is even more uncomfortable when we keep in mind that ethics asks those profound and disruptive questions about the things we most value in life. Is a meat-based diet ethical given its impact on the environment and on animals? Is the capitalist system in which we are embedded the better way of facilitating human development? Is monogamy and serial monogamy, as widely socially practiced and legally enforced in the Western world, the ideal conjugal life? Should medical aid in dying for patients with mental illness be offered to people who live with profound suffering?

Asking theoretical questions about these matters is already something uncomfortable. But dealing concretely with situations involving them, situations that shake our most profound values is another, let alone trying to change situations and evaluate those changes honestly and humbly. Hence, ethics is a disruptive field—a profoundly disruptive field—where common morality is challenged and reconstructed based on knowledge, that is, critical thinking and reflection on experience and observation.[2] Ethics inquiry can generate frontal challenges to different forms of shared human habits and customs about all aspects of human lives. Certainly, there is personal difficulty in maintaining flexibility in our habits such that we question ourselves about what guides us and what we want. But this is made more difficult when the inertia in our habits is reinforced by social, political, and economic forces that promote a certain understanding of human well-being and flourishing (e.g., based on capitalist productivity, based on monogamous relationships). Political and economic regimes, as thinkers have encountered across history, have their own vested interest in promoting, implicitly or explicitly, forms of thinking that legitimize and support ideas that allow a political system to sustain itself and preserve its own legitimacy (Capps 2003; Hollinger 1980). This is a problem that historical pragmatism ran into as it was replaced by a more docile form of analytic philosophy to avoid raising questions that could have brought suspicion of its complicity with communist and socio-democratic political philosophies (Capps 2003).

In sum, ethics can create trouble. Given that the current proposal is to bring ethics to bear on human flourishing, I recognize that evaluating enactments is a conceptually and methodologically challenging task, but it is also a kind of inquiry that raises questions about social order and ways of doing things in the spheres of law, work, education, sexuality, politics, and economics, notably.[3] As Rousseau wrote, "Repose and liberty seem to me to be incompatible; it is necessary to choose between them" (Rousseau 1819, p. 359, personal translation). To use our intelligence and our freedom to scrutinize and question our habits, and to change them afterward, requires effort and courage. Accordingly, ethics—if not critical and aware of itself such that it considers its own outcomes and implications—can embody a form of intellectual laziness and internalize and perpetuate the alienation of thought from experience and action, and experience and action from thought. In this respect, evaluating outcomes of ethics-oriented changes is therefore of crucial importance. Otherwise, we have no way of gauging if we have really done anything better in practice and in given situations despite our best intentions. We would be, so to speak, a car without a steering wheel or driver, a ship without keels or a compass, because in the end what is sought is an expanded experience of the world and a deeper eudaimonic experience of it, not simply claims that this is the case without proof. But to be committed to evaluating outcomes means that we also reflect carefully on the kinds of outcomes that are meaningful in a given situation. Indeed, when realizing that a given act is a means, the question quickly irrupts about the adequacy of that means to what end. And, moreover, we start asking about the kinds of outcomes that are relevant from an ethical standpoint to avoid falling prey to thoughtlessly using existing outcome measures that may be partially or completely inadequate to evaluate the ethical meaningfulness of a given change.

A Proposal for Guideposts for the Evaluation of Enactments

According to the vision proposed by the theory of deliberative wisdom, the resolution of a moral problem should foster the growth of the people involved such that, through the discussion and engagement with a specific moral matter, their vision of the world and their ability to experience deeper connection with other human beings expands. If an act is considered ethical insofar as it facilitates human flourishing, this orientation toward flourishing—although flexible and deliberately open-ended—can serve as a focal point, a cardinal participatory orientation, for the evaluation of outcomes. Dewey, who formulated such an ideal, also keenly noted that there are limitations to such an experience- and growth-oriented approach because it can veer into essentializing the ethicality of an act or idea narrowly, universally, or authoritatively such that a single account of flourishing can be taken as applicable to all human beings in all settings at all times. He thus hesitated to formulate growth as a cardinal orientation or a kind of categorical imperative, even inviting us to rebuke the search for a general formula.[4]

There is a fair share of wisdom in this self-imposed reserve since defining human flourishing authoritatively is an invitation to intellectual and existential laziness where the ideals of some are imposed on others. The current proposal must be understood in keeping with this open-mindedness about what a meaningful life is.

Despite these reservations, I propose to build from other components of the theory of deliberative wisdom, notably the proposal to adopt an open-ended and participatory account of human flourishing as embodied in the deliberative account of wisdom (chapter 2) and the ability to articulate ethical analysis in terms of criteria of ethical acceptability and ethical praiseworthiness (chapter 5). Illustrating what these general dimensions represent in a given context is a matter of interpretive inquiry and calls for thoughtful engagement with the situation, that is, the agents involved, the context, the lived experience. The bottom line is that the criterion of ethical acceptability requires that little or no infringement is committed to key dimensions of human flourishing, while the criterion of ethical praiseworthiness requires a more active development in key eudaimonic dimensions (see chapters 2 and 11). Clearly, not all resolutions will conserve or promote all dimensions of eudaimonic well-being all at once. For example, a given situation may be problematic with respect to autonomy or mostly with respect to positive relationships or combinations of different dimensions of flourishing. A good evaluation needs to be open to accepting this complexity and handling it in a meaningful and thoughtful way such that the focus is calibrated to what is at stake.

An orientation toward human flourishing has great appeal, but can it be further specified in order to yield practical guidance? What are the features of a successful and meaningful evaluation of an ethical enactment? Since processes of enactments can imperil ends pursued, are there features of processes of evaluation and general orientations that are more consistent with an open-ended and generous understanding of human flourishing? In an effort to advance these questions, table 13.1 describes that a given enactment can be evaluated in terms of its outcomes related to human flourishing, notably based on important dimensions such as autonomy and positive relationships with others (as reviewed in chapter 2). A qualitative gradation of evaluative stances (i.e., ethically condemnable, ethically unacceptable, ethically acceptable, ethically praiseworthy) is also provided based on respective contributions to flourishing (i.e., major encroachment, infringement, preservation, and growth). However, these substantive guideposts are insufficient because they could be taken too easily as essences or abstract constructs in ways that do not capture how people actually experience a given outcome. Accordingly, table 13.2 presents more methodologically oriented procedural guideposts consistent with the current theory's emphasis on the deliberative nature of ethical wisdom. These are presented in no particular order of priority, although they make sense together as reflecting a cohesive and coherent approach to the evaluation of ethical enactments. In this regard, I show that each individual guidepost helps avoid specific kinds of problems, while reliance on a single procedural guidepost is perilous. Cursory definitions and illustrations of the guidepost at work are provided.

In short, as presented in figure 13.1, table 13.1 describes the kinds of outcomes valued as a result of enactments based on different dimensions of human flourishing (e.g., personal growth, positive relationships with others, meaning and purpose in life). Table 13.2 explains how the eventual assessment of certain valued dimensions of human flourishing should be carried out from a more methodologically deliberative perspective. These procedural guideposts prevent the essentialization of outcomes of enactments based merely on the dimensions of human flourishing while also stressing the importance of the process of evaluation itself. For example, for actions aiming to promote autonomy, it would be important to assess them holistically (e.g., examine outcomes beyond consequences), adopt forms of nonreductionist inquiry and so forth (see below for procedural guideposts), with autonomy specifically in mind. For example, is autonomy something actually meaningful (procedural guidepost of narrativity and meaningfulness) to the person in question since autonomy is sometimes not the most valued aspect of certain situations? I recognize that this first tentative list (table 13.2) could be eventually refined and reworked based on further experimentation and trialing, as well as maturation of judgments about how to best do this. It would be encouraging in fact that progress be made in that direction in light of the pursuit of open-ended experience and existential evaluation of ethical enactments in different contexts. For the sake of providing clear explanations, the proposed procedural guideposts are explained and exemplified in isolation of the substantive guideposts (which are well covered in other chapters) but the reality is that they are interwoven such that the procedural guideposts are used to specify how the substantive guideposts are operationalized and used.

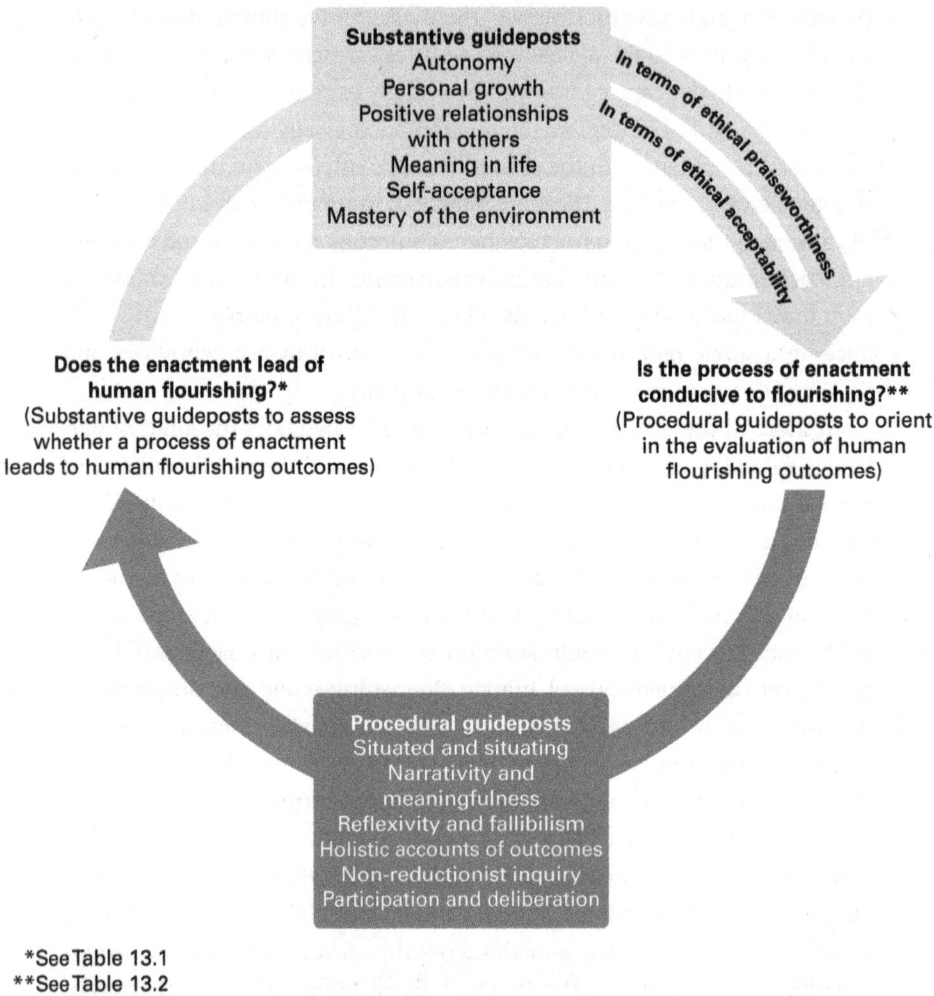

Figure 13.1
Guideposts for the ethical evaluation of enactments.

Situated and Situating

Consistent with the situated account of moral and ethical agency (chapter 1) and the nature of situated moral experience (chapters 3–8), outcomes of enactment need to be evaluated in light of how they return to experience. To take the example of medical aid in dying (a.k.a. physician-assisted suicide or active euthanasia, depending on the country in which this takes place), in some societies, discussion of this situation cannot be fathomed because

Table 13.1

Substantive guideposts to assess whether a process of enactment leads to human flourishing outcomes

	Ethically condemnable	Ethically unacceptable	Ethically acceptable	Ethically praiseworthy
	Major encroachment	**Infringement**	**Preservation**	**Growth**
Autonomy	Destroys the autonomy of individuals or communities	Impedes the autonomy of individuals or communities	Does not impede or has a minimal impact on the autonomy of individuals or communities	Promotes the autonomy of individuals or communities
Personal growth	Abolishes the growth of individuals or communities	Diminishes the growth of individuals or communities	Does not diminish or preserves the growth of individuals or communities	Fosters the growth of individuals or communities
Positive relationships with others	Destroys engagement and relationships	Imperils engagement and relationships	Does not imperil or maintains current engagement and relationships	Deepens engagement and relationships
Meaning in life	Invalidates the meaning of people's lives	Impoverishes the meaning of people's lives	Does not impoverish or invalidate the meaning of people's lives	Promotes genuinely authentic meaning making
Self-acceptance	Invalidates the identity of individuals or communities	Challenges the identity of individuals or communities	Does not challenge the identity of individuals or communities	Promotes identity development and self-acceptance within individuals or communities
Mastery of the environment	Takes away the ability of a person to control their environment	Undermines a person's ability to control their environment	Does not undermine a person's ability to control their environment	Empowers a person to act upon their environment

Table 13.2

Procedural guideposts to orient in the evaluation of assess of human flourishing outcomes

Guidepost	Description	Problem(s) avoided	Illustration of questioning related to guidepost	Pitfall(s) if guidepost were to be used alone
Situated and situating	Assessing flourishing outcomes in context-sensitive ways. Avoiding the reduction of flourishing outcomes to personal or individual characteristics.	Universalism	Is there an effort to understand the different aspects of the situation at stake, notably agential factors, contextual factors that shape the given situation, and the lived experience of the situation? Is the problem understood in relationship to the different aspects of the situation?	Relativism. Reduction of flourishing outcomes to untestable hypotheses because too heavily context dependent.
Narrativity and meaningfulness	Assessing flourishing outcomes such that the role of events is situated within lived human life and existential meaning.	Reduction of human flourishing to impersonal, depersonalized, and meaningless outcome measures. Abstraction and essentialism (existential insignificance or personal irrelevance in assessing outcomes, e.g., using allegedly objective measures).	Is the evaluation done such that the outcomes are considered within the broader context of people's lives and not only one's own life or a narrow understanding of other people's desires, choices, and projects? Is there a way of approaching outcomes that grants room for the narrative and life trajectory of the person and the meaning of specific events, choices, and situations to them?	Reduction of flourishing outcomes to mere intentions guiding the act. Reduction of ethical outcomes to biographical and personal meaning making. Reduction of flourishing to person-related outcomes without consideration of the impact on relationships with others and the connections between one's flourishing and that of others.
Reflexivity and fallibilism	Assessing flourishing outcomes in ways that provide for ongoing learning and that are open to invalidation.	Reduction of flourishing outcomes to authoritative measures. Uniformity, conformism, and dogmatism.	Is the solution envisioned as being fallible such that insights gleaned from more general understandings of human flourishing can be evoked and question the proposed resolution and the understanding of its relationship to the original problem situation? Is the outcome considered with an open mind?	Subjectivism and the ineffability of flourishing if confined to intractable knowledge.

Holistic accounts of outcomes	Assessing flourishing outcomes beyond immediate or simplified consequences narrowly construed to capture existential and narrative aspects of outcomes.	Narrow consequentialism (reducing outcomes to certain types of consequences, e.g., utility, or only to those consequences related to actions and not the agents themselves).	Is the evaluation carried out in a way that avoids only measuring what can be easily counted or described? Is there an open mind about what counts as a satisfying or even laudable outcome, inclusive of the intents of agents and the meaning of the deeds involved? Is there an open mind about the way of envisioning outcomes (beyond consequences *sensu stricto*) such that implications on moral agents themselves are considered?	Failing to take into account agent-related outcomes such that effects on the person as an ethical agent are left out given an overriding focus on holistic outcomes.
Nonreductionist inquiry	Assessing flourishing outcomes using inquiries and knowledge without using science in an authoritative and dismissive way.	Closure to learning from others and to questioning dogmas that can be supported by authoritative accounts of science. Obscurantism related to the inability of reporting how knowledge is generated and mobilized.	Is there openness to cross-check one's understanding of the situation with that of others and with existing knowledge about human flourishing?	Scientism and positivism, reduction of flourishing outcomes to existing scientific understandings and available validated measures and phenomena.
Participation and deliberation	Assessing flourishing outcomes in ways that are participatory and that engage stakeholders. Assessing flourishing outcomes while accepting that experiences vary between persons and that this impacts the dimensions of human flourishing.	Nonparticipation and authoritarianism in imposing a set views on what flourishing means. Individualism and solipsism. Closure to learning from life and from others.	Are open-ended processes followed for understanding and resolving the situation also reflected in how the outcome is evaluated? Does the evaluation of the resolution take into account the experiences and perspectives of different stakeholders involved in the situation?	Reduction of ethics to participatory processes without an eye to participation as a process of self-actualization and enactment of what counts as meaningful. Reduction of ethics to formal consultation and discussion procedures.

the moral structure of the society preempts any discussion about inducing death in an ethical way. This was the case not too long ago in Canada where more traditional morality—heavily shaped by rather rigid customs and religious beliefs and a certain understanding of the value and meaning of life—precluded any discussion of this matter. There is a way of interpreting this change as an "evolution" for the better, a progression with all nuances, limitations, and qualifications that this interpretation summons. However, we should be careful to understand the nature of this evolution and how it represents progress. The broader framework to recognize is that both agents and contexts shape situations and that situations occur within social networks. This means that the search for human flourishing cannot be abstracted from situations and that situations cannot be abstracted from the broader environments in which they occur. This goes back to the example of intergenerational change (see chapter 4) and social change (chapter 5 on how situations evolve) discussed previously.

Human flourishing needs to be envisioned within situations and within the contexts in which experiences occur. For example, the desire to end one's life may be understandable in a society where means of prolonging suffering and pain are a collateral effect of advanced treatments and care. It may also be that a given society is mostly guided by a form of inflexible morality that does not offer any kind of leeway to raise questions about such situations despite being confronted with the effects of advanced technology. For example, some societies are rigid and morally conservative yet are nevertheless avid users of very advanced technology (e.g., Saudi Arabia). This creates contexts where moral thinking is confronted with a challenging new reality, made even more challenging by this rigidity. For example, religious-based conservatism in Saudi Arabia is defied by the many advances touching on women's health (Muaygil 2018), while a whole range of medical practices challenge practitioners who must wrestle with opposing forces of change and status quo (Al Sayyari, Hejaili, and Shaheen 2016). The distinction between morality and ethics becomes useful to thinking about what human practices are amenable to the critical lens of human flourishing and, in contrast, what proportion is excluded from open questioning by taboos or implicitly held or unquestioned or unquestionable practices.

The emphasis on the situated nature of the enactment and its evaluation does not lead to pure relativism. This methodological guidepost should not prevent engaging with comparisons of which ways of life make human beings happier and better through the lens of human flourishing. However, given the complexity of comparing situations across cultures and contexts, this likely paves the way for mutual learning rather than brute evaluative comparisons of cultures and contexts (for a discussion of this idea in the context of discussion about ethical implications of neuroscience in an international perspective, see Racine and Senghor [2023]). Of course, stating that every human being can cherish the goal of flourishing—and not only a political or economic elite—is an important and subversive commitment. It implies that all aspects of our lives—what we eat, how we relate to each other, how we relate to the environment, how we love and engage in sexual relationships—should all be considered critically in light of the well-being of ourselves and others. In other words, social structures and human habits should be in service of human flourishing rather than human beings acting in service of structures and habits. It is important to keep in mind—given the example I used above—that Western societies have their own implicitly held and problematic assumptions about a number of fundamental aspects that may not be in service of human flourishing. These include exploitative and irreverent relationships with the environment; rigid ways of envisioning intimate relationships; the impoverished place granted to friendship; modes of economic production that are often alienating and leave little to no pride or contentment to the person doing the work. There is nothing set in stone to say that human flourishing is best pursued based on the ideas of the Western world, notably how we engage with ourselves, each other, and the world; all of this should be evaluated contextually. By that I also mean that some of the dominant ways of thinking should be recognized for what they are, and critical thinking be exercised toward ideas that we seem to take for granted and see acontextually as being universally valid or true or relevant. As transacting beings, our environments shape us such that we are tributaries of our contexts in ways that make it impossible to extirpate ourselves from our own contexts, but we have the ability to engage in discussions about the differing effects of contexts in the evaluation of outcomes. Doing so can nurture our ethical imagination and on ability to act differently.

Narrativity and Meaningfulness

The corollary to understanding the situated aspects of the problematic situation and of understanding consequences holistically is the need to also understand holistically the meaning of the outcomes of enactments. The evaluation of ethics-oriented change needs to capture the specificities of a given morally problematic situation and the rationale behind the selection of a given preferred response-scenario and not only its consequences in a narrow sense. This is another reason why ready-made measures and tightly defined theoretical models of outcome measurement can fall short of the goal of measuring true ethical outcomes, namely human flourishing. This can only be done if the evaluation of the outcome takes into consideration the connection to the intent of the agents and the interpretation of the situation as per the nature of situations (see chapter 3). It also needs to pay heed to the meaning-making contribution of morality to our existence as human beings (chapter 1).

Consider again the example of medical aid in dying. Facilitating someone's death, taken out of context, could represent a deeply problematic action and outcome. Moreover, the intent of the people involved (to help someone end their life) could also appear highly problematic. However, in this example, understanding the motivations of those involved (e.g., a compassionate gesture) and ensuring a process wherein these motivations are screened (e.g., to ensure that there are no conflicting interests and motivations, and that the decision is informed and made freely) is crucial because of the enormous stakes. If the meaning of the gesture is disregarded, the evaluation of what was intended will fall short of grasping the full significance of the situation.

There are ways that the intents and the meaning of enactments can be evaluated. For example, if compassionate care and a humane death are the supreme goal of medical aid in dying, then follow-ups to evaluate the outcomes with families and those concerned directly throughout this process will help ensure that this goal is achieved and that the processes in place do not disorient the unfolding scenario from its chief motivation. Some of these important aspects for outcome evaluation could eventually be simplified such that a short interview with family members would get to some of the more basic points (e.g., "Did you feel that your loved one received compassionate care?" "Did you feel they were respected throughout the process?"). However, in any case, some tailoring of the outcomes is required because simply relying on predetermined outcomes will not unravel the narrative and meaning that structures a given action-scenario and will not be sensitive to it. This is because each event has meaning in the context of a lived life, and there is no substitute for grasping what something means in one's life to understand its impact. This does not mean that the meaning we attribute to certain events or phenomena is static or cannot be revised. It simply means that meaning is inherently tied to the narrative structure of our existence and that the ethical evaluation of enactments must take it into account. We cannot presume that common dimensions of human flourishing such as the pursuit of autonomy or of positive relationships will lead to meaningful and narratively coherent outcomes all the time for everyone.

Reflexivity and Fallibilism

Ethics, as described and proposed in this book, constitutes a deliberative process of inquiry, a journey we actively undertake more than a fixed destination. Accordingly, leaving room for learning and questioning means that we can ask whether the outcomes of an ethical enactment help nurture ethics development or if they actually foreclose further discussion and learning. The very way that an ethical enactment is structured and guided—that is, as an open process of engagement, of understanding the situation, of reconstructing the problem situation, and of deliberating about solutions—should make it clear that the outcomes are bound to be partial and fallible. This means that if the initial morally problematic situation was rooted in moral convictions about what were the most desirable and least desirable options, the related enactment should have shaken those convictions to show their limitations and their fallibility. As a result of an ethics-guided process, stakeholders—even if this is painful and confronting—should be more willing to learn because they are now more familiar with the temporal and fallible nature of their moral convictions. If the outcome is seen as a simple victory or defeat, the process has likely gone wrong such that participants have somewhat framed ethics as a process able to determine who is or was right or wrong with a sense of certainty that does not really exist. The outcome of ethical enactments should be learning and fostering the willingness to learn because participants will have recognized that, in a situation, most often no one is or was totally right or wrong with absolute certainty. Because of the existential and experiential nature of moral situations (see chapter 6), the importance of moral awareness (see chapter 8) and its grounding in the experiential, and the process of collaboratively envisioning and constructing the understanding of the problem situation and its resolution (see chapter 10), the evaluation of ethical enactments is preferably carried out following methods that respect the importance of the procedural aspects of deliberation and open processes of understanding and resolution. Reinforcing this process supports the scientific nature of ethics as an open-ended and fallible inquiry through which we can question and enrich our moral existence.

Procedurally, because of the evolving nature of human realities and the evolving perspectives and experiences of agents, there needs to be a recognition of the provisional nature of the evaluations we can make of a given ethical enactment to stand as a genuine re-situation, a dynamic state of transactional equilibrium and fulfillment (see chapter 12). The lack of closure or consensus on the best ways for human beings to flourish does not mean that conclusions cannot be drawn. Rather, it means that we should be humble about learning more and enriching our understanding of the situation and about the ability of a resolution to be conducive to human flourishing. Taking again the example of medical aid in dying, if it is learned later down the road that patients are somehow coerced in the process, then clearly the resolution would need to be called into question. Or, if we realize that the demands of patients are triggered by the lack of appropriate palliative care, as some worry (Gallagher and Passmore 2020), then again this would call into question the satisfactory nature of the resolution. This means that we need to avoid triumphalism and illusions about the unbounded nature of human action and human reasoning (for a seminal exposé on the bounded nature of human rationality, see Simon [1955]). This can be done by recognizing the provisional nature of even our best attempts to resolve difficult moral matters. We should remain cognizant of the tension(s) that generated the problem situation and recognize its complexity and the stakes that were involved in the first place. This is akin to acknowledging the existence of moral residue (Epstein and Hamric 2009), the uneasiness felt after being distressed by a given moral situation and having attempted a solution that does not squarely resolve the situation. In chapter 10, I alluded to abortion as an example of a situation, understood here as a broader societal phenomenon that is addressed in a way that the majority of, for example, Canadian women are comfortable with, at least legally (since its decriminalization in 1988) because access is still geographically limited in some areas of the country. Broader Canadian society also sides with pro-choice orientations, but there is still a small but significant proportion who does not (Weeks 2022). So even though an issue could be seen as "resolved" it can still raise important questions, both for those who are pro-choice (e.g., because of issues with access to abortion in Canada) and those who are pro-life (Bowden 2022). In other words, there is still room for improvement and learning about the foundations of these points of view and the questions raised about human life in terms of ideals and well-being because there are moral residues even though we seem to have a satisfactory and livable re-situation.

Holistic Accounts of Outcomes

An important additional procedural consideration is that ethics outcome evaluation cannot be reduced to a single set of considerations such as consequences in a narrow sense as potentially implied in some simplified forms of consequentialism. Nor for that matter can it be limited to the intents or acts of the agent or their character. When understandings of morality and moral matters are reduced to being about singular components (e.g., agents, deeds, consequences), they can set the stage for forms of ethics that deploy a revised morality according to a thinner account, urging us, for example, to become strict consequentialists or stringent deontologists. This use of ethics theories falls under the "foundations metaphor," a kind of pitfall to which the feminist bioethicist Sherwin contrasts the "lens" metaphor to describe another, more constructive and less authoritative role of ethics theory (Sherwin 1999). What this means is that to evaluate outcomes, we cannot solely rely on the consequences in the narrow *sensu stricto* because the implications and meaningfulness of actions is much more extensive than consequences.

Understanding what a given act does within a situation requires a holistic approach to assessing outcomes. Consider a child asking about lying to their younger sibling about the imminent death of a grandparent: "Dad, should we tell (a younger sibling), that grandpapa is dying?" Guiding the older child in making this decision—and deciding with them to use this as a learning experience—requires attention to the sadness that it will provoke. Telling the truth also matters. But is not telling the truth (immediately) lying, the child may ask? When and how that truth will be told also matters. Providing cultural and intellectual resources to make sense of life and death is also part of the equation, given the brutality of the announcement of imminent death. As this simple example shows, quotidian ways of dealing with the assessment of the ethical appropriateness of our acts fly in the face of single-component and overly rationalist ways of approaching ethical decisions and behaviors where only certain kinds of outcomes defined a priori would matter (Fesmire 2003). We naturally guide ourselves relying on a broad take on outcomes, the understanding of which is deepened by the observation of the consequences of our acts but cannot be reduced to it.

Nonreductionist Inquiry

To properly evaluate outcomes of enactments, we need research and data-gathering efforts. But knowledge and what is "scientific" does not equate to hard sciences, reductionism, or to the scientific views of those who hold political and economic power. "Scientific" here—and throughout this book—designates the frame of mind of accepting the fallible nature of our ideas and the willingness to test them in real-word settings to learn from such tests. It is not science as a specific and established set of methods, results, or institutions and, even more so, of the authority of a certain set of people: scientists and academics. A commitment to the scientific mindset is an engagement to see vexing moral situations as problems that we can overcome and learn from, and experiment in trying to address them. It is also tied to the understanding that these problems are rooted in everyday experience and that therefore the resolution of moral matters is also rooted in our existence and not simply in ideas. Within the problematization of this experience comes the need to find resolutions through inquiries. Once inquiries are viewed as efforts to reconstruct a deeper understanding of the situation and to find a meaningful response to the situation via open deliberation, the eventual resolution must be understood within the confines of the processes that generated its development. In other words, it is a fallible process in which we try to generate, for example, the best possible but imperfect understandings of the situation, best possible but imperfect attempts to engage in open deliberation to identify a preferable action-scenario, and best possible but imperfect attempts to enact a given response-scenario given the circumstances. A scientific approach to ethics does not judge from the outside, since this would reduce the situation to its context, but remains insightful about the intents and internal logic and experience of a given situation. It is connected to the experience of the kind of problems that moral matters are, that is, not biological problems, but rather social and relational problems that mobilize our biology, among other things, and whose nature makes sense when we recognize the existence of meaning and intentions (Dennett 1981) and their connection to our existence as irreducible aspects in scientific inquiry into human enactments.

Participation and Deliberation

Given the situated nature of moral matters, which I have described as a complex of agent(s), contextual factors, and lived morally problematic experience (see chapters 3 and 4), the evaluation of outcomes of a resolution needs to reflect the situation as a lived experience involving, most often, several agents. Each agent experiences life from their own valuations and interests. This means that the evaluation will not only need to be *subjective*[5] but *intersubjective*, where the view of different agents is apprehended as constitutive of the lived situation. More subjective evaluations can be used perhaps as a shortcut or a surrogate in some exceptional circumstances where there is clear and convincing evidence to think that another party's evaluation is representative or that only a single agent is substantially involved, for instance in the case of a fundamentally personal dilemma or problem.

Concretely speaking, this intersubjective understanding of the nature of problem situations calls for methods that allow us to grasp the perspective and experience of those affected by a given resolution and follow up with them in a meaningful (not tokenistic) way. Because calibrating a given evaluation to a given resolution of a given situation implies variance and tailoring to specificities, possible indicators or measures of a successful resolution should be discussed in advance. When a preferred response-scenario is identified (see chapter 12), and the underpinning values or principles justifying the scenario also explored, this should be an opportunity to evaluate whether this given justification or rationale is explicitly guiding the scenario in real-world settings. For example, if compassionate care and respect for patient preferences (respect for autonomy) would be the key motivation for a medical aid in dying policy as alluded to above, then the debriefing evaluation of each situation should assess whether the patient (perhaps in their last moments), their relatives, and surrounding healthcare team felt like these two motivating forces were present. If the motivating justifications are infringed upon (e.g., patients did not feel like they could decide, clinicians felt patients opted for medical aid in dying because of suboptimal palliative care), then at least two types of correctives could be envisioned. One is to ensure that the enactment of the response-scenario is aligned with the initial justifying motivations such that the response-scenario is truly able to embody compassionate care and autonomy. Furthermore, there is always a possibility for those involved in situations to live a given response as a new morally problematic situation (or relive the initial situation). In this case, there is perhaps a fundamental problem with the response that was envisioned such that the motivating justifications would not be acceptable or praiseworthy, or that the justifications envisioned were acceptable and/or praiseworthy but simply cannot be properly enacted.

A second is that there is also a possibility that a series of amendments could be made, and that the resolution is progressively attuned to the lived worries of those concerned such that greater harmony and richer experiences (see chapter 2) are progressively enabled by the enactment of a scenario as a participatory process. There is also a possibility that a given scenario may be found in the end to have been a false good idea. The overarching point here is that the evaluation cannot abstract from the fact that the situation is fundamentally a social reality involving different people with different values and valuations. Participatory and deliberative evaluation methods recognize and reflect this.

Formal and Informal Methods of Evaluating Ethical Enactments

The critical assessment of ethical enactments needs to take various forms; this is why I have not elaborated on which specific methods should be used, especially given the more tentative and exploratory nature of the ideas shared in this chapter. The proposed substantive and procedural guideposts should make clear that a wide array of methods can be used with flexibility, creativity, and ingenuity, because contexts in which ethics can be used vary tremendously. For example, evaluating how one deals with an urgent situation with limited time and resources to respond is very different than evaluating the outcomes of a formal ethics initiative within an institution (e.g., implementation of a formal ethics service, of a code of ethics) where there are more time and resources available to deploy more comprehensive forms of evaluation. Some informal evaluation of ethical outcomes is routinely carried out when, as individuals, we follow up on situations to assess whether a problem is resolved or not. To do so, we also rely heavily on more intuitive processes and perceptions of social realities (e.g., observing people's behaviors, emotions, attitudes) to evaluate whether a given action has managed to satisfactorily resolve a problem situation. More formal and extensive mechanisms to evaluate outcomes build and elaborate on these processes, which are part of the very nature of our existence as social and moral beings.

Informal Ethical Enactment Evaluation

There are times when informal assessment of ethics outcomes is the preferable route for practical reasons, or because no relevant research-based ethics outcome method is available. Here I have in mind situations (e.g., emergency situations) where one wishes to reflect lucidly on a situation and examine whether the resulting re-situation satisfactorily responds to the initial morally problematic situation. Also, there are situations where there is no comprehensive support available to engage in a thorough evaluation of ethics outcomes. For example, situations where no research is available to guide the structure of ethics evaluation, where opportunities to engage in open discussions about meaningful resolution are limited, or where there are limited time and resources to engage in a more thorough or institutionalized form of ethics outcome evaluation. Accordingly, the points presented and discussed as substantive and procedural guideposts in this chapter can be used to self-assess, or assess in small groups, whether a given resolution provided a meaningful resolution that fosters human flourishing.

Formal Ethical Enactment Evaluation

There could be healthy skepticism about the possibility of evaluating ethical enactments beyond informal processes. Establishing what is ethical and most likely to promote human flourishing in our present actions is at times very challenging. It is not uncommon to see professional ethicists become very uncomfortable and hesitant when asked whether a certain action or outcome is ethical. This uneasiness is sometimes caused by lack of knowledge of a given situation, but there is also the unease created by being asked to assess without really having a ruler with which to measure the outcome. While we often agree loosely on what is in our best interest and that of others (e.g., in our personal lives, in our professional lives, even as a collective) such that we get by without encountering moral problems frequently, ethics steps in when problems arise therein. Thus, ethics appears to be the domain of complex and even unresolvable problems. Is it possible to evaluate the outcomes of enactments on which we may have only partial agreement? There are potentially more generic forms of ethics evaluation and more targeted human flourishing outcome evaluation, as well as participatory forms for the evaluation of ethical enactments that can help here.

Generic ethics outcome evaluation The field of ethics outcome evaluation is nascent, with only a few attempts to evaluate the outcomes of an ethics-oriented practice change. For example, there are published studies of randomized controlled trials of measures such as advanced care planning and practices to seek informed consent (Sugarman et al. 2005; Weathers et al. 2016) or of ethics services (e.g., ethics consultation) (Chen and Chen 2008). These protocols have the virtues of being very tight and formalized and using standardized instruments. However, they may lack the ability to capture lived experience and may not measure the outcomes of any single given situation, a situation here being a combination of agents, contexts, and actual lived experience.

Other efforts to evaluate ethics outcomes have focused on institutional ethics-oriented change (i.e., "policy"). An obvious example here is the impact of introducing a code of ethics in an institution. There are several ways in which a code of ethics can be developed, ranging from the most top-down to the more inclusive and participatory (Martineau, Johnson, and Pauchant 2017). Top-down enactment and evaluation methods may be considered to be more effective in carrying out a specific kind of outcome such as compliance (Adam and Rachman-Moore 2004). However, they may also fail to engage important dimensions of flourishing and promote mere compliance, without helping to tackle processes of dehumanization within organizations (Craze 2020). In contrast, more participatory methods of developing and implementing a code of ethics may preserve and nurture connections with the actual situations that led to its need being felt (Nijhof et al. 2003). Some strategies to develop codes of ethics have used a combination of more top-down and bottom-up approaches to assess compliance with a code of ethics, but also solicit perspectives on the meaningfulness of the values promoted in a code of ethics (Peguero et al. 2015). There is also noteworthy literature on the value of ethics consultation services, which have used more quantitative and sometimes more participatory methods (Linkeviciute et al. 2016; Metselaar et al. 2017; Racine 2007; Robles 1999; Slowther et al. 2001; Slowther 2000; Tulsky and Fox 1996; White, Dunn, and Homer 1997).

However—and I do recognize the value of these efforts—many of these lack the ability to inform us of whether the desired and evaluated ethics outcome make sense in terms of human flourishing. For example, granted that a given concise process of seeking informed consent is more effective (than a longer process) in the context of, let's say, acute stroke urgent intervention, does that mean that the care provided to the patient and family is more ethical by virtue of using this adapted consent process? On the surface, yes, because consent is presumed to serve important ethical functions (Dickert et al. 2017), functions that are meaningful and relate to autonomy and to flourishing. Yet, we may not really know if that's the case. What if this concise content process actually increased anxiety or appeared to induce insensitive legalism in an acute situation because of its emphasis on being informed and on making a quick decision?[6] What if it produced these outcomes for persons who have been marginalized by the healthcare system and therefore mistrust healthcare professionals? The same goes for different forms of ethics consultation services. Do they make things better for patients? Do they improve work climates in ways that are consistent with human flourishing? Do they actually help healthcare professionals and patients (McClung et al. 1996)? Are they only dealing with ethical crises or tackling the more substantial causes of ethics problems (Beyleveld, Brownsword, and Wallace 2002)? There are risks that such ethics consultation services usurp existential moments of questioning to offer experts leeway in becoming authoritarian (Parizeau 1995). I doubt this is the (general) case, but how can we know if we do not ask? Participatory methods of evaluation have been proposed such that some of these worries are offset (Metselaar et al. 2017), but there remains some flux over what is supposed to be accomplished and why. To tackle such a question, more clarity and precision in what is pursued needs to be provided, even if a worthwhile participatory and open-ended process is adopted.

Human flourishing outcome evaluation A number of studies have already attempted to measure the success of interventions in improving some morally problematic aspects of a situation in light of flourishing or at least some key dimensions of human flourishing. It is important to acknowledge these efforts and to understand how they relate to the criteria of evaluation for deliberative wisdom. For example, there is considerable literature on the development of empathy and of measures that could augment empathy, notably in healthcare professionals (Kiosses, Karathanos, and Tatsioni 2016). Another relevant example is research on stigma reduction, which tries to avoid harm to people done by demeaning them and impoverishing their prospect for well-being (Corrigan 2012). There are also personhood-enhancing psychosocial interventions (Johnston et al. 2016) and dignity-promoting interventions (e.g., dignity therapy) (Chochinov 2012), as well as interventions aiming to increase autonomy and self-determination (Algozzine et al. 2001). Several scales have also been developed to explicitly measure human flourishing, and their use has generated a remarkable body of evidence (Ryff 2014a; VanderWeele 2017). Incorporating such measures into participatory approaches could be a way to specify the evaluation of outcomes.

There are also risks that this strategy overspecifies the evaluation such that only what can be counted counts, while what counts sometimes cannot be easily counted, to paraphrase Einstein cited earlier, and that these measures send the wrong message that evaluations can be acontextual. However, these efforts have characteristics in common such as considerable efforts to refine the construct to be improved (e.g., empathy, stigma and self-acceptance, dignity, autonomy and self-determination). These constructs have clear relevance to a eudaimonic theory of ethics such as the theory of deliberative wisdom, which proposes that the ability to engage with others, to feel valued, to be autonomous, are key features of the flourishing person (Ryff 2014a). However, keeping in mind the methodological orientations articulated above, these efforts also exhibit some limitations that are less salient than, in my opinion, the need to expand these orientations to ensure their validity, that is, their ability to capture meaningful flourishing for those concerned. First, the use of established measures and associated constructs makes it difficult to relate a given concept (e.g., empathy, stigma, self-determination) to the experiences of individuals. These kinds of measures will seldom grasp the interpretation that people have of their own situation and their own narrative, as well as the complexity of their intents. Moreover, as a second point, established measures and associated constructs tend to be acontextual such that, in principle, the increase of autonomy, diminishment of stigma, and so on will solve some morally problematic situations (or lead to avoiding some), but they are unlikely to be useful to fully resolve a given situation in its concrete and contextual form. Increasing a trait or feature of a person (e.g., autonomy) will not necessarily lead to the actual response-scenario to be enacted in a given situation This is because a situation will rarely be about only a given personality trait or a given kind of intervention. Established measures of eudaimonic outcomes are therefore useful in generating general knowledge and expanding the toolbox. But research on such outcomes may also benefit from keeping an open door to unexpected findings that do not fit the given outcome measure, from an ethical perspective at least; the gap between knowledge of various situations and the actual understanding and resolution of a given problematic situation is not a simple implementation exercise. Thus, there is a significant amount to learn from this emerging literature despite the urge to adopt at the same time open-ended and participatory evaluation approaches such that our understandings of flourishing, dignity, autonomy, and other meaningful ethical constructs do not become essentialized.

Toward situational, realistic, and participatory evaluation of ethical enactments Some potential pitfalls of more closed-ended and predefined measures could be avoided, especially if closed-ended measures are supplemented or enshrined within a participatory mindset that embodies a view of the person as an agent. Participatory methods for understanding human flourishing can also move away from the more typical dualistic object-subject spectator theory of knowledge and instill a more pragmatist agent-as-knower and agent-as-actor paradigm that is much closer to actual human agency (Kramer-Roy 2015; Willen et al. 2022). It is also closer to the spirit of ethics as a field interested in using insights about human flourishing, which means much more than just "applying" knowledge; it means trying to understand what flourishing is from the standpoint of those concerned, including its cultural diversity (for an example of how this can be approached, see Cele et al. [2021]). Although participatory methods are household methods in research and in various domains such as social work and education, their use in the study of human flourishing has been more limited, at least under the umbrella of "flourishing." The critical perspective at the heart of many participatory methods may have led these approaches to be more interested in related concepts like inequalities, empowerment, exclusion, and so on. In a sense, these are the flipside of flourishing, since equality and inclusion are important ingredients in social contexts that support flourishing.

Some researchers influenced by pragmatist philosophy and versed in qualitative and participatory methods are in the process of developing tools to assess the impact of deliberative approaches to tackling ethics problems using such participatory methods (Hartman et al. 2019; Inguaggiato et al. 2019a; Metselaar et al. 2017). For example, they have developed an instrument, the European Moral Case Deliberation Outcomes Instrument (Euro-MCD) (Svantesson et al. 2014) and a participatory method to evaluate ethics consultation services (Metselaar et al. 2017). Likewise, others, also pulling from pragmatism's emphasis on growth, have described the impact of ethics consultation as the co-creation of norms or as moral learning for participants (Aiguier and Cobbaut 2016; Aiguier et al. 2012; Aiguier and Loute 2016; Inguaggiato et al. 2019a). These efforts point in a promising direction of greater contextual sensitivity to the nature of moral matters in the process of understanding them using engaging methods, resolving them using open deliberation, and assessing their outcomes in terms of the growth of those concerned (Inguaggiato et al. 2019a).

In all these efforts, we must always keep in mind the dynamic, the evolving, and our growing understanding of human flourishing. More formal research-based ethics outcome evaluation can yield more structured information because of its increased thoroughness. However, less structured evaluation that is more naturally connected to and reflective of the actual situation has advantages over a formal and apparently rigorous evaluation that perpetuates reductionism and closure about human flourishing.

Conclusion

This chapter argued that, consistent with commitments to valuing the experiential aspects of moral situations and deliberative strategies of moral problem resolution, ethics outcome evaluation is unavoidable, especially for a theory like the theory of deliberative wisdom; and I would venture this is ultimately the case for any meaningful understanding of ethics. This process represents a genuine opportunity for ethics to grow as a science and a practical field of knowledge. Without a commitment to evaluation, ethics fails to open itself to understanding new human experiences and situations, and to assessing its ability to guide human beings toward human flourishing. Ethics can become one's individual fancy about the best life, a potential projection of biases and assumptions, a rigid system of beliefs and of behavioral conduct that is not guided by flourishing but by conformity to established moral standards that remain unquestioned, unchecked, and unrelated to human experiences. I hope this more tentative chapter opens the discussion and shared some useful ideas, and that others will be enticed to bring their energy, intelligence, and collaborative spirit toward the much-needed evaluative task of ethics theory. There is a real opportunity for progress and development, to test ideas, to learn from the process of (in)validating, unstiffen our theories and habits, and learn from failures in the field of ethics such that the result will be a more flexible ethics better suited to the future for growth and human flourishing.

Conclusion

Moral matters are important existential problems that concern the value and meaning of human experience and human life. They are rooted in who we are and concern who we want to be. When left unaddressed, they take a significant toll on our welfare and our growth as human beings, because we are struck by moral distress or become stalled in the face of uncertainty, especially uncertainty about what matters and about the humanistic and social implications of what we do. These problems are, in their most concrete and experienced forms, vexing situations that tend to overwhelm our abilities to find clarity, direction, and meaning in what we should do. Ethics theory and methodology are accountable for their ability to assist human beings from all walks of life in making sense of such situations. In other words, ethics theory is only as good as its ability to accompany people in understanding such situations, create spaces for dialogue where people can talk about these problems and envision solutions to them, and support all of us in evaluating honestly and with humility what we have done and whether and how we have done better or worse in light of human flourishing.

The theory of deliberative wisdom is a vision (a *theoria*) of what ethics is and should be, based on an ever-changing scientific understanding of human morality, which itself is a moving and plural phenomenon. Despite the plurality of moral goods and valuations, the theory of deliberative wisdom is a eudaimonic theory not afraid to put forth—following Aristotle and other early philosophers, and more recently, Dewey—the ideal of human flourishing and development (or growth) as focal points of human existence. This flourishing or growth—although at face value, a grandiose project—simply designates the process through which deeper meanings of our lives are acquired and expanded by approaching new situations, attempting to overcome difficulties, and reflecting on and learning from these. Ethics is sometimes taught like this and likely even more often practiced like this, as a body of knowledge that helps humans understand, learn from, and constructively resolve difficult morally problematic situations. Yet, the theoretical approaches and vocabularies at our disposal are often plagued by static metaphysics and sterile collateral debates that are hard to reconcile with the vision of ethics as a practical and experiential field. This led me, after many years of research, stakeholder engagement, discussion, and training of students in ethics, to more seriously explore some of the insights of the pragmatist tradition, as well as recent scholarship in different fields. I undertook this work to scaffold a vision that would connect the dots between what I have experienced—as an academic but also as a person having to deal with difficult personal and interpersonal situations—as key moments in the use of ethics as a tool for growth. The chapters of this book present distinct attempts at making sense of these moments (recognizing the situated nature of moral agency and of the ethical life, opting for a flourishing orientation, cultivating moral and ethical awareness, developing situational insight of moral problems, fostering ethical deliberation and imagination, pursuing ethical enactments and our evaluations of them). Thus, I hope that this book is useful in describing these moments and making connections between them. I also hope that it encourages the recognition of the moral dimension of life as an unavoidable component of human existence and a part of the very fabric of our individual and collective existences, independent of how dominant discourses of organized religion and politics make sense of it. I also hope that this theory encourages a focus on human flourishing and an orientation toward life as a positive and open-ended project in various fields such as ethics, as well as other more theoretical and more practical disciplines that can support this project. Whether it is in psychology, medicine, business, or nursing, engagement with various aspects of reality and with others can be a learning opportunity. I hope that this synthesis facilitates learning about ourselves, about others, and about the world we inhabit. Despite the book's title, I do not pretend to be the bearer of more profound truth than others have already put forth. What I do hope is that the work of researching and writing this book helps to assemble a number of rather disparate ideas and insights for the sake of the search for coherence and meaning. I realize that much work lays ahead to refine many key concepts at the core of the theory of deliberative wisdom.

From the perspective of the development of deliberative wisdom theory and practice, there are several crucial matters. Education and investments therein are central to ethical change. In our contemporary societies that are showing unprecedented signs of openness, there are also powerful counterforces (e.g., social media, entrenched identarian discourses) that seriously impede the ability to have dialogue with and learn from each other. The environmental crisis we are living in calls for much greater awareness of moral matters since this crisis is partly a moral crisis and calls on human beings to organize individual and collective life according to certain values. Not unrelatedly, capitalism and its full impact on human beings needs rethinking given the incredible amount of harm and alienation caused by the ambient status quo in how we organize production and share wealth. Stubborn adhesion to old ideas prevents us from imagining other forms of human associations held together by more fulfilling values. In all matters, we will need to strengthen and cultivate our abilities to live in a plural world by finding common goals and values. Also, concentration of political power leads to the possibility of pitting human populations against each other in ways that too easily disregard what ordinary people have in common. To mend disagreements and foster open deliberation in a pluralistic world, we must avoid situations wherein the meaning of the words we learn and use do the thinking for us and imprison us in the meanings of past experiences. We must tap into creativity to ensure that our vocabularies genuinely connect to our experiences but also reflect the meaning of our aspirations. Therein lies much future work for deliberative wisdom.

Notes

Introduction

1. "Testing" here refers not only to scientific research, but also to informal experimentation.

2. The proposed theory describes a vision of ethics, including how ethics can contribute to making decisions in the broad sense of the term because making decisions is one of many important processes involved. This practical orientation of the theory clearly implies a methodology on how to understand, discuss, imagine solutions, and so forth, to moral problems. At the same time, the procedural aspects of ethics embody a search for greater ethical wisdom, that is, knowledge about human flourishing. As proposed, this search is deliberative and participatory and thus mimics a research process itself because ethics is a form of experiential and inquiry-based learning. In this light, the proposed theory does not constitute a research methodology per se, although it could help guide ethics research to certain specific topics and mechanisms central to deliberative wisdom.

3. There are numerous examples of the progress made in the increased awareness of the needs and interests of others in matters of life decisions, sexuality, religious beliefs, and so on. For example, morality has been revised critically such that racism is now widely viewed as unethical, while it may at some point have been viewed as a morally acceptable attitude in certain societies. Indeed, political and religious authorities in the US, for example, encouraged racial segregation as being the right thing to do based on a sense of superiority of one race over another. Likewise, in Canada, beliefs of racial superiority fueled colonialism and systematic attempts to eradicate Indigenous peoples, their knowledge, and their culture. A similar analysis could be undertaken for sexism and homophobia. Critical reflection on the values at the basis of such behaviors and ways of thinking—often initially forced by vehement protest of the groups concerned and not readily accepted by those favored by a moral status quo—has led progressively to the reconsideration of certain aspects of a common, shared, and accepted morality based on its impact on others and their well-being. In

other words, a form of moral awareness has been *ethically revised*; it has been revised critically in light of attitudes and behaviors (e.g., beliefs, policies) favoring greater human flourishing.

Chapter 1

1. In ancient Greek, *theoria* designates a vision, a view, a contemplation. I sometimes use the term vision to reflect how theories provide a view, a lens, or a perspective.

2. This is a common positivist fallacy in fields like economics and psychology where knowledge is viewed as being objective and separable from contexts such that universal "laws" can be extracted. These trends share a common faith when ethics follows a staunch deductive model of reasoning and of rationality (Fesmire 2003; Hoffmaster 2018; Lekan 2006).

3. Dewey criticized, especially in the later stages of his career, customary ways of envisioning the mind as being located only within the organism and the organism as completely distinct from its environment (Shusterman 2008).

4. The notion of transaction has the unfortunate connotation of reductive financial exchanges. Clearly this is not the meaning of the term here.

5. Social psychology, in the spirit of Mead's social thinking, has now developed sophisticated (methodological and statistical) interactionist models (Rauthmann and Sherman 2019) that capture how the meaning of situations is shaped by a series of ever-evolving interactions that need to be taken into account to explain subtle human behaviors such as social cognition (Kihlstrom 2013).

6. Peirce's formulation of this logical maxim is the following (and is made in the context of the presentation and comparison of three grades of clarity of ideas): "It appears, then, that the rule for attaining the third grade of clearness of apprehension is as follows: Consider what effects, that might conceivably have practical bearings, we conceive the object of our conception to have. Then, our conception of these effects is the whole of our conception of the object."

7. My use of the word "science" is generous such that I equate it with knowledge and the process of gaining knowledge about the world. Experimental sciences structure inquiries in ways where the parameters of action and observations are controlled for. A scientific investigation represents a kind of pinpointed and specified transaction to test for the accuracy and predictability of certain ideas. For example, if I observe this kind of cell prepared in this way using this microscope with these specifications under these conditions, I observe the following biological structures. If I interview participants with these kinds of life experiences, using this interview protocol, in this kind of setting, they report their lived experiences. Science is a way to unpack details of transactions such that we can share and eventually agree on the conditions of observation and on the observations themselves. Science makes the transactional nature of

our ideas explicit such that ideas become fallible tools based on the intentional test of their implications.

8. The fact that human existence ceases is unequivocal, as proponents of higher brain death argue (Veatch 2019). Their argument is that meaningful and conscious human life, that is, human existence, ceases and that only some biological life remains, thus signaling the end of the human being as an existing human being and not only as a (biologically) living human being. No jurisdiction has adopted the higher-brain formulation of death determination (Bernat 2006b), but it is debated in the context of organ donation because some argue that the dead donor rule requiring that biological life has ended should be abandoned (Miller, Truog, and Brock 2010).

9. In contrast to irreversible coma (or brain death), typical coma reverts after a month. Typical coma patients with severe brain injury tend to transition to a state of unresponsive wakefulness syndrome (UWS, previously called a vegetative state) (Laureys et al. 2010) or a minimally conscious state (Giacino et al. 2002), and this state can last for months and even years with sometimes rather basic life support such as artificial nutrition and hydration (Bernat 2009).

10. Relationships are a specific and very important subset of transactions since environments are frequently social in nature and human beings are greatly impacted by other human beings.

11. Symptoms of depression (e.g., hopelessness, pessimism, sense of failure, dislike of self, feelings of shame, suicidal ideation, worthlessness) are captured in the commonly used Beck Depression Inventory (BDI I, BDI II) (Beck, Steer, and Brown 1996; Beck et al. 1961).

12. Interestingly, the early pragmatists like Peirce and Dewey described how the habits that make up who we are the product of our interpretation of the world and that thus, the self is a kind of matrix that sums up our past and provides an orientation to the future (Crick and Bodie 2016).

13. Dewey described "consummatory experiences" as experiences in which an individual is immersed, enacting their abilities, and transforming situations (Dennis and Powers 1974). This is very much congruent with how modern psychologists describe "flow" or being "in zone" (Csikszentmihalyi 1990). Although some take flow to be a rather narrow construct dealing with performance, it has actually been developed by Csikszentmihalyi as an account of human flourishing.

14. While biological death designates the inability to transact with the environment from a physiological point of view, social death is the inability to transact productively with other human beings (Card 2003; Patterson 1985).

15. For a discussion on the relational and ontological aspects of personhood attribution for this clinical situation, see a paper coauthored with Athena Detmertzi and Steven Laureys' group (Demertzi et al. 2012).

16. In past research, I have shown that there is widespread confusion about this in public controversies about cases of patients suffering from an unresponsive wakefulness syndrome, such as the Schiavo case in the US (Racine et al. 2008; Racine et al. 2010). There is also confusion among healthcare professionals in different countries, including specialty physicians (Kuehlmeyer et al. 2014; Kuehlmeyer et al. 2012. There is much perplexity in families who struggle to make sense of biological life without human existence, notably in the interpretation of patients' preserved cranial or postural reflexes, depending on their condition (Huber 2011; Macdonald et al. 2008). The question of whether it is ethical to remove life support from biological life can thus be reframed: Is a biological life without human existence worth living? Is such a state meaningful, and is it ethical to maintain a person, a loved one, once the patient is in such a state?

17. "Through the social activity of appraisal or evaluation, private preferences, or what Dewey terms 'prizings' (i.e., what is subjectively valued or desired), are converted into publicly shared values (i.e., what is objectively valuable or desirable)" (Ralston 2010, p. 26).

18. What I designate as "scientific studies of human moral behavior and psychology" is distinct from empirical ethics as a hybrid and more encompassing field. See also table 5.2. where I differentiate the empirical study of human morality from empirical ethics.

19. Briefly put, a moral agent is someone who acts and has responsibilities whereas a moral patient is someone who is the object of concerns from other moral agents but who does not have this ability and related expectations.

20. James's and Dewey's accounts of agency (and consciousness) across time are often neglected, but are deep philosophical insights of the pragmatist tradition (see Fesmire [2003, pp. 42–44, 50–51]). Dewey's account of deliberation as "dramatic rehearsal" in particular emphasizes the story-telling structure of human existence and human thinking (Fesmire 2003, p. 79). In subsequent chapters, I explain how action-scenarios generated as the outcome of deliberation and moral imagination are possible "episodes" that human beings could "enact."

21. Deliberation is participation applied to thinking, that is, thinking together, in dialogue. Deliberation could be a separate point, but its inclusion with participative strategies reflects the social nature of intelligence and the fact that human beings—as social beings—appreciate being able to think with others and to imagine their futures with others.

Chapter 2

1. Originally published in French as "Or la tâche essentielle de l'État consiste bien en ceci: Protéger l'individu, lui offrir la possibilité de se réaliser en tant que personne

humaine créatrice" (Einstein, À propos de la conférence du désarmement en 1932, *Comment je vois le monde*, Einstein [1979/1997, p. 65]). In the French version, there is an emphasis is on the creative human person and less narrowly on the creative personality, a nuance important to the topic of this chapter.

2. Berlin's work follows the tradition of Kant. Kant discussed how our positive rights are weaker than our negative rights. For example, the right not to be harmed entails a stronger obligation (perfect duties) than the right to pursue certain goals, orientations, and choices (imperfect duties) (Kant 1785/1998). This has been extensively discussed by contemporary theories of rights and responsibilities, and the stronger obligations associated with perfect duties may be due to the fact that harms are more readily felt and recognized than goods.

3. Ancient Greek moral philosophies point out the existential and holistic goal of eudaimonia. Eudaimonic well-being or human flourishing is frequently distinguished from hedonic well-being, or "pleasure." Contemporary psychology supports this distinction (Baumeister et al. 2013; Ryff 1989; Ryff and Singer 2008).

4. The Western intellectual tradition is not a monolithic one and has certainly been influenced by other major traditions in its inception and afterwards. One of the famous maxims of ancient Western philosophy, "know thyself," seen in the Delphic maxims inscribed in the forecourt of the Temple of Apollo at Delphi, is believed to have come from the city of Luxor in ancient Egypt. It is believed that Pythagoras traveled to the near East and Egypt where he was exposed to other philosophies (Riedweg 2002). Thus, even the early—not to say anything of the contemporary—Western tradition is multifaceted and pluralistic.

5. This is what I have described elsewhere as a process of synergetic enrichment between concepts and experience as a way of describing positive and enriching transactions (Racine 2017). This aligns with the idea that ethical wisdom must be deliberative, as ancient thinkers recognized (Gadamer 1991).

6. I attribute the maternity of the model to Carol Ryff even though her work is often published in collaboration, including the 2008 paper (Ryff and Singer 2008) I use to expose her model.

7. As remarked by Hoffmaster and Lekan, scholarly ethics sometimes becomes synonymous with a quest for semantic clarity and purity, for perfect justification, or for universal principles in the form of coherent theories (for critical reviews see Hoffmaster [2018] and Lekan [2006]). There are clear reasons for these pursuits because providing justifications for our choices, identifying and defending principles that guide them, and seeking coherence between them make sense in trying to reflect upon human morality and the tensions we experience therein. But problems surface when we lose sight of why we embark on such inquiries and what our purposes are. We should question whether pursuing these goals actually allows us to make better, more ethical decisions and live better lives (Stengel 2009). Are our ethical methods effective in

bringing about what we seek? Is it worthwhile to pursue theoretical ethics and isolate ethics from the consideration of its outcomes? One manifestation of the general tendency to avoid questions concerning the good and flourishing life is finding refuge in narrow, argumentative, impersonal, abstract, and rationalistic concerns to move away from the messiness and complexity of experience. This is a frequent move made in thinking and writing about ethics; see Parry (2021) for an overview.

8. For a more extensive discussion on autonomy discussed as being an instrument in a flourishing life, see Racine et al. (2021).

9. One of the positive contributions of thin and procedural forms of ethics is that they have brought worthwhile attention to the open-endedness of the domain of ethics. For example, modern liberal societies have secured a level of open-endedness for individuals to choose whether life should be ended based on someone's wishes or not, or whether the development of a fetus should be terminated. Modern societies have also progressively limited the domain of traditional morality where there exist strong and explicit prescriptions about what is right and wrong in a whole range of matters, in turn creating room in the domain of ethics for explicit and conscious answers to be deliberated on, envisioned, and proposed. However, it must be noted that a defensible and worthwhile process (e.g., to enunciate the conditions of ethical deliberations or the conditions of a just society) itself does not define the good; a process can be a crucial aspect of the former but must also deal with ends pursued and their subsequent outcomes.

10. In French, they referred to "*mieux-être*" (better-being) instead of "*bien-être*" (well-being), thereby articulating that human flourishing is aspirational and a process of becoming better and of trying to do better.

11. The tension between current moral experience and idealized future moral experience is at the core of ethics as a discipline (Racine and Cascio 2020). Ethics builds from current moral experiences as we live them such that there is no real staunch dichotomy between morality and ethics—but ethics examines morality critically to see what courses of action could be more conducive of growth and flourishing. Ethics attempts to reconstruct morality in the face of new situations where moral insights are challenged; it helps repurpose behaviors and thinking toward the betterment of human existence with the help of knowledge and inquiry. Ethics also helps break vicious circles of problematic moralities by pointing to questions about their impact on well-being.

12. Despite the seemingly Western origins of philosophical ethics and of this discussion on ethical wisdom and its relationship to flourishing, I—like many others—have observed through personal experiences, conversations, and friendships with persons of different nationalities and cultural backgrounds that there is often rather considerable agreement about what makes a good and meaningful life, as opposed to the discord that theoretical disagreements in academic ethics would suggest. When starting

from everyday experience and opting for an open mind, there is much room for embracing existential diversity and mutual learning between people. There is hope for constructing a common notion of the good dialogically. Important areas of convergence on key aspects of human flourishing are evidenced the literature (VanderWeele et al. 2022) and the insights of figures such as those of Mandela, Confucius, or Gandhi speak to humanity.

13. The ancient word "*aretê*" designates forms of human excellence that are broader than "virtues." "Virtue" here, as usually conceived, designates a settled disposition to act in a certain way, involving practical knowledge and the motivation to act consistently.

14. Sometimes, these concepts are respectively translated as practical wisdom and transcendental wisdom (Trowbridge and Ferrari 2011, p. 91).

15. He also put forward this existential focus before Heidegger, Sartre, and Gadamer developed their existential and hermeneutic thinking and in ways that are often much more concrete, interdisciplinary, and workable (for comparisons see Higgins [2010] and Kestenbaum [1992]). This is perhaps because Dewey was a man of general science, as well as a man of action—not only an academic philosopher *sensu stricto*, but also a public intellectual in a crucial formative period of American intellectual and political history. The return to experience was more than mere written doctrine as Dewey and other leading pragmatists were heavily involved in social reforms of the time. Dewey was also heavily influenced by Peirce and James who themselves influenced Husserl (e.g., Peirce developed his own account of phenomenology, while James provided an account of consciousness and horizons) (Fesmire 2003).

16. Some of Dewey's ideas on human flourishing were evidently in the spirit of the time. Russell described how the goal of the good life—individual and collective—is to allow the creative (and constructive) impulse to take maximum precedence over the impulse for possession (Russell 2011). Einstein, as cited at the beginning of this chapter, urged that the state be in service of the individual as a creative person (Einstein 1979/1997). Adler promoted a positive orientation of the individual toward the "useful side of life" (Dwornik 2003). Existentialism, which came just a few decades after, likewise promotes self-creation and the perpetuation of ontological freedom of human beings, encouraging actions that foster greater liberties, possibilities, and opportunities for others (de Beauvoir 1947/2011; Sartre 1946/1996). Despite being embedded in a context wherein positive life orientations were discussed, Dewey scaffolded the basis of an experience-based ethics that foreshadowed the importance of deliberation and open-ended trialing of ethical enactments.

17. Dewey's ideal of growth is reminiscent of the concept of praxis or formative action described in the writings of Hegel and Marx (Pekarsky 1990, p. 291); see notably Pekarsky's excellent footnote explanation (p. 291).

18. Note here that Dewey tends to use the term "morals" to designate the science of morality, which is what I refer to as "ethics." Sometimes, Dewey alludes to a reflective

morality; I am unable to state whether or not he systematically differentiates between morality and ethics throughout his vast writings.

19. In the words of Dewey:

> When we observe that morals is at home wherever considerations of the worse and better are involved, we are committed to noting that morality is a continuing process, not a fixed achievement. Morals means growth of conduct in meaning; at least it means that kind of expansion in meaning which is consequent upon observations of the conditions and outcome of conduct. It is all one with growing. Growing and growth are the same fact expanded in actuality or telescoped in thought. In the largest sense of the word, morals is education. It is learning the meaning of what we are about and employing that meaning in action. The good, satisfaction, "end," of growth of present action in shades and scope of meaning is the only good within our control, and the only one, accordingly, for which responsibility exists. (Dewey 1922, pp. 280–281)

For a recent proposal for a living ethics, see Racine et al. (2024)

20. The value of this model of human flourishing is further illustrated in subsequent discussions about moral situations where it serves to unpack the "morally problematic" nature of situations (chapter 5) and to describe ethical deliberation, where it provides a guidepost to evaluate the ethical merits of action-scenarios (chapter 11).

21. There are numerous sayings capturing this experiential wisdom: "For it is in giving that we receive" (Saint Francis of Assisi); "The sole meaning of life is to serve humanity" (Leo Tolstoy); "Making money is a happiness; making other people happy is a superhappiness" (Muhammad Yunus); "Giving back is as good for you as it is for those you are helping, because giving gives you purpose. When you have a purpose-driven life, you're a happier person" (Goldie Hawn); and as a Chinese saying goes, "If you want happiness for an hour, take a nap. If you want happiness for a day, go fishing. If you want happiness for a year, inherit a fortune. If you want happiness for a lifetime, help somebody" (Santi 2017).

22. It is important to note that Mill argued, in collaboration with Harriet Taylor Mill, in *The Subjugation of Women* (written in 1861, published in 1869):

> The object of this Essay is to explain as clearly as I am able grounds of an opinion which I have held from the very earliest period when I had formed any opinions at all on social political matters, and which, instead of being weakened or modified, has been constantly growing stronger by the progress reflection and the experience of life. That the principle which regulates the existing social relations between the two sexes—the legal subordination of one sex to the other—is wrong itself, and now one of the chief hindrances to human improvement; and that it ought to be replaced by a principle of perfect equality, admitting no power or privilege on the one side, nor disability on the other. (Mill 1869)

23. Interestingly translated as "The Ethics of Ambiguity" in English. The English translation would be coherent with the ethic/morality distinction proposed in this book.

24. In his *Pensées sur l'interprétation de la nature*, Didérot writes in the *dedicace*:

> Young man, take and read. If you get go to the end of this book, you will not be unable to read a better one. As I proposed myself less to instruct you than to exercise you, it matters

> little to me whether you adopt my ideas or reject them, provided that they employ all your attention. A more skillful author will teach you to know the forces of nature; it will be enough for me to have you try yours. (Diderot 1754/2005, personal translation)

25. Aristotle states:

> The happy person needs to have goods of the body and external goods added [to good activities], and needs fortune also, so that he will not be impeded in these ways. Some maintain, on the contrary, that we are happy when we are broken on the wheel, or fall into terrible misfortunes, provided that we are good. Whether they mean to or not, these people are talking nonsense. (Aristotle 1999, pp. 116–117).

26. More than ever, failures of liberal economic ideologies to promote human welfare are clear given the expanding disparities in the wealth of people, as well as marked inability of these ideologies to envision human beings as existing within broader ecosystems. There is a clear impetus to orient our societies toward other, more mature and substantive goals for human growth and welfare. We are seeing encouraging signs that broader thinking is making its way to higher levels of particular societies and groups (e.g., the Wellbeing Budget in New Zealand [Roy 2019]) where obstacles to human flourishing are becoming targets of action. Ethics has an important role to play in shifting attention toward the broadest and deepest impacts of policies of all sorts (e.g., health, economic, educational) on human development and flourishing.

Chapter 3

1. It is hard to be systematic with such uses, but I tend to use the concept of environment to designate the general surroundings in which agents are and evolve while the concept of context is used to describe the narrower circumstances experienced by agents. For example, I tend to allude to more general transactions between organisms and their environments while I tend to talk about the context that is experienced by agents as part of situations.

2. Dewey writes very eloquently about the interdisciplinary nature of ethics:

> Morals is the most humane of all subjects. It is that which is closest to human nature; it is ineradicably empirical, not theological nor metaphysical nor mathematical. Since it directly concerns human nature, everything that can be known of the human mind and body in physiology, medicine, anthropology, and psychology is pertinent to moral inquiry. Human nature exists and operates in an environment . . . Moral science is not something with a separate province. It is physical, biological and historic knowledge placed in a human context where it illuminates and guides the activities of men. (Dewey 1922, pp. 295–296)

3. A thorough review of the basic properties of human social and psychological situations cannot be undertaken here (for general overviews see Kihlstrom [2013] and Rauthmann et al. [2015]).

4. Understanding agency, and notably moral agency, as situated builds directly from Dewey's moral and social philosophy, including his attempt to explain human moral behavior using the tools of moral and social psychology of the time (Dewey 1922), a project perhaps first fully articulated by Hume in a more primitive form (Hume

1739/1975) that is central to Smith's philosophy (Müller 2016) and at work in the writings of Mill (1863/2001). As I mentioned in chapter 1, Dewey proposed that habits (i.e., functional dispositions to act, react, think, and so on in certain ways) are best understood in relationship to environments or more precisely *as transactions with* the environment.

5. I recognize the confusion generated in my own work in developing, for example, a theory of contextualized autonomy (Bogossian, Majnemer, and Racine 2020; Racine et al. 2017b; Racine and Dubljević 2016), which could have rather been named a theory of situated autonomy (see note about this in Racine et al. [2021]).

6. Here, I side resolutely with anthropocentrists in terms of how I understand human valuations as coming from human interests and activities. However, these valuations may need to adopt a strong ecocentrist orientation to recognize the value of ecosystems. Humans' best interests may rely to some extent on their ability to take distance from their own interests. However, I hesitate to formulate this as pure disinterest because there are compelling and admirable motivations to this recalibration of understanding humans within a broader worldview and the interconnections of the welfare of human beings and of their natural environments. I think that intrinsic motivation to adopt ecocentric habits of thought and behavior are rooted in human values that we must recognize as such, that is, as being human and tied to human agency, but at the same time appreciating mutual links between organisms and environments is crucial.

7. The authors of this review paper remark that "little over one-tenth of the study (*n* = 117; 11.6 percent) used more than one tool to collect data, for example, interviews and focus groups" (Wangmo et al. 2018, p. 5).

8. However, with greater effort to categorize morally problematic situations, perhaps in certain kinds of contexts (e.g., healthcare) and in light of dimensions of human flourishing (see table 2.1 in chapter 2 on wisdom and flourishing), some patterns might potentially be identified to predict how situations evolve (e.g., their problematic nature either diminishing or augmenting) and perhaps the kind of key factors typically involved.

9. The current and future work that would allow us to better understand how lived moral situations are remembered and how that experience helps or hinders us in projecting scenarios that promise to bring about greater human flourishing is of great importance for ethics as laid out in this book.

10. As Fesmire explains, "Reasoning, then, is a story-structured, aesthetically funded capacity. In Paul Ricoeur's and Alasdair MacIntyre's terms, it has a narrative structure" (Fesmire 2003, p. 51).

11. For example, William explains that:

> If a man wakes up some fine day unable to recall any of his past experiences, so that he has to learn his biography afresh, or if he only recalls the facts of it in a cold abstract way, as things

that he is sure once happened, or if, without this loss of memory, his bodily and spiritual habits all change during the night, each organ giving a different tone, and the act of thought becoming aware of itself in a different way, he feels and he says that he is a changed person. (James [1890/1910] as cited in Schacter, Chiao, and Mitchell [2003])

Chapter 4

1. Dewey wrote that a situation was a combination of objective and internal factors: "Any normal experience is an interplay of these two sets of conditions. Taken together, they form what we call a situation" (Dewey [1938/1986] as cited in Pekarsky [1990, p. 284]).

2. The authors of these seminal studies were rather mindful of the limitations therein, contrary to some commentators. For example, Milgram describes at length the limitations of his experimental setup and how the situation may have featured physical (e.g., novel environment, limited time for reflection) and interpersonal (e.g., obligations to the researcher, ambiguity of the situation) factors that may explain the results. Darley and Batson's study also stresses the impact of the hurried situation on the cognitive ability to attend carefully to the bystander victim and to interpret the situation.

3. Credit goes to Dr. Ariane Quintal and her pioneering work on morally problematic situations in the context of rare diseases (Quintal 2023) for suggesting this name (moral matters) to designate the broad range of moral preoccupations encountered in moral life (Quintal 2023).

Chapter 5

1. Dewey formulated the following instructive criticism, which is still relevant given the impetus encountered in ethics theory in defining the moral domain monolithically:

> In the face of the role played by the real conflict of forces in moral situations and the manifest uncertainty about which side to take, I am inclined to think that one of the causes of the inefficiency of moral theories resides in their attachment to the unitary concept, which has led them to simplify moral life excessively. The result is an abyss between the involved realities of practice and the abstract forms of the system. A moral philosophy which frankly recognizes the impossibility of reducing all the elements of moral situations to one single principle, one which would admit that every human being can only do his best to shift for himself among the disparate forces, would throw light on our real difficulties and would help us make a more accurate estimate of competing factors. It would be necessary to sacrifice the idea that there exists, theoretically and beforehand, a unique and ideally correct solution for every difficulty into which a person will be thrown. Personally, I believe that this sacrifice, far from being a loss, would be a gain. By turning our attention from rules and rigid standards, it would lead us to take fuller consideration of the specific elements which necessarily enter into every situation where we must act. (Dewey 1966, p. 204)

2. Level 1: Preconventional morality (Stage 1: Punishment and obedience orientation, and Stage 2: Individualism and exchange). Level 2: Conventional morality (Stage 3: Good interpersonal relationships, and Stage 4: Maintaining the social order). Level 3:

Postconventional morality (Stage 5: Social contract and individual rights, and Stage 6: Universal principles). For a quick review, see Kohlberg (1968).

3. This is what the former president of UNESCO's International Bioethics Committee, Michèle Stanton-Jean, described as the trap of words (Stanton-Jean 2017) in trying to achieve global consensus in the development of the Universal Declaration on Bioethics and Human Rights (UNESCO 2005).

4. This issue of vagueness was already pointed out by James, who stressed that some vagueness about human morality is inescapable because experience precedes language and thus the words we use can sometimes lead to the tendency to "explain away" aspects of experience that do not conform to our existing categories and concepts (Fesmire 2003).

5. For example, eating meat is becoming increasingly considered an unhealthy tradition that raises questions about the impact on the environment and animal welfare of nonvegetarian/nonvegan diets (Dombrowski 1984). Accordingly, eating meat can gain moral significance insofar as it raises questions about human flourishing (and is not only an issue of taste preference), notably the ethical praiseworthiness of perpetuating health and environmental conditions that challenge basic conditions for humans (and other beings). This question also has implications for healthcare professionals and health authorities who are engaged in promoting dietary regimes. Nutrition is but one of the many examples of how a given behavior or habit can come to be seen as moral. Interestingly, as I noted in chapter 2, ancient ethics tackled issues related to entrenched behaviors such as nutrition and sexuality in ways that raise questions about the worthiness and justification of habits in light of human flourishing because of its commitment to philosophy as a way of life (Hadot 2002; Racine and Prud'homme 2019).

6. There are longstanding debates about the relationship between ethics and empirical work to understand human morality, such as illustrated in the strikingly different orientations of Hume's and Kant's work, notably Hume (1739/1975 and Kant (1785/1998). Discussions have tended to become mired in two oversimplified options (see Fesmire [2003, pp. 28–30] for an analysis): either (1) the justification of moral rules and the designation of the morally desirable comes from experience and there is thus no authoritative appeal for moral justification (e.g., Hume 1739/1975); or (2) moral rules stem from pure disincarnated reason that offers the benediction to moral rules based on their alleged ability to be universalized, applied to everyone in all contexts (e.g., Kant 1785/1998). The first stance is rightfully grounded in everyday experiences since everything about human morality is connected to our lives and experiences of the world. However, it fails to provide argumentation for why some experiences are preferable to others and appear circular. Yes, a person can point to, for example, appreciating the ability to exercise their autonomy (over paternalism) in a given situation, but what kind of argument is offered in support of this? The second stance offers a solution to ethical justification of moral rules, but it is often challenged by the complexity and diversity of human experiences.

7. There is a very real sense that foundational views—when taken literally—are contrary to ethics itself because ethics is not about defending this or that position, but about ensuring that human habits are constructed with an orientation toward flourishing. The foundational view of ethics theory and ethical principles is incompatible with an open-ended view of morally problematic situations because moral valuations are manifold such that no theory of ethics can provide an encompassing set of general principles that exhausts the reality of the moral domain and the scope of human aspirations for the future. Taking away the ambition of providing foundations is very much like saying that the emperor has no more clothes such that the garments of authority and power that ethics can wear are brought to their blunt truth, which is human experience. In that light, the dream of foundations is more aligned with the goal of dogmatizing certain moralities in ways that provide moral certainty. The only real way of determining the value of ethical ideas is through human experience, which changes across time, space, and between persons. Thus, ethics principles can point to and express the meaning of experiences, yet they fall short of justifying the value of this or that experience, which calls for broader appeals to human experience.

8. For example, when deliberating about projected response-scenarios, ethical principles facilitate the critical analysis of different courses of action in terms of our commitment to values, our departure from our values, and our ability to promote these values. In chapter 10 on ethical deliberation, I come back to the role that ethical principles play in structuring the analysis of strengths and weaknesses of competing action-scenarios.

Chapter 6

1. Take the example of lobotomies. Although this unfortunate episode in medicine is often considered an example of clearly unethical behavior to which modern psychosurgery is compared and contrasted (Fins, Rezai, and Greenberg 2006), it is important to examine the standards and contexts at the time when such procedures occurred. This scientific procedure seemed highly promising based on the Nobel Prize awarded to Egas Moniz for a reportedly successful clinical series: no pharmaceutical treatment for severe psychiatric illness was available at the time, the understandings of such conditions were rudimentary, and societies did not uphold the rights of individuals with psychiatric illnesses as we do today. Though morally wrong according to today's standards, they may have had a stronger justification in prior times that explains how protagonists of lobotomies like Walter Freeman justified their actions. The progressive recognition of their clinical failure coupled with the emergence of movements for the rights of (psychiatric) patients provided and welcomed another interpretation of the situation based on its consequences and its motivations, judged to be increasingly problematic and out of control (Gostin 1980).

2. I leave aside a discussion of the experience of the world of other sentient beings, though it is clear that other beings do have some kind of experience of the world, as evidenced by the ability to understand other animals and to communicate with

some of them, let alone their ability to communicate between themselves (Allen and Bekoff 1999; Griffin 1978).

3. James' notions of "horizon," "penumbra," and "fringe" influenced Husserl and Gadamer, as acknowledged by Gadamer in *Truth and Method* (Fesmire 2003, pp. 43–44, 138). The impact of pragmatism on phenomenology and hermeneutics is underappreciated. Fuller recognition would bring additional insights to these intellectual traditions, especially in their ability to be used as powerful tools for human betterment (e.g., Gallagher 2022).

4. This could be explained more simply with reference to values and ethical principles, but the term "valuations"—as explained earlier—indicates that values are not static and do not exist independently of the object of valuation.

5. The study of some facts is more inherently embedded in valuations (e.g., the study of empathic responses) and call for an immediate recognition of how values are instilled in their very objects, while others (e.g., basic research oriented toward basic discoveries of mechanisms) may involve valuations insofar as the human agents undertaking this research are guided by values and interests, as well as those of the institutions they work in and the contexts they live in. Science can be rather objective yet still value-laden because values guide the conduct of science in the pursuit of knowledge. Recognizing the existence of values does not at all undermine the epistemic worth of highly objective research. It simply adds context to understand it from a social science and humanities standpoint.

6. Ethics does not promise the continued comfort of being secure in one's own experiences or one's interpretations of situations. Sometimes we are unsettled, and rightfully so, because there is harm to someone's flourishing that we have disregarded; sometimes we are disturbed by different ways of thinking to which we should be open because they have supported greater flourishing or show promise of doing so. However, in current debates, the pain of growth toward freedom *from* religion is often underappreciated by those who advocate for a more radically secular society, but at the same time, the mere sanctification of freedom *of* religion cannot replace a commitment to engage in dialogue on the true value of religious beliefs in fostering human flourishing. Ethics cannot start or end with the claim that religious beliefs have some kind of special authority that protects them from critical scrutiny, but ethics cannot discount human experiences rooted in religious faith. I give this example here because it exemplifies well how different worldviews, including my own (as an atheist, I do not believe in God), must be fundamentally understood in their diversity before deciding what the best policy option is. In cases involving wearing religious symbols, I can partly understand and appreciate—as someone who has been religious in his early life—the profound connection of religious belief to one's identity, but I also see the roadblock in the entrenched, static worldviews on which these beliefs can sometimes rely given the lack of commitment to the fallibility of many religious teachings and some of their more general problems, such as the pervasive sexism of several major

religious institutions. Fallible beliefs about human life and the moral way of living—based on religious or for that matter on other kinds of foundational political, social, and economic beliefs—can easily be used to reinforce desires for segregation and even hatred toward others because the truth of those beliefs relies on authority or lack of willingness to criticize beliefs based on their impact on others. Thus, squarely questioning moral beliefs cemented by religious faith can be difficult and lead to irascibility given the stakes at hand, for the reason that these beliefs are typically not the result of experimental inquiry but of faith, which can be in tension with experimental inquiry (Peirce 1877).

7. He wrote that:

> No subject would seem to me more worthy of investigation than what I will call the "ordinary language of moral thinking and discourse." Most people do not talk about their ethical problems in the language of philosophers. And I have yet to meet one professional ethicist who, when dealing with his own personal moral dilemmas, talks the language of his professional writings; he talks like everyone else, and presumably he is thinking through his own problems in banal everyday language like everyone else. (Callahan 1973, p. 69)

8. We can predict—although this would need to be tested—that the expected growth of someone is proportional to the magnitude of the distance between one's vision and that of another insofar as developing this shared experiential understanding will only be accommodated by a more generous and encompassing understanding of a shared human world and the possibilities it can offer to those involved in a given situation. Consider the example of Nelson Mandela refusing to promote revenge for Apartheid, endorsing instead reconciliation and remembrance, as an example of the magnanimity and sense of justice required to envision and encourage in others such a path.

Chapter 7

1. The corollary of this understanding of the role of ethics in moral situations—and this is found in more detail in the chapters on moral growth and flourishing (see chapter 2) and ethical deliberation (see chapter 10)—is that it offers a "starting point" for moral experience that helps describe ethics as an effort to resolve situations. It stresses how agents should be actively engaged in envisioning action-scenarios, as well as how they can learn from experience and engagement with difficult situations. For example, dealing with morally distressing situations—although not sought after actively—can be part of a "crucial life process of moral maturation" (Tigard 2019, p. 601). Likewise, viewing a situation as problematic already mobilizes a certain initial human effort toward its resolution because of the need to first reckon with the problematic nature of the situation. This is why many ethicists have declared each moral matter as its own solution (Jonsen 1991; Jonsen, Siegler, and Winslade 2010; Roy, Williams, and Dickens 1994) or, as described when discussing Kohlberg's work (see chapter 8 on moral awareness), that the awareness of the problematic nature of a situation "portends action" (Reynolds 2008, p. 1036). This idea requires that different components

of ethics theory (e.g., the process of understanding moral matters, the process of generating solutions to moral matters) be understood in an agent-consistent way that is participatory and actionable, because this is the epistemological standpoint required to enact human flourishing.

2. There is a long tradition of scholarship in sociology based on such intents. However, the normative stance of this work is often implicit. It criticizes society for failing to do such and such without providing a vision for what should be pursued. Here, I leave aside the logical possibility and coherence of critical social sciences that do not commit to some vision of human well-being and flourishing and the undergirding issue of crypto-normativity in the social sciences (e.g., Habermas' critique of Foucault in Habermas [1990]).

3. Following the writings and thinking on ethical analysis by my mentor and bioethicist Hubert Doucet (Dion-Labrie 2009; Doucet 2004), I adopt a broader unit of analysis for ethics wherein different action-scenarios are the focal integrative point of discussion through which problematic situations are converted, via analysis and discussion, into possible responses. These possible responses, being potentialities, are explored by those engaged in ethical deliberation, as scenarios before their actual implementation.

4. The following explanation helps understand these Heideggerian concepts:

> Thrownness (German: *Geworfenhei*) for *Dasein* is not just experienced at the inception of Being, for the reason that this past way of being "thrown" subsequently echoes into every moment thereafter. It would follow that the notion of thrownness can be understood, to an extent, as a "past-for-the-present" mood that contingently allows for, and shapes interpretation. Subsequently, thrown-projection, or *Entwurf*, within the Heideggerian canon, is best placed by acknowledging the etymological translated of this term as "a sketched project, yet to be carried through." As such, it can be suggested that *Dasein* forms an intelligibility of the world within this thrownness by way of projecting "itself in terms of possibilities." (Yarrow n.d.)

Chapter 8

1. One exception is Reynolds and Miller who distinguish moral sensitivity from moral awareness insofar as moral awareness designates "an individual's determination that a single situation contains moral content and legitimately can be considered from a moral point of view," whereas moral (or ethical) sensitivity would designate the "broader cognizance of moral issues" in response to a wide variety of moral matters.

2. Other terms used include (see Miller, Rodgers, and Bingham [2014] and for an instructive review, see their table 1.1) cognitive recognition (of moral matters), affective recognition (of moral matters), cognitive perspective taking, affective perspective taking, ascription of importance, unconscious awareness, moral attentiveness, moral construal, moral imagination, moral perception, implicit assumptions and associations, moral disengagement, ethical fading, moral neutrality, and bounded ethicality.

3. These processes are sometimes underrecognized in ethics theory, which often starts with cases that are predefined as moral matters from the point of view of the writer or teacher. This way of learning and teaching ethics does not reflect the psychosocial complexity involved in how a person or group of persons recognizes, or fails to recognize, problems as moral.

4. The other components are moral judgment, moral motivation and commitment, and moral character and competence. For Rest, these four components interact with each other and should not be viewed necessarily as separate building blocks of human reality but perhaps more as a heuristic strategy to organize a scholarly understanding of human morality (Rest 1986).

Chapter 9

1. Celebrated American football coach.

2. Miller and colleagues describe five defining components of moral awareness: (1) cognitive recognition, (2) affective recognition, (3) cognitive perspective-taking, (4) affective perspective-taking, and (5) ascription of importance (Miller, Rodgers, and Bingham 2014). In this chapter, components 1 and 3 of this classification fall under "cognitive barriers" while components 2 and 4 fall under "affective barriers." Ascription of importance is implied in all forms of moral awareness. I further submit that relational and social barriers need to be taken into account. Again, the differences in presentations partly concern the fact that I am using knowledge about human morality (and moral awareness) to inform a model of ethics that brings a specific goal to the forefront, that is, overcoming moral unawareness to grow ethical awareness as part of the pursuit of deliberative ethical wisdom. I am approaching this literature within the context of an ethics dedicated to human flourishing; knowledge about moral awareness is instrumentalized toward this goal and follows a specific account of ethical agency (chapter 1).

3. In addition to its role in supporting moral and ethical awareness, empathy in clinical professions is attributed a role in (1) understanding the experience of patients (Halpern 1993, p. 167), (2) understanding patient history (Halpern 1993, p. 167), (3) showing interest in the patient (Halpern 1993, p. 167), (4) accessing the meaning of the life of patients (Charon 1993, p. 148), (5) offering empathic care, (6) understanding patient suffering, (7) recognizing the motivations of patients, and (8) giving the chance to patients to relate their narrative (Charon 1993, p. 150). These roles intersect with good clinical practices and speak to the broader role of empathy in morality and in clinical practice. Likewise, in the context of business ethics, empathy has been attributed different positive functions such as (1) allowing for more informed and grounded decision-making in contexts of complexity; (2) representing a rampart against cold, impersonal, and dehumanized business; and (3) being an essential aspect of efficient organizations (Martineau, Decety, and Racine 2020).

4. Explicit attitudes are "beliefs and opinions developed through cognitive deliberation" of which the moral agent is aware and can thus self-report about (Reynolds, Leavitt, and DeCelles 2010, p. 753).

5. Herein lies one of the pitfalls of moral awareness, which is to reduce the domain of morality to a preferred "variable" and not cultivate a richer understanding of the former in which values of different moral perceptions are accounted for in the construction of a reality of a given situation (Reynolds 2008). The implications of a restricted moral awareness (to one or a few sets of considerations) often lies at the root of certain important mishaps. Reynolds gives the example of major business scandals in which an attraction to arguably worthwhile goals (focused on purportedly self-sufficient ends) prevented business managers from seeing the importance of basic rules and norms (Reynolds 2006). He suggests that if managers had a broader understanding of the moral domain and recognized the importance of norms, they may have avoided generating "moral catastrophes."

6. In this literature, teleological or ends-based decision-making (*telos* comes from the Greek "ends") describes frameworks that focus "attention on the consequences of a situation and argues that the moral act is that which optimizes or creates the greatest good" (Brady [1985] as cited in Reynolds [2006, p. 234]). Teleological reasoning, in this sense, is often equated with utilitarianism, perhaps one of the best know forms of teleological moral theories. In contrast, deontological moral decision-making "emphasizes the importance of patterns, rules of behaviors, or some other formal features of ethics to determine moral behavior" (Brady and Wheeler [1996] and Honderich [1995] as cited in Reynolds and Ceranic [2007, p. 1611]). It is sometimes referred to as "formalism."

7. This interpretation would be consistent with developmental models of moral reasoning pioneered by Piaget and Kohlberg (Kohlberg 1968; Piaget 1932; Tanner 2016), which have been heavily criticized, but which may capture some real progressive development in moral awareness and development of the moral point of view. For the sake of this chapter, this explanation is not needed, and a multifactorial and nonhierarchical (nondevelopmental or at least nonlinear developmental) view of moral awareness will be encouraged as one that does not imply any a priori superiority of one predisposition over another, although this should be explored further.

8. The main formulations of the categorical imperative are:

1. "Act only in accordance with that maxim through which you can at the same time will that it become a universal law." (Kant 1781/1787/1996, p. 31)
2. "So act that you use humanity, whether in your own person or in the person of any other, always at the same time as an end, never merely as a means." (p. 38)
3. "From this there follows now the third practical principle of the will, as supreme condition of its harmony with universal practical reason, the idea of the will of every rational being as a will giving universal law." (p. 39)
4. "Consequently, every rational being must act as if he were by his maxims at all times a lawgiving member of the universal kingdom of ends. The formal principle of these

maxims is, act as if your maxims were to serve at the same time as a universal law (for all rational beings)." (p. 45)

9. I was actually interested in doing this research in younger individuals, but because of local regulation we would have faced additional obstacles to conducting a pediatric study, which we did not have the means to do; this speaks to the very issue at stake, namely the difficulty of allowing people with disability to be active and heard agents.

10. This proposal is made in the context of the development of a "person-oriented research ethics" (Cascio and Racine 2018), an adaptive form of ethics responsive to the interpersonal moral aspects of research that are somewhat left aside by more regulatory conventional research ethics approaches.

11. The six factors are magnitude of consequences, social consensus, probability of effects, temporal immediacy, proximity, and concentration of effect.

12. However, I do not imply that freedom of speech is without any limits since some speech can be harmful and needs to be considered in the broader scheme of things. However, most democratic jurisdictions have tended to be careful in imposing such limits because censorship would otherwise jeopardize the ability to seek the truth. For example, if offensive or disruptive speech becomes censored, there is a risk that certain historical or contemporary realities will become taboos and not be expressed in their wholeness.

13. Gender is defined as "women" in the cited paper, that is, as biological sex and not really as gender in ways that would account for social construction of gender identification (Reynolds and Miller 2015). The paper confuses social (gender) and biological (sex) categories.

14. As Kekes puts it, "This kind of depth must be achieved. Conventional morality and breadth can be taught, but depth cannot be. It requires the constant imaginative effort to try to see moral situations as a person would who approximated more closely than one persistently can to the ideals to which he aspires. This imaginative effort is the driving force of morality" (Kekes 1984, p. 9).

Chapter 10

1. It is important to distinguish between agents who are in a prereflective mode (*moral agents* dealing with situations without being confronted with any notably difficult moral matter) and those in a reflective mode (*ethical agents* who are engaged in a deliberate, cognitively, and socially effortful process). This distinction is similar to the one I made regarding moral awareness (as a rather instinctive capacity) versus ethical awareness (as a cultivated and nurtured capacity). Accordingly, the learning moral agent is, in a sense, an ethical agent, an agent engaged in an inquiry process.

2. Or a dialogical mode of being. Transactions occur between organism (as Dewey puts it) and environment. Human thinking, because of its purpose-oriented nature,

allows human beings to make sense of themselves and their experiences, ensuring the transition from the past to the present. In this sense, following James, Gerald Edelman, and others, human consciousness is a remembered present where memory plays a key role (Edelman 1989).

3. Hoffmaster notes this regarding bioethics, but this comment could apply to other areas of ethics: "As long as bioethics remain mired in the philosophical conception of morality as a moral theory comprised of principles, bioethics will be applied ethics" (Hoffmaster 2018, p. 119). Here the term "applied ethics" is used to designate a common (and rather unhelpful) way of considering ethics as a process of applying more general, even universal, ethical principles following the canons of deductive reasoning. In such a view, deliberation plays little role as a process that underpins the development of a reasonable judgment because the principles in themselves represent the standard by which actions are guided. However, Hoffmaster explains that in concrete cases of moral problem resolution, deliberation is what serves to validate, as a process, the judgment to which moral agents come (Hoffmaster 2018). He identifies a deeply entrenched rift that plagues much of the empirical work conducted in bioethics where researchers make sense of the contribution of empirical research to ethics theory with difficulty. It is my hope that this chapter—based on previous work on deliberation theory and pragmatist ethics—offers a justification for the essential role that deliberation plays in ethics, conceived here as a form of deliberative wisdom.

4. On the empirical basis of valuations, Czeżowski states very tellingly:

> Primary evaluations are, as we have said, subject to generalisation; and it is owing to the generalisation that evaluational criteria are established. General propositions in which criteria of ethical evaluation are established, are not other than our ethical principles. As all generalisations, they too are uncertain and variable; i.e., it may happen that a criterion, established with regard to a certain scope of moral experience, will prove insufficient when the scope is enlarged and when new cases of primary evaluations are incapable of being subordinated to that criterion. Another reason for the variability of ethical principles lies in the fact that general terms, used for their formulation, are themselves subject to change; terms such as fidelity, courage, heroism, falsehood, and treason, vary their meanings unceasingly, along with changing customs and standards of morality. As a result, the meaning of propositions expressing ethical principles undergoes changes, and the ethical principles require new formulations. And finally, changing social relations also demand that evaluational criteria be adapted to them, if the latter are not sufficiently differentiated to embrace the variety of situations . . . The role of foundations in ethics is played not by a priori ethical principles but by empirical primary evaluations. (Czeżowski 1953, pp. 167–168)

5. The last residential school (Kivalliq Hall in Rankin Inlet) closed in 1997 (Buscemi 2019). Residential schools "had the nominal objective of educating Indigenous children but also the more damaging and equally explicit objectives of indoctrinating them into Euro-Canadian and Christian ways of living and assimilating them into mainstream white Canadian society" (Hanson, Gamez, and Manuel 2020).

6. Pragmatism's commitment to dialogical approaches may resemble some features of "two-eyed seeing" (*Etuaptmumk*), first advanced by Mi'kmaq Elder Albert Marshall

(from the community of Eskasoni, Nova Scotia, Traditional Territory of *Mi'kma'ki*). This idea is to "learn to see from your one eye with the best or the strengths in the Indigenous knowledges and ways of knowing . . . and learn to see from your other eye with the best or the strengths in the mainstream (Western or Eurocentric) knowledges and ways of knowing . . . but most importantly, learn to see with both these eyes together, for the benefit of all" (Elder Albert Marshall as cited in Bartlett [2017, p. 1]; see also Bartlett et al. [2012]; Bartlett et al. [2015]; Marshall et al. [2015]).

7. Smith also explains that when a society settles on an imaginary, that imaginary becomes a normative, standard, third-person-like perspective (Müller 2016).

8. The Doucet method is structured in eight steps: (1) list at least three scenarios; (2) further examine in greater detail each scenario, including its consequences and implications; (3) discuss/deliberate on each scenario to identify the moral aspects (e.g., values, principles) advanced and neglected for each scenario; (4) name the moral grounding or foundation that would allow you to choose the scenario; (5) choose the better option; (6) enact the chosen scenario; (7) communicate the decision; and (8) evaluate the decision taken.

9. In contrast to scenarios, moral situations are the actual or past problem situations in need of further understanding and eventual resolution.

10. This contrast between more conventional accounts of moral judgments and scenario-based deliberation guides the following explanations. In other words, ethics needs to be useful in resolving lived situations in meaningful ways. This is partly its cash value, to speak the language of James, granted that meaningful resolution does not equate to mere problem fixing. Resolutions can also imply profound self-reflection of the moral agent because the real outcome of ethics is partly the resolution of the situation, but more importantly, ethical learning.

11. Accordingly, moral deliberation is somewhat of a contradiction (given my previous differentiation of morality and ethics), though I would not want to overdo the difference and distance between morality and ethics.

12. Casuistic theory describes common aspects of cases, the "loci," the disputes, the circumstances of cases (Jonsen 1995). However, casuistry has the tendency to generate conservatism inasmuch as precedent decisions (like jurisprudence) set a standard, unless the standards are also granted the ability to evolve and be enriched based on knowledge and experience. Furthermore, it tends to adopt the view of the expert who sees in situations "cases" in their profession and the problematic aspects of the case from the standpoint of the authoritative decision-maker (Kopelman 1994). This is not wrong but simply limited in scope.

13. In an analogy to a mathematical function, action serves as a transformative operation between the complexity of the input (the situation) and the requirement of an output (a response to the situation). This analogy can lead to

oversimplification—mathematic function theory is complex—but it helps to convey the idea that regardless of the situational complexity and its multidimensional aspects and variables, a response representing the full range of human behavior will have to be generated.

14. In an early paper in the history of bioethics, Callahan states very clearly that most philosophers do not even use the theories they teach to solve their own moral quandaries (Callahan 1973). In the view of many ethicists (Miller, Fins, and Bacchetta 1996)—and this is very much my own experience too, notably in the deliberative and qualitative research I have undertaken (e.g., Racine 2008a; Racine et al. 2017a)—richer understandings of the perspectives of moral agents, the interpretation and description of situations, and an openness to various moral insights are key.

15. In Gadamerian hermeneutics, a situation designates what is within reach of the agent, but a situation is not isolated or isolating because it occurs within a "horizon." A horizon represents the perspective from which we can look at the situation. Everyone is situated differently and operates from a perspective; the fusion of horizon represents the possibility for agents to foresee a common path and understanding (Gadamer 1989). Much like in the pragmatist tradition, situations occur in relationship to our history such that an interpretative consciousness (hermeneutical consciousness) designates the awareness of how our understanding is shaped by history. The parallels between the pragmatist and hermeneutic traditions are striking, though rarely recognized. Fesmire explains well how Husserl's account of consciousness is shaped by James' psychology (Fesmire 2003).

16. When an important compromise takes place, it can yield what is sometimes described as "moral residue" (Epstein and Hamric 2009); see the explanation in chapter 7.

17. In brief, rationality and reasonableness are not equivalent; while the former suggests an effort to remove human understanding from its existential context, the latter calls for its inclusion.

Chapter 11

1. This statement does not imply relativism since conditions for assessing whether meanings are warranted will be part of a discussion as protagonists try to assess the relevance and worthiness of different interpretations of the situation (Prairat 2014). Inquiring about the nature of the situation will thus be a dialogical process (Crick and Bodie 2016).

2. Fesmire notes that Dewey identifies four ways in which people will deliberate and engage in dramatic rehearsal: (1) "Some people deliberate by dialogue," (2) "Others visualize certain results," (3) "Others rather take the motor imagery and imagine themselves doing a thing," (4) "Others imagine a thing done and then imagine someone else commenting upon it" (Fesmire 2003, p. 74).

3. These were described as *moral* acceptability and *moral* praiseworthiness in my previous work (see chapter 6 in Racine [2010]) before I started to adopt a more rigorous conceptual distinction between the moral and the ethical, such as is required by the theory of deliberative wisdom.

4. In other words, *ago, ergo sum* because existence occurs in transactions.

5. Accordingly, some faith in the ability to make a given scenario possible must be granted. Some charity principle (or rational accommodation) as discussed by Quine, Davidson, and others, must be accorded to different options again to avoid dismissing different scenarios that are not in line with one's initial preference or otherwise.

6. Here, I mean that human flourishing is not a preset *telos*; it is based on orientations that we give to ourselves, which are ends, but not ordained or preset ends.

7. The concepts of ethical acceptability and ethical praiseworthiness that I am elucidating are reminiscent of Berlin's distinction between negative liberty ("freedom from") and positive liberty ("freedom to") (Berlin 2002). Just like Berlin, I intend the concepts to be complementary. Contrary to the sometimes acrimonious tone of some ethics debates, both ethical acceptability and ethical praiseworthiness are genuine views that together help express the complex nature of moral thought and behavior (see table 11.1).

8. Kant uses the terminology of perfect (e.g., refraining from suicide, not lying and telling the truth) and imperfect duties (e.g., improving oneself, aiding others) to describe roughly negative and positive obligations (Kant 1785/1998).

9. The application of a criterion (or criteria) of ethical praiseworthiness may be contingent on the possibility of sharing a common worldview, a common life, on a more personal level such that there is considerable shared experiences and agreement on their value. The application of ethical acceptability may require commitment to a thinner concept of the good (not doing harm). There is perhaps a sense according to which these criteria are inversely proportional such that the more intension ethics has (in the logical sense), the less extension is has, and vice versa. However, the theory of deliberative wisdom seeks to avoid this purely conflicting and dualistic dynamic by arguing for the complementary nature of both standards.

10. This terminology is analogous to the triadic distinction between the concept of death, criteria for death determination, and tests to ascertain that death determination criteria have been fulfilled (Bernat 2006b; Bernat, Culver, and Gert 1981).

11. It is possible that tightly woven societies with more limited diversity of lifestyle and values feature more consensus on the criteria for ethical praiseworthiness such that a certain account of human flourishing is enforced more strongly throughout social institutions. Social cohesion around a standard of ethical praiseworthiness is a project that is partly defined by the shared and communicable experiences between members of a society, but also, importantly, as an ongoing task in recognition that traditional standards can simply not account for all modern situations where new

kinds of goods surface, involving new preferences and valuations. In *Pragmatic Neuroethics*, I describe this as requiring a creative effort rather than simply falling back on tradition (e.g., cultural, religious, customs, or other). What traditional societies had to offer (clear and strong views on ethical praiseworthiness) now needs to be re-created to some extent, based on experience and knowledge. This commitment to the evaluation of outcomes based on experience was likely at the root of the traditions that different cultures adopted. But the initial situational reasons for why something became preferable have been lost due to the time past and the adoption of these traditions into religious or dogmatic and unquestionable discourse. Moreover, the slower evolution of social and physical environments prior to capitalism and industrialization did not put as much at the forefront the need for ethical innovation, as now needed by modern societies given the quick and significant changes in technologies and values.

12. See also other annual editions of the world happiness report at https://worldhappiness.report/.

13. In doing so, it builds from previous scholarship and furthers the empirical turn in ethics (Borry, Schotsmans, and Dierickx 2005) and pragmatist theories of ethical analysis (Martela 2015, 2017; Racine et al. 2019) and deliberation (Caspary 1991; Gouinlock 1978; Lekan 2006; Ralston 2010).

14. When different ethical theories are instrumentalized as lenses (Sherwin 1999), they allow us to see valuable outcomes and experiences, but they also need to be cross-checked with each other to ensure that (1) no false negatives occur (e.g., the inability to see a valuable experience or a valuable good) and (2) no false positive occurs such that a theory brings idiosyncrasies that are inconsistent with a broader integrative understanding of the domain of morality. In this sense, ethics theories are used to explore the ethical worth of different action-scenarios and are also used to evaluate their merits based on the importance of the values they point to and the coherence of the analysis based on the values expressed in different theories. This kind of use of ethics theories differs from envisioning them as pure foundations of moral judgment insofar as the process of generating a resolution to an ethical problem is more holistic and involves other considerations in order to support the analysis of different actions. It would be ideal for ethics to become more scientific and allow the evaluation of the actual eudaimonic contribution of different courses of actions based on their outcomes (Martela 2015).

Chapter 12

1. As Dewey argued against James' account of will as a purely mental entity, there is no need to consider the will as being removed from causality and from context to make sense of human behavior as voluntary. For a review and discussion of this point, see Shusterman (2008).

2. I do not claim here that ownership of the world by humans is the end goal, but rather that human existence be greater expressed in its full depth and diversity.

3. The ability of that response to bring about a *resolution* or *re-situation* to the moral matter is a complex matter and is discussed more extensively in the next chapter. A resolution implies an element of quality that needs to be carefully assessed given the rather definitive connotation. This is not an easy task, but not insurmountable, and we should find the courage to undertake the better characterization of what counts as a successful/less successful/unsuccessful resolution in a problem situation without feeling self-defeated, that is, believing that moral matters are unsolvable. In human experience, we have understandings of when situations are resolved, pacified, as well as when they are not. Although these understandings are difficult to articulate, we can find ways to characterize them.

4. Nietzsche explains in the *On the Genealogy of Morality: A Polemic* that true reaction is action (Nietzsche 1887/1971) and that resentful human beings are human beings deprived of the ability to act; they do not truly re-act, or act in response to situations, but fall prey to imaginary revenge and resent instead.

5. Other possible response-scenarios could also be to seize this opportunity to enact major changes or to formulate a clear rejection.

6. As Kant wrote about the good will, "This will need not, because of this, be the sole and complete good, but it must still be the highest good and the condition of every other, even of all demands for happiness" (Kant 1785/1998, p. 10).

Chapter 13

1. This is what Dewey called the philosophical fallacy.

2. There have been quite appropriate criticisms of bioethics when it is used to reconstruct purpose without questioning the more foundational goals of why changes based on scientific development and technology are desirable (see notably Hedgecoe [2010] and Turner [2004]). This is the problem of co-option of bioethics to the technological and scientific agenda. This is not the whole story of bioethics, but it has certainly been part of it.

3. Perhaps this explains why ethics was pursued in small schools in antiquity with limited general impact on the public life of the city-state. Discussing provocative issues publicly was not necessarily prudent, a thesis elaborated on by Strauss (1952). Socrates was accused of corrupting the youth with his teachings. Descartes feared sanctions by religious authorities, as did Spinoza. Dewey's pragmatism was overshadowed by McCarthyism and conservatism, which gained traction in postwar America (Capps 2003; McCumber 2001; Menand 2001; West 1989), and its preference for philosophy that stayed away from asking the big, disruptive questions about the meaning of life and its unavoidable social and political implications.

4. Dewey suggests that everyone should "act as to increase the meaning of present experience," (Dewey 1922, p. 283) and that this statement could amount to a reformulation of Kant's categorical imperative. However, Dewey himself discourages the

use of his own reformulation of the categorical imperative in terms of present meaning of actions on the grounds that universalizable maxims empty statements of their meaning. Rather, we need to understand how our needs and the possible responses to our needs lie within unique and localized situations (Dewey 1922). Dewey explains:

> Progress means increase of present meaning, which involves multiplication of sensed distinctions as well as harmony, unification. This statement may, perhaps, be made generally, in application to the experience of humanity. If history shows progress it can hardly be found elsewhere than in this complication and extension of the significance found within experience. It is clear that such progress brings no surcease, no immunity from perplexity and trouble. If we wished to transmute this generalization into a categorical imperative we should say: "So act as to increase the meaning of present experience." But even then in order to get instruction about the concrete quality of such increased meaning we should have to run away from the law and study the needs and alternative possibilities lying within a unique and localized situation. The imperative, like everything absolute, is sterile. Till men give up the search for a general formula of progress they will not know where to look to find it. (Dewey 1922, p. 283)

5. Here I do not intend to use the term "subjective" to designate a somehow biased or erroneous kind of evaluation but simply one that takes as its primary referent the subject, that is, the agent. Subjective in this sense is not really opposed to objective because subjects do apprehend a world external to them. However, other agents do too, hence the ideal of a more intersubjective account of the objective and subjective domains, on a spectrum.

6. This is an outcome of autonomy-emphasizing approaches in neonatal care (e.g., Orfali 2004; Orfali and Gordon 2004).

References

2004. *Loi encadrant le port de signes ou de tenues manifestant une appartenance religieuse dans les écoles, collèges et lycées publics*. 2004–228. France. L.141-5-1.

2014. *Loi concernant les soins de fin de vie*. Québec. S-32.0001

2019. *Loi sur la laïcité de l'État*. Québec. L-0.3.

Abbasi, Mahmoud, Nasrin Nejadsarvari, Mehrzad Kiani, Fariba Borhani, Shabnam Bazmi, Saeid Nazari Tavaokkoli, and Hamidreza Rasouli. 2014. "Moral Distress in Physicians Practicing in Hospitals Affiliated to Medical Sciences Universities". *Iranian Red Crescent Medical Journal* 16 (10): e18797.

Abma, Tineke A., Yolande Voskes, and Guy Widdershoven. 2017. "Participatory Bioethics Research and Its Social Impact: The Case of Coercion Reduction in Psychiatry." *Bioethics* 31 (2): 144–152.

Adam, Avshalom M., and Dalia Rachman-Moore. 2004. "The Methods Used to Implement an Ethical Code of Conduct and Employee Attitudes." *Journal of Business Ethics* 54 (3): 225–244.

Aiguier, Grégory, and Alain Loute. 2016. "L'intervention éthique en santé: Un apprentissage collectif. " *Nouvelles Pratiques Sociales* 28 (2): 158–172.

Aiguier, Grégory, and Jean-Philippe Cobbaut. 2016. "Le tournant pragmatique de L'éthique en santé: enjeux et perspectives pour la formation. " *Journal international de bioéthique* 27 (1–2): 17–40

Aiguier, Grégory, Rozenn le Berre, Dominique Vanpee, Jean-Philippe Cobbaut. 2012. "Chapitre 8. Du pragmatique aupragmatisme: quels enjeux pour la formation à l'éthique?" *Journal international de bioéthique* 23: 123–148.

Al Sayyari, Abdulla, Fayez Hejaili, and Faissal Shaheen. 2016. "Interaction between Society and Medical Ethics in Saudi Arabia." *Anthropology of the Middle East* 11 (2): 60–75.

Alexander, Thomas. 1993. "John Dewey and the Moral Imagination: Beyond Putnam and Rorty Toward a Postmodern Ethics." *Transactions of the Charles S. Peirce Society* 29 (3): 369–400.

Algozzine, Bob, Diane Browder, Meagan Karvonen, David W. Test, and Wendy M. Wood. 2001. "Effects of Interventions to Promote Self-Determination for Individuals with Disabilities." *Review of Educational Research* 71 (2): 219–277.

Alkire, Sabina. 2002. *Valuing Freedoms: Sen's Capability Approach and Poverty Reduction.* New York, NY: Oxford University Press.

Allen, Collin, and Marc Bekoff. 1999. *Species of Mind: The Philosophy and Biology of Cognitive Ethology.* Cambridge, MA: MIT Press.

Altman, Andrew. 1983. "Pragmatism and Applied Ethics." *American Philosophical Quarterly* 20 (2): 227–235.

Ameen, Elsie C., Daryl M. Guffey, and Jeffrey J. McMillan. 1996. "Gender Differences in Determining the Ethical Sensitivity of Future Accounting Professionals." *Journal of Business Ethics* 15 (5): 591–597.

American Academy of Neurology. 1989. "Position of the American Academy of Neurology on Certain Aspects of the Care and Management of the Persistent Vegetative State Patient." *Neurology* 39 (1): 125–126.

American Academy of Pediatrics, American Academy of Family Physicians, and American College of Physicians-American Society of Internal Medicine. 2002. "A Consensus Statement on Health Care Transitions for Young Adults with Special Health Care Needs." *Pediatrics* 110 (6): 1304–1306.

Andre, Judith. 2004. *Bioethics as a Practice.* Chapel Hill: The University of North Carolina Press.

Anscombe, Gertrude E. M. 1958. "Modern Moral Philosophy." *Philosophy* 33 (124): 1–19.

Apfel, Lauren J. 2011. *The Advent of Pluralism: Diversity and Conflict in the Age of Sophocles.* Oxford, UK: Oxford University Press.

Apperly, Ian A. 2012. "What is "Theory of Mind?" Concepts, Cognitive Processes and Individual Differences." *Quarterly Journal of Experimental Psychology* 65 (5): 825–839.

Ardelt, Monika. 2003. "Empirical Assessment of a Three-Dimensional Wisdom Scale." *Research on Aging* 25 (3): 275–324.

Aristotle. 1999. *Nicomachean Ethics.* Indianapolis, IL: Hackett Publishing Company.

Arminjon, Mathieu, François Ansermet, and Pierre Magistretti. 2010. "The Homeostatic Psyche: Freudian Theory and Somatic Markers." *Journal of Physiology–Paris* 104 (5): 272–278.

Arras, John D. 2002. "Pragmatism in Bioethics: Been There, Done That." *Social Philosophy and Policy* 19 (2): 29–58.

Assemblée Nationale du Québec. 2021. "Commission Spéciale sur L'évolution de la Loi Concernant les Soins de Fin de Vie." *Assemblée Nationale du Québec*, December 8, 2021. http://www.assnat.qc.ca/fr/travaux-parlementaires/commissions/cssfv-42-1/index.html.

Azikiwe, Ben N. 1931. "Ethics of Colonial Imperialism." *Journal of Negro History* 16 (3): 287–308.

Baertschi, Bernard. 1998. Les méthodes de résolution de cas. *Bioethica Forum* 26: 4–11.

Baier, Kurt. 1958. *The Moral Point of View: A Rational Basis for Ethics*. Ithaca, NY: Cornell University Press.

Bailey, Zinzi D., Nancy Krieger, Madina Agénor, Jasmine Graves, Natalia Linos, and Mary T. Bassett. 2017. "Structural Racism and Health Inequities in the USA: Evidence and Interventions." *The Lancet* 389 (10077): 1453–1463.

Bakracheva, Margarita. 2019. "Coping Effects on Life Meaning, Basic Psychological Needs and Well-Being." *Psychology, Health & Medicine* 10: 1375–1395.

Baloyannis, Stavros J. 2019. "Is Alzheimer's Disease an Existential Fragmentation or an Imprisonment of Mind?" *Journal of Neurology & Stroke* 9 (3): 155–157.

Baltes, Paul B., and Ursula M. Staudinger. 2000. "Wisdom: A Metaheuristic (Pragmatic) to Orchestrate Mind and Virtue Toward Excellence." *American Psychologist* 55 (1): 122.

Baltes, Paul B., Ursula M. Staudinger, Andreas Maercker, and Jacqui Smith. 1995. "People Nominated as Wise: A Comparative Study of Wisdom-Related Knowledge." *Psychology and Aging* 10 (2): 155–166.

Bandini, Julia, Christine Mitchell, Zachary D. Epstein-Peterson, Ada Amobi, Jonathan Cahill, John Peteet, Tracy Balboni, and Michael J. Balboni. 2017. "Student and Faculty Reflections of the Hidden Curriculum." *American Journal of Hospice and Palliative Medicine* 34 (1): 57–63.

Bangen, Katherine J., Thomas W. Meeks, and Dilip V. Jeste. 2013. "Defining and Assessing Wisdom: A Review of the Literature." *American Journal of Geriatric Psychiatry* 21 (12): 1254–1266.

Bargh, John A., and Tanya L. Chartrand. 1999. "On the Unberable Automaticity of Being." *Scientific American* 54 (7): 462–479.

Barned, Claudia, Corinne Lajoie, and Eric Racine. 2019. "Addressing the Practical Implications of Intersectionality in Clinical Medicine: Ethical, Embodied and Institutional Dimensions." *American Journal of Bioethics* 19 (2): 27–29.

Baron-Cohen, Simon, Alan M. Leslie, and Uta Frith. 1985. "Does the Autistic Child Have a Theory of Mind?" *Cognition* 21 (1): 37–46.

Bartlett, Cheryl. 2017. "Two-Eyed Seeing: Elder Albert's Marshall's Guiding Principle for Inter-Cultural Collaboration." http://www.integrativescience.ca/uploads/files/Two-Eyed%20Seeing-AMarshall-Thinkers%20Lodge2017(1).pdf.

Bartlett, Cheryl, Murdena Marshall, and Albert Marshall. 2012. "Two-Eyed Seeing and Other Lessons Learned Within a Co-Learning Journey of Bringing Together Indigenous and Mainstream Knowledges and Ways of Knowing." *Journal of Environmental Studies and Sciences* 2: 331–340.

Bartlett, Cheryl, Murdena Marshall, Albert Marshall, and Marilyn Iwama. 2015. "Integrative Science and Two-Eyed Seeing: Enriching the Discussion Framework for Healthy Communities." In *Ecosystems, Society and Health: Pathways through Diversity, Convergence and Integration*, edited by Lars K. Hallström, Nicholas P. Guehlstorf, and Margot W. Parkes, 280–326. McGill-Queen's University Press, Montreal.

Batson, C. Daniel, Shannon Early, and Giovanni Salvarani. 1997. "Perspective Taking: Imagining How Another Feels versus Imaging How You Would Feel." *Personality and Social Psychology Bulletin* 23 (7): 751–758.

Batson, C. Daniel, Tricia R. Klein, Lori Highberger, and Laura L. Shaw. 1995. "Immorality from Empathy-Induced Altruism: When Compassion and Justice Conflict." *Journal of Personality and Social Psychology* 68 (6): 1042–1054.

Bauer, Mark S., Laura Damschroder, Hildi Hagedorn, Jeffrey Smith, and Amy M. Kilbourne. 2015. "An Introduction to Implementation Science for the Non-Specialist." *BMC Psychology* 3 (1): 32.

Bauman, Zygmunt, and Leonidas Donskis. 2013. *Moral Blindness: The Loss of Sensitivity in Liquid Modernity*. Cambridge, UK: Polity Press.

Baumeister, Roy F. 1991. *Meanings of Life*. New York, NY: Guilford Press.

Baumeister, Roy F., Kathleen D. Vohs, Jennifer L. Aaker, and Emily N. Garbinsky. 2013. "Some Key Differences between a Happy Life and a Meaningful Life." *Journal of Positive Psychology* 8 (6): 505–516.

Beauchamp, Tom, and James Childress. 2001. *Principles of Biomedical Ethics*. Oxford, UK: Oxford University Press.

Beck, Aaron T., Clyde H. Ward, Myer Mendelson, John Mock, and J. K. Erbaugh. 1961. "An Inventory for Measuring Depression." *Archives of General Psychiatry* 4 (6): 561–571.

Beck, Aaron T., John Rush, Brian F. Shaw, and Gary Emery. 1979. *Cognitive Therapy of Depression*. New York, NY: Guilford Press.

Beck, Aaron T., Robert A. Steer, and Gregory K. Brown. 1996. Beck Depression Inventory—II. *APA PsycTests*.

Bell, Derrick. 1995. "Who's Afraid of Critical Race Theory?" *University of Illinois Law Review* (4): 893–910.

Bell, Emily, Veljko Dubljević, and Eric Racine. 2013. "Nudging without Ethical Fudging: Clarifying Physician Obligations to Avoid Ethical Compromise." *American Journal of Bioethics* 13 (6): 18–19.

Berkeley, George. 1734/1957. *A Treatise Concerning the Principles of Human Knowledge.* New York, NY: Liberal Arts Press.

Berlin, Isaiah. 2002. *Liberty: Incorporating Four Essays on Liberty.* New York, NY: Oxford University Press.

Bernat, James L. 2006a. "Chronic Disorders of Consciousness." *Lancet* 367 (9517): 1181–1192.

Bernat, James L. 2006b. "The Whole-Brain Concept of Death Remains Optimum Public Policy." *Journal of Law Medicine and Ethics* 34 (1): 35–43, 33.

Bernat, James L. 2009. "Chronic Consciousness Disorders." *Annual Review of Medicine* 60: 381–392.

Bernat, James L., Charles M. Culver, and Bernart Gert. 1981. "On the Definition and Criterion of Death." *Annals of Internal Medicine* 94 (3): 389–394.

Bernstein, Amit, Yuval Hadash, Yael Lichtash, Galia Tanay, Kathrine Shepherd, and David M. Fresco. 2015. "Decentering and Related Constructs: A Critical Review and Metacognitive Processes Model." *Perspectives on Psychological Science* 10 (5): 599–617.

Beveridge, Madeleine, and Martin Pickering. 2013. "Perspective Taking in Language: Integrating the Spatial and Action Domains." *Frontiers in Human Neuroscience* 7: 577.

Beyleveld, Deryck, Roger Brownsword, and Susan Wallace. 2002. "Clinical Ethics Committees: Clinician Support or Crisis Management?" *HEC Forum* 14 (1): 13–25.

Biggs, Antonia, Katherine Brown, and Diane G. Foster. 2020. "Perceived Abortion Stigma and Psychological Well-Being over Five Years after Receiving or Being Denied an Abortion." *PLoS One* 15 (1): e0226417.

Biggs, M. Antonia, Ushma D. Upadhyay, Charles E. McCulloch, and Diana G. Foster. 2017. Women's Mental Health and Well-Being 5 Years after Receiving or Being Denied an Abortion: A Prospective, Longitudinal Cohort Study. *JAMA Psychiatry* 74 (2): 169–178.

Billman, George E. 2020. "Homeostasis: The Underappreciated and Far too Often Ignored Central Organizing Principle of Physiology." *Frontiers in Physiology* 11: 200.

Bilodeau, Antoine, Luc Turgeon, Stephen White, and Ailsa Henderson. 2018. "Strange Bedfellows? Attitudes Toward Minority and Majority Religious Symbols in the Public Sphere." *Politics and Religion* 11 (2): 309–333.

Blain-Moraes, Stefanie, Eric Racine, and George A. Mashour. 2018. "Consciousness and Personhood in Medical Care." *Frontiers in Human Neuroscience* 12: 306.

Blair, Robert J. R. 1995. "A Cognitive Developmental Approach to Morality: Investigating the Psychopath." *Cognition* 57 (1): 1–29.

Blodgett, Jeffrey G., Long-Chuan Lu, Gregory M. Rose, and Scott J. Vitell. 2001. "Ethical Sensitivity to Stakeholder Interests: A Cross-Cultural Comparison." *Journal of the Academy of Marketing Science* 29 (2): 190–202.

Bloom, Paul. 2016. *Against Empathy: The Case for Rational Compassion*. New York, NY: Ecco Books.

Blumer, Herbert. 1969. *Symbolic Interactionism: Perspective and Method*. Berkeley, CA: University of California Press.

Bogossian, Aline, Annette Majnemer, and Eric Racine. 2020. "Contextualized Autonomy in Transitional Care for Youth with Neurologic Conditions: The Role of the Pediatric Neurologist." *Journal of Child Neurology* 35 (8): 536–542.

Bogossian, Aline, Jan W. Gorter, and Eric Racine. 2018. "Protocol for a Scoping Review about Ethics in Transition Programmes for Adolescents and Young Adults with Neurodisabilities." *BMJ Open* 8 (8): e020914.

Bollard, Mara. 2013. "Psychopathy, Autism and Questions of Moral Agency." In *Ethics and Neurodiversity*, edited by Chris D. Herrera and Alexander Perry, 238–259. Newcastle upon Tyne: Cambridge Scholars Publishing.

Borry, Pascal, Paul Schotsmans, and Kris Dierickx. 2005. "The Birth of the Empirical Turn in Bioethics." *Bioethics* 19 (1): 49–71.

Bouchard, Gérard, and Charles Taylor. 2008. *Commission de consultation sur les pratiques d'accommodement reliées aux différences culturelles*. Québec, Canada: Gouvernement du Québec.

Bouchier-Hayes, David. 2005. "'Life Can Only Be Understood by Looking Backward; But It Must Be Lived Looking Forward'—Soren Kierkegaard (1813–1855)." *Irish Journal of Medical Science* 174 (1): 4–4.

Bourdieu, Pierre. 1977. *Outline of a Theory of Practice*. Cambridge, UK: Cambridge University Press.

Bowden, Olivia. 2022. "Here's the Reality of Abortion Access in Canada." *Toronto Star*, May 3, 2022. https://www.thestar.com/news/canada/here-s-the-reality-of-abortion-access-in-canada/article_51ef7ec4-4ec7-5b54-89e7-e6cbb34d61d9.html.

Boyer, Yvonne. 2017. "Healing Racism in Canadian Health Care." *CMAJ* 189 (46): E1408–E1409.

Boyer, Yvonne, and Judith Bartlett. 2017. "External Review: Tubal Ligation in the Saskatoon Health Region: The lived Experience of Aboriginal Women." Saskatoon, Canada: Saskatoon Health Region. https://senatorboyer.ca/wp-content/uploads/2021/09/Tubal-Ligation-in-the-Saskatoon-Health-Region-the-Lived-Experience-of-Aboriginal-Women-Boyer-and-Bartlett-July-11-2017.pdf.

Brady, F. Neil. 1985. "A Janus-Headed Model of Ethical Theory: Looking Two Ways at Business/Society Issues." *Academy of Management Review* 10 (3): 568–579.

Brady, F. Neil, and Gloria E. Wheeler. 1996. "An Empirical Study of Ethical Predispositions." *Journal of Business Ethics* 15 (9): 927–940.

Brinkmann, Svend. 2011. "Dewey's Neglected Psychology: Rediscovering His Transactional Approach." *Theory & Psychology* 21 (3): 298–317.

Brown, James M., Alison L. Kitson, and Terence J. McKnight. 1992. "Moral Deliberation." In *Challenges in Caring: Explorations in Nursing and Ethics,* edited by James M. Brown, Alison L. Kitson, and Terrence J. McKnight, 13–29. New York, NY: Chapman & Hall/Springer Science+Business Media.

Brown, Scott C., and Jeffrey A. Greene. 2006. "The Wisdom Development Scale: Translating the Conceptual to the Concrete." *Journal of College Student Development* 47 (1): 1–19.

Brown, Todd A., John A. Sautter, Levente Littvay, Alberta C. Sautter, and Brennen Bearnes. 2010. "Ethics and Personality: Empathy and Narcissism as Moderators of Ethical Decision Making in Business Students." *Journal of Education for Business* 85 (4): 203–208.

Browne, Annette J., Emily Hill, Barry Lavallee, Josée Lavoie, and Mary J. Logan Mcallum. 2017. Out of Sight. *Brian Sinclair Working Group,* September 2017. https://professionals.wrha.mb.ca/old/professionals/files/OutOfSight.pdf

Bryden, Joan. 2021. "Government Agrees Mentally Ill Should Have Access to Assisted Dying—in 2 Years." *CBC News,* February 23, 2021. https://www.cbc.ca/news/politics/maid-bill-senate-amendments-1.5924163.

Budig, Kirsten, Julia Diez, Paloma Conde, Marta Sastre, Mariano Hernán, and Manuel Franco. 2018. "Photovoice and Empowerment: Evaluating the Transformative Potential of a Participatory Action Research Project." *BMC Public Health* 18 (1): 432.

Bury, Michael. 1982. "Chronic Illness as Biographical Disruption." *Sociology of Health & Illness* 4 (2): 167–182.

Buscemi, Franco. 2019. "Kivalliq Hall Recognized as a Residential School." *Nunavut Tunngavik,* April 15, 2019. https://www.tunngavik.com/2019/04/15/kivalliq-hall-recognized-as-a-residential-school/.

Butterfield, Kenneth, Linda K. Trevino, and Gary R. Weaver. 2000. "Moral Awareness in Business Organizations: Influences of Issue-Related and Social context Factors." *Human Relations* 53 (7): 981–1018.

Callahan, Daniel. 1973. "Bioethics as a Discipline." *Hastings Center Studies* 1 (1): 66–73.

Callahan, Daniel. 1994. "Bioethics: Private Choice and Common Good." *Hastings Center Report* 24 (3): 28–31.

Callahan, Daniel. 2005a. "Bioethics and the Culture Wars." *Cambridge Quarterly of Healthcare Ethics* 14 (4): 424–431.

Callahan, Daniel. 2005b. "Conservatives, Liberals, and Medical Progress." *New Atlantis* 10: 3–16.

Campbell, James. 1988. "Dewey's Understanding of Marx and Marxism." In *Context Over Foundation: Dewey and Marx*, edited by William J. Gavin. Dordrecht, 119–145. Netherlands: Springer.

Campbell, Robert L., and John Chambers Christopher. 1996. "Moral Development Theory: A Critique of Its Kantian Presuppositions." *Developmental Review* 16 (1): 1–47.

Canadian Association of Pediatric Health Centres (CAPHC), National Transitions Community of Practice. 2016. *A Guideline for Transition from Paediatric to Adult Health Care for Youth with Special Health Care Needs: A National Approach*. Ottawa, Canada: Canadian Association of Pediatric Health Centres. https://www.childhealthbc.ca/sites/default/files/caphc_transition_to_adult_health_care_guideline_may_2017.pdf.

Caplan, Arthur L. 1980. "Ethical Engineers Need not Apply: The State of Applied Ethics Today." *Science, Technology, & Human Values* 5 (4): 24–32.

Capps, John. 2003. "Pragmatism and the McCarthy Era." *Transactions of the Charles S. Peirce Society* 39 (1): 61–76.

Card, Claudia. 2003. "Genocide and Social Death." *Hypatia* 18 (1): 63–79.

Carden, Stephen. 2006. *Virtue Ethics: Dewey and MacIntyre*. London, UK: Continuum.

Carter, Adrian, Perry Bartlett, and Wayne Hall. 2009. "Scare-Mongering and the Anticipatory Ethics of Experimental Technologies." *American Journal of Bioethics* 9 (5): 47–48.

Cascio, M. Ariel, and Eric Racine. 2018. "Person-Oriented Research Ethics: Integrating Relational and Everyday Ethics in Research." *Accountability in Research* 25 (3): 170–197.

Cascio, M. Ariel, and Eric Racine. 2019. *Research Involving Participants with Cognitive Disability and Differences: Ethics, Autonomy, Inclusion, and Innovation*. New York, NY: Oxford University Press.

Cascio, M. Ariel, Jonathan A. Weiss, and Eric Racine. 2020. "Empowerment in Decision-Making for Autistic People in Research." *Disability & Society* 36 (1): 100–144.

Caspary, William R. 1991. "Ethical Deliberation as Dramatic Rehearsal: John Dewey's Theory." *Educational Theory* 41 (2): 175–188.

Cele, Lindile, Sarah S. Willen, Maydha Dhanuka, and Emily Mendenhall. 2021. "Ukuphumelela: Flourishing and the Pursuit of a Good Life, and Good Health, in Soweto, South Africa." *SSM—Mental Health* 1: 100022.

Chambliss, William J. 1993. "Common Ground in Aristotle's and Dewey's Theories of Conduct." *Educational Theory* 43 (3): 249–260.

Changeux, Jean-Pierre. 1983. *L'homme neuronal*. Paris, France: Hachette.

Charon, Rita. 1993. "The Narrative Road to Empathy." In *Empathy and the Practice of Medicine: Beyond Pills and the Scalpel*, edited by Howard M. Spiro, Enid Peschel, and Deborah St. James, 147–159. New Haven, CT: Yale University Press.

Charon, Rita. 2001. "Narrative Medicine: A Model for Empathy, Reflection, Profession, and Trust." *JAMA* 286 (15): 1897–1902.

Charon, Rita, and Martha Montello. 1999. "Framing the Case: Narrative Approaches for Healthcare Ethics Committees." *HEC Forum* 11 (1): 6–15.

Chen, Yen-Yuan, and Yu-Chieh Chen. 2008. "Evaluating Ethics Consultation: Randomised Controlled Trial is not the Right Tool." *Journal of Medical Ethics* 34 (8): 594–597.

Chevalier, Jacques M., and Daniel J. Buckles. 2013. *Participatory Action Research: Theory and Methods for Engaged Inquiry*. London, UK: Routledge.

Chirkov, Valery, Richard M. Ryan, Youngmee Kim, and Ulas Kaplan. 2003. "Differentiating Autonomy from Individualism and Independence: A Self-Determination Theory Perspective on Internalization of Cultural Orientations and Well-Being." *Journal of Personality and Social Psychology* 84 (1): 97–110.

Chochinov, Harvey M. 2012. *Dignity Therapy: Final Words for Final Days*. Oxford, UK: Oxford University Press.

Chomsky, Noam. 2012. *Occupy*. London, UK: Penguin.

Cialdini, Robert B., Stephanie L. Brown, Brian P. Lewis, C. Luce, and Steven L. Neuberg. 1997. "Reinterpreting the Empathy-Altruism Relationship: When One into One Equals Oneness." *Journal of Personality and Social Psychology* 73 (3): 481–494.

Cimpian, Andrei, and Erika Salomon. 2014. "The Inherence Heuristic: An Intuitive Means of Making Sense of the World, and a Potential Precursor to Psychological Essentialism." *Behavioral and Brain Sciences* 37 (5): 461–480.

Clark, David A., Mario Biggeri, and Alexandre A. Frediani. 2019. *The Capability Approach, Empowerment and Participation: Concepts, Methods and Applications*. London, UK: Palgrave-Macmillan.

Clouser, Danner K. 1995. "Common Morality as an Alternative to Principlism." *Kennedy Institute of Ethics Journal* 5 (3): 219–236.

Cohan, Audrey, and Charles F. Howlett. 2017. "John Dewey and His Evolving Perceptions of Race Issues in American Democracy." *Faculty Works: Education* 17 (1): 16–22.

Cohen, Eric. 2006. "Conservative Bioethics and the Search for Wisdom." *Hastings Center Report* 36 (1): 44–56.

Cohen, Shlomo. 2013. "Nudging and Informed Consent." *American Journal of Bioethics* 13 (6): 3–11.

Collège des médecins du Québec. 2017. *L'aide médicale à mourir: guide d'exercice et lignes directrices pharmacologiques*. Montréal, Canada: Collège des médecins du Québec.

Conway, Jane R., Caroline Catmur, and Geoffrey Bird. 2019. "Understanding Individual Differences in Theory of Mind via Representation of Minds, not Mental States." *Psychonomic Bulleting and Review* 26 (3): 798–812.

Conway, Martin A. 2009. "Episodic Memories." *Neuropsychologia* 47 (11): 2305–2313.

Corrigan, Oonagh. 2003. "Empty Ethics: The Problem with Informed Consent." *Sociology of Health and Illness* 25 (7): 768–792.

Corrigan, Patrick. 2012. *A Toolkit for Evaluating Programs Meant to Erase the Stigma of Mental Illness*. Chicago, IL: Illinois Institute of Technology. https://www.montefiore.org/documents/Evaluating-Programs-Meant-to-Erase-the-Stigma-of-Mental-Illness.pdf

Cotkin, George. 1985. "William James and the Cash-Value Metaphor." *ETC: A Review of General Semantics* 42 (1): 37–46.

Craze, Gareth. 2020. "Corporate Social Responsibility and Dehumanization." In *Organizational Neuroethics*, edited by Joé T. Martineau and Eric Racine, 131–146. Cham, Switzerland: Springer.

Crespo, Ricardo F., and Belén Mesurado. 2015. "Happiness Economics, Eudaimonia and Positive Psychology: From Happiness Economics to Flourishing Economics." *Journal of Happiness Studies* 16 (4): 941–946.

Crick, Nathan, and Graham D. Bodie. 2016. ""I Says to Myself, Says I": Charles Sanders Peirce on the Components of Dialogue." *Communication Theory* 26 (3): 273–289.

Csikszentmihalyi, Mihaly. 1975. *Beyond Boredom and Anxiety: Experiencing Flow in Work and Play*. San Francisco, CA: Jossey-Bass.

Csikszentmihalyi, Mihaly. 1990. *Flow: The Psychology of Optimal Experience*. New York, NY: Harper and Row.

Czeżowski, Tadeusz. 1953. "Ethics as an Empirical Science." *Philosophy and Phenomenological Research* 14 (2): 163–171.

Dahl, Robert A., and Ian Shapiro. 2015. *On Democracy*. New Haven, CT: Yale University Press.

Dalley, Tessa. 1984. *Art As Therapy: An Introduction to the Use of Art as a Therapeutic Technique*. New York, NY: Routledge.

Damasio, Antonio R. 1994. *Descartes' Error*. New York, NY: Penguin Putnam.

Darley, John M., and Daniel C. Batson. 1973. ""From Jerusalem to Jericho": A Study of Situational and Dispositional Variables in Helping Behavior." *Journal of Personality and Social Psychology* 27 (1): 100–108.

Darwall, Stephen L. 1983. *Impartial Reason*. Ithaca, NY: Cornell University Press.

Darwin, Charles. 1859. *On the Origin of Species by Means of Natural Selection, or the Preservation of Favoured Races in the Struggle for Life*. London, UK: John Murray.

Darwin, Charles. 1872. *The Expression of the Emotions in Man and Animals*. London, UK: John Murray.

Davies, Rachel, Jonathan Ives, and Michael Dunn. 2015. "A Systematic Review of Empirical Bioethics Methodologies." *BMC Medical Ethics* 16 (15). https://doi.org/10.1186/s12910-015-0010-3.

Davies, Robert D., and Madeline E. Davies. 2020. "The (Slow) Depathologizing of Gender Incongruence." *Journal of Nervous and Mental Disease* 208 (2): 152–154.

Davis, Dena S. 2014. "Alzheimer Disease and Pre-emptive Suicide." *Journal of Medical Ethics* 40 (8): 543–549.

Davis, Mortimer, and Caroline Tait. 2003. *Fetal Alcohol Syndrome Among Aboriginal People in Canada: Review and Analysis of the Intergenerational Links to Residential Schools*. Ottawa, Canada: Aboriginal Healing Foundation. https://www.publicsafety.gc.ca/lbrr/archives/cn76093876-eng.pdf.

de Beauvoir, Simone. 1947/2011. *Pour une morale de l'ambiguïté*. Paris, France: Gallimard.

De Jaegher, Hanne, and Ezequiel Di Paolo. 2007. "Participatory Sense-Making: An Enactive Approach to Social Cognition." *Phenomenology and the Cognitive Sciences* 6 (4): 485–507.

Decety Jean, and Jason M. Cowell. 2014. "The Complex Relation between Morality and Empathy." *Trends in Cognitive Science* 18 (7): 337–339.

Decety, Jean, and Jason M. Cowell. 2015. "Empathy, Justice, and Moral Behavior." *AJOB Neuroscience* 6 (3): 3–14.

Deci, Edward L., and Richard M. Ryan. 1985. *Intrinsic Motivation and Self-Determination in Human Behavior*. New York, NY: Springer.

Deci, Edward L., and Richard M. Ryan. 2000. "The "What" and "Why" of Goal Pursuits: Human Needs and Self-Determination Theory in Human Behavior." *Psychological Inquiry* 11 (4): 227–268.

Demertzi, Athena, Eric Racine, Marie-Aurélie Bruno, Didier Ledoux, Olivia Gosseries, Audrey Vanhaudenhuyse, Marie Thonnard, Andrea Soddu, Gustave Moonen, and Steven Laureys. 2012. "Pain Perception in Disorders of Consciousness: Neuroscience, Clinical Care, and Ethics in Dialogue." *Neuroethics* 6 (1): 37–50.

Dennett, Daniel. 1981. *Brainstorms: Philosophical Essays on Mind and Psychology*. Cambridge, MA: MIT Press.

Dennis, Lawrence J., and Francis Powers. 1974. "Dewey, Maslow, and Consummatory Experience." *Journal of Aesthetic Education* 8 (4): 51–63.

DeRenzo, Evan G., and Michelle Strauss. 1997. "A Feminist Model for Clinical Ethics Consultation: Increasing Attention to Context and Narrative." *HEC Forum* 9 (3): 212–227.

Dewey, John. 1896. "The Reflex Arc Concept in Psychology." *Psychological Review* 3 (4): 357–370.

Dewey, John. 1922. *Human Nature and Conduct: An Introduction to Social Psychology*. New York, NY: Holt.

Dewey, John. 1925. *Experience and Nature*. Chicago, IL: Open Court.

Dewey, John. 1934/1958. *Art as Experience*. New York, NY: Capricorn Books.

Dewey, John. 1938/1986. *Experience and Education*. New York, NY: Taylor & Francis.

Dewey, John. 1941. "Propositions, Warranted Assertability, and Truth." *Journal of Philosophy* 38 (7): 169–186.

Dewey, John. 1948. "Common Sense and Science: Their Respective Frames of Reference." *Journal of Philosophy* 45 (8): 197–208.

Dewey, John. 1966. "Three Independent Factors in Morals." *Educational Theory* 16 (1): 198–209.

Dewey, John. 1983. "Human Nature and Conduct." In *John Dewey: The Middle Works*, edited by Jo Boydston, 146–153. Carbondale, IL: Southern Illinois University Press.

Dewey, John. 1985. "Ethics." In *John Dewey: The Later Works*, edited by Jo Boydston, 262–310. Carbondale, IL: Southern Illinois University Press.

Dickert, Neal, Nir Eyal, Sara F Goldkind, Christine Grady, Steven Joffe, Bernard Lo, Franklin G Miller, Rebecca D Pentz, Robert Silbergleit, Kevin P Weinfurt, David Wendler, Scott Y H Kim 2017. "Re-framing Consent for Clinical Research: A Function-Based Approach." *American Journal of Bioethics* 17 (12): 3–11.

Diderot, Denis. 1754/2005. *Pensées sur L'interprétation de la nature*. Paris, France: GF Flammarion.

Dion-Labrie, Marianne. 2009. "La méthode des scénarios." *Comité provincial d'éthique clinique en réadaptation de l'Association des établissement de réadaptation en déficience physique du Québec (AERDPQ)*. https://docplayer.fr/12354362-La-methode-des-scenarios.html

Dombrowski, Daniel. 1984. *The Philosophy of Vegetarianism*. Amherst: University of Massachusetts Press.

Doris, John, Stephen Stich, Jonathan Philips, and Lachlan Walmsley. 2020. "Moral Psychology: Empirical Approaches." *Stanford Encyclopedia of Philosophy Spring*, April 19. https://plato.stanford.edu/archives/spr2020/entries/moral-psych-emp/.

Doucet, Hubert. 1994. Le développement des morales, des législations et des codes, garder le dialogue ouvert et la conscience inquiète. In *Élargir les horizons: Perspectives scientifiques sur l'intégration sociale*, edited by Office des personnes handicapées du Québec, 135–141. Sainte Foy, Canada: Éditions Multimondes.

Doucet, Hubert. 2001. "Historical Context: Deliberation and Methodology in Bioethics." In *Ethical Deliberation in Multiprofessional Health Care Teams*, edited by Hubert Doucet, Jean-Marc Larouche, and Kenneth R. Melchin, 13–34. Ottawa, Canada: Ottawa University Press.

Doucet, Hubert. 2004. L'analyse de cas: Une pluralité d'approches pour une diversité de situations. Colloque de bioéthique de l'Université de Montréal, May 7, 2004.

Doucet, Hubert. 2015. *La mort médicale: est-ce humain?* Montréal, Canada: Médiaspaul.

Doucet, Hubert, Jean-Marc Larouche, and Kenneth R. Melchin. 2001. *Ethical Deliberation in Multiprofessional Health Care Teams*. Ottawa, Canada: University of Ottawa Press.

Drescher, Jack. 2015. "Out of DSM: Depathologizing Homosexuality." *Behavioral Sciences* 5 (4): 565–575.

Druckman, James N. 2001. "Using Credible Advice to Overcome Framing Effects." *Journal of Law, Economics, and Organization* 17 (1): 62–82.

Dubljević, Veljko, and Eric Racine. 2013. "Judging Deeds, not Psychopaths." *AJOB Neuroscience* 4 (2): 33–34.

Dubljević, Veljko, and Eric Racine. 2014. "The ADC of Moral Judgment: Opening the Black Box of Moral Intuitions with Heuristics about Agents, Deeds, and Consequences." *AJOB Neuroscience* 5 (4): 3–20.

Dubljević, Veljko, Sebastian Sattler, and Eric Racine. 2018a. "Deciphering Moral Intuition: How Agents, Deeds, and Consequences Influence Moral Judgment." *PLoS One* 13 (10): e0204631.

Dubljević, Veljko, Sebastian Sattler, and Eric Racine. 2018b. "Preferences for Precepts Implied in Moral Theories Instrument (PPIMT)" [Database record]. *APA PsycTests.*

Duffy, Bobby, and Gideon Skinner. 2021. "Culture Wars Around the World: How Countries Perceive Divisions." June 2021. https://www.kcl.ac.uk/policy-institute/assets/culture-wars-around-the-world-how-countries-perceive-divisions.pdf.

Duplantie, Andrée, Guy Durand, Yvon Laroche, and Danielle Laudy. 2000. *Histoire de l'éthique médicale et infirmière.* Montréal, Canada: Presses de l'Université de Montréal.

Durand, Guy. 1999. *Introduction générale à la bioéthique: Histoire, concepts et outils.* Montréal, Canada: Fides-Cerf.

Durkheim, Émile. 1893. *De la division du travail social.* Paris, France: Librairie Félix Alcan.

Durkheim, Émile. 1925. *L'éducation morale.* Paris, France: Librairie Félix Alcan.

Dutton, Jane, Laura M. Roberts, and Jeffrey Bednar. 2010. Prosocial Practices, Positive Identity, and Flourishing at Work. In *Applied Positive Psychology: Improving Everyday Life, Schools, Work, Health, and Society*, edited by Stewart Donaldson, Mihaly Csikszentmihalyi, and Jeanne Nakamura, 155–170. New York, NY: Routledge.

DuVal, Gordon, Leah Sartorius, Brian Clarridge, Gary Gensler, and Marion Danis. 2001. "What Triggers Requests for Ethics Consultations?" *Western Medical Journal* 175 (9): 24–30.

Dworkin, Gerald. 1988. *The Theory and Practice of Autonomy.* Cambridge, UK: Cambridge University Press.

Dworkin, Ronald. 1982. "Law as Interpretation." *Critical Inquiry* 9 (1): 179–200.

Dwornik, Isabel A. 2003. "A Comparison of Educational Theory: Alfred Adler and John Dewey." *The Journal of Individual Psychology* 59 (1): 52–71.

Edelman, Gerald M. 1989. *Remembered Present: A Biological Theory of Consciousness.* New York, NY: Basic Books.

Effing, M. 2009. The Origin and Development of Self-Help Literature in the United States: The Concept of Success and Happiness, an Overview." *Atlantis* 31 (2): 125–141.

Einstein, Albert. 1979/1997. *Comment je vois le monde.* Manchecourt, France: Champs Flammarion.

Eisenberg, Nancy, and Janet Strayer. 1987. *Empathy and its Development.* Cambridge, UK: Cambridge University Press.

El Haj, Mohamad, Karim Gallouj, and Pascal Antoine. 2019. "Autobiographical Recall as a Tool to Enhance the Sense of Self in Alzheimer's Disease." *Archives of Gerontology and Geriatrics* 82 (1): 28–34.

Emmons, Robert A. 2003. "Personal Goals, Life Meaning, and Virtue: Wellsprings of a Positive Life." In *Flourishing: Positive Psychology and the Life Well-Lived*, edited by Corey Keyes and Jonathan Haidt, 105–128. Washington, DC: American Psychological Association.

Epstein, Elizabeth G., and Ann B. Hamric. 2009. "Moral Distress, Moral Residue, and the Crescendo Effect." *Journal of Clinical Ethics* 20 (4): 330–342.

Erickson, Erik. 1968. *Identity: Youth and Crisis*. New York, NY: Norton.

Fahey, Gregory M. 2010. "The Idea of the Good in John Dewey and Aristotle." *Pragmatism and Neopragmatism* 3 (2): 201–226.

Fang, Zhuo, Wi Hoon Jung, Marc Korczykowski, Lijuan Luo, Kristin Prehn, Sihua Xu, John A. Detre, Joseph W. Kable, Diana C. Robertson, and Hengyi Rao. 2017. "Post-Conventional Moral Reasoning is Associated with Increased Ventral Striatal Activity at Rest and During Task." *Scientific Reports* 7 (1): 7105.

Faucher, Luc. 2016. "Revisionism and Moral Responsibility for Implicit Attitudes." In *Implicit Bias and Philosophy*, edited by Michael Brownstein and Jennifer Saul, 115–144. Oxford, UK: Oxford University Press.

Fennell, M. 2012. "Cognitive Behaviour Therapy for Depressive Disorders." In *New Oxford Textbook of Psychiatry*, edited by John R. Geddes, Nancy C. Andreasen, and Guy M. Goodwin, 1304–1312. New York, NY: Oxford University Press.

Fenton, Margaret. 1988. "Moral Distress in Clinical Practice: Implications for the Nurse Administrator." *Canadian Journal of Nursing Administration* 1 (3): 8–11.

Fesmire, Steven. 2003. *John Dewey and Moral Imagination*. Indianapolis, IN: Indiana University Press.

Fiester, Autumn M. 2015. "Weaponizing Principles: Clinical Ethics Consultations & the Plight of the Morally Vulnerable." *Bioethics* 29 (5): 309–315.

Fins, Joseph J. 2008. "A Leg to Stand On: Sir William Osler and Wilder Penfield's 'Neuroethics.'" *American Journal of Bioethics* 8 (1): 37–46.

Fins, Joseph J., Ali R. Rezai, and Benjamin D. Greenberg. 2006. "Psychosurgery: Avoiding an Ethical Redux While Advancing a Therapeutic Future." *Neurosurgery* 59 (4): 713–716.

Fischer, Peter, Joachim I. Krueger, Tobias Greitemeyer, Claudia Vogrincic, Andreas Kastenmüller, Dieter Frey, Moritz Heene, Magdalena Wicher, and Martina Kainbacher. 2011. "The Bystander-Effect: A Meta-Analytic Review on Bystander Intervention in Dangerous and Non-Dangerous Emergencies." *Psychological Bulletin* 137 (4): 517–537.

Fish, Stanley. 2015. "It's Not About Free Speech or Academic Freedom." *Huffington Post*, November 23 2015. https://www.huffpost.com/entry/its-not-about-free-speech_b_8634022.

Fishkin, James S. 1991. *Democracy and Deliberation: New Directions for Democratic Reform*. New Haven, CT: Yale University Press.

FitzGerald, Chloë, Angela Martin, Delphine Berner, and Samia Hurst. 2019. "Interventions Designed to Reduce Implicit Prejudices and Implicit Stereotypes in Real World Contexts: A Systematic Review." *BMC psychology* 7 (1): 1–12.

FitzGerald, Chloë, and Samia Hurst. 2017. "Implicit Bias in Healthcare Professionals: A Systematic Review." *BMC Medical Ethics* 18 (1): 19.

Flanagan, Owen. 1991. *Varieties of Moral Personality: Ethics and Psychological Realism*. Cambridge, MA: Harvard University Press.

Flanagan, Owen. 2009. Moral Science? Still Metaphysical After All These Years. In *Moral Personality, Identity and Character: Explorations in Moral Psychology*, edited by Darcia Narvaez and Daniel K. Lapsle, 52–78. Cambridge, UK: Cambridge University Press.

Fletcher, Joseph. 1966. *Situation Ethics: The New Morality*. Philadelphia: Westminster Press.

Flett, Gordon L., Kirk R. Blankstein, and Thomas R. Martin. 1995. "Procrastination, Negative Self-Evaluation, and Stress in Depression and Anxiety." In *Procrastination and Task Avoidance: Theory, Research, and Treatment*, edited by Joseph R. Ferrari, Judith L. Johnson and William G. McCown, 137–167. Boston: Springer US.

Foucart, Jean. 2012/2013. "Pragmatisme et transaction. La perspective de John Dewey." *Pensée Plurielle* 33–34 (2): 73–84.

Fouke, Daniel. 2009. "Democratic Deliberation and Moral Awareness." *Journal of Public Deliberation* 5 (1): Article 10.

Fowler, Marsha D. 2017. "Why the History of Nursing Ethics Matters." *Nursing Ethics* 24 (3): 292–304.

Frankl, Viktor E. 1966. "Self-Transcendence as a Human Phenomenon." *Journal of Humanistic Psychology* 6 (2): 97–106.

Frankl, Viktor E. 1985. *Man's Search for Meaning*. New York, NY: Washington Square Press/Pocket Books.

Friedman, Elliot M., and Carol D. Ryff. 2012. "Living Well with Medical Comorbidities: A Biopsychosocial Perspective." *Journals of Gerontology: Series B* 67 (5): 535–544.

Furnham, Adrian, Jean Ann Graham, and Michael Argyle. 1981. *Social Situations*. Cambridge, UK: Cambridge University Press.

Gadamer, Hans-Georg. 1989. *Truth and Method*. London, UK: Sheed and Ward.

Gadamer, Hans-Georg. 1991. *Plato's Dialectical Ethics: Phenomenological Interpretations Relating to the "Philebus"*. New Haven, CT: Yale University Press.

Galinsky, Adam D., and Gordon B. Moskowitz. 2000. "Perspective-Taking: Decreasing Stereotype Expression, Stereotype Accessibility, and In-Group Favoritism." *Journal of Personality and Social Psychology* 78 (4): 708–724.

Gallagher, Romayne, and Michael J. Passmore. 2020. "Canada Needs Equitable, Earlier Access to Palliative Care." *CMAJ* 192 (20): E559.

Gallagher, Shaun. 2014. "Pragmatic Interventions into Enactive and Extended Conceptions of Cognition." *Philosophical Issues* 24 (1): 110–126.

Gallagher, Shaun. 2022. "Phenomenology and Pragmatism: From the End to the Beginning." *European Journal of Pragmatism and American Philosophy* 4 (2).

Gaudine, Alice, and Linda Thorne. 2001. "Emotion and Ethical Decision-Making in Organizations." *Journal of Business Ethics* 31 (2): 175–187.

Geary, Aidan. 2017. "Ignored to Death: Brian Sinclair's Death Caused by Racism, Inquest Inadequate, Group Says." *CBC News*, September 18 2017. https://www.cbc.ca/news/canada/manitoba/winnipeg-brian-sinclair-report-1.4295996.

Gee, Gilbert C., and Chandra L. Ford. 2011. "Structural Racism and Health Inequities: Old Issues, New Directions." *Du Bois Review* 8 (1): 115–132.

Geertz, Clifford. 1973. "Thick Description: Toward an Interpretive Theory of Culture." In *The Interpretation of Cultures: Selected Essays*, 3–30. New York, NY: Basic Books.

Gelenberg, Alan J. 2010. "American Psychiatric Association Practice Guidelines for the Treatment of Patients with Major Depressive Disorder." *American Journal of Psychiatry* 167 (Suppl): 9.

Giacino, Joseph T., Stephen Ashwal, Nancy Childs, Ronald Cranford, Bryan Jennett, Douglas I. Katz, James P. Kelly, Jay H. Rosenberg, John Whyte, Ross Zafonte, and Nathan D. Zasler. 2002. "The Minimally Conscious State: Definition and Diagnostic Criteria." *Neurology* 58 (3): 349–353.

Gibson, James J. 1977. "The Theory of Affordances." In *Perceiving, Acting, and Knowing: Toward an Ecological Psychology*, edited by Robert Shaw and John Bransford, 67–82. Hillsdale, NJ: Lawrence Erlbaum.

Gilligan, Carol. 2001. *In a Different Voice. Psychological Theory and Women's Development.* Cambridge, MA: Harvard University Press.

Glannon, Walter. 2019. *The Neuroethics of Memory: From Total Recall to Oblivion.* New York, NY: Cambridge University Press.

Gostin, Larry O. 1980. "Ethical Considerations of Psychosurgery: The Unhappy Legacy of the Prefrontal Lobotomy." *Journal of Medical Ethics* 6 (1): 149–156.

Gouinlock, James. 1975. "John Dewey's Philosophy of Value." *Revue philosophique de la France et de l'étranger* 165 (3):367.

Gouinlock, James. 1978. "Dewey's Theory of Moral Deliberation." *Ethics* 88 (3): 218–228.

Gouinlock, James. 2002. *The Moral Writings of John Dewey*. Amherst, NY: Prometheus Books.

Government of Canada. 2021. *Canada's New Medical Assistance in Dying (MAID) Law*. Ottawa, Canada: Department of Justice Canada. https://www.canada.ca/en/department-justice/news/2021/03/new-medical-assistance-in-dying-legislation-becomes-law.html.

Gracia, Diego. 2001. "Moral Deliberation: The Role of Methodologies in Clinical Ethics." *Medicine, Health Care and Philosophy* 4 (2): 223–232.

Gracia, Diego. 2003. "Ethical Case Deliberation and Decision Making." *Medicine, Health Care and Philosophy* 6 (3): 227–233.

Grant, Colin. 2002. "Whistle Blowers: Saints of Secular Culture." *Journal of Business Ethics* 39 (4): 391–399.

Gray, Kurt, Anne T. Knickman, and Daniel M. Wegner. 2011. "More Dead than Dead: Perceptions of Persons in the Vegetative State." *Cognition* 121 (1): 275–280.

Greene, Jeffrey A., and Scott C. Brown. 2009. "The Wisdom Development Scale: Further Validity Investigations." *International Journal of Aging and Human Development* 68 (4): 289–320.

Griffin, Donald R. 1978. "Prospects for a Cognitive Ethology." *Behavioral and Brain Sciences* 4 (1): 527–538.

Griswold, Charles L., Jr. 1999. *Adam Smith and the Virtues of Enlightenment*. Cambridge, UK: Cambridge University Press.

Gross, Neil. 2009. "A Pragmatist Theory of Social Mechanisms." *American Sociological Review* 74 (3): 358–379.

Gross, Olivia, Rémi Gagnayre, and Pierre Lombrail. 2020. "L'évaluation des projets d'accompagnement à l'autonomie en santé: pourquoi privilégier une approche pragmatiste?" *Santé publique* 32 (4): 301–313.

Habermas, Jürgen. 1973. *La technique et la science comme "idéologie."* Paris, France: Gallimard.

Habermas, Jürgen. 1990. *The Philosophical Discourse of Modernity: Twelve Lectures*. Cambridge, MA: MIT Press.

Habermas, Jürgen. 1991. *Moral Consciousness and Communicative Action*. Cambridge, MA: MIT Press.

Habermas, Jürgen. 1997. *Between Facts and Norms: Contributions to a Discourse Theory of Law and Democracy*. Cambridge, MA: MIT Press.

Habermas, Jürgen. 1999. *De l'éthique de la discussion*. Paris, France: Flammarion.

Hadot, Pierre. 2002. *La philosophie comme manière de vivre*. Paris, France: Albin-Michel.

Hagner, Michael. 2001. "Cultivating the Cortex in German Neuroanatomy." *Science in Context* 14 (4): 541–563.

Haidt, Jonathan. 2001. "The Emotional Dog and its Rational Tail: A Social Intuitionist Approach to Moral Judgment." *Psychological Review* 108 (4): 814–834.

Halberstam, Joshua. 1988. "From Kant to Auschwitz." *Social Theory and Practice* 14 (1): 41–54.

Hall, Joanne M., and Becky Fields. 2015. ""It's Killing Us!" Narratives of Black Adults about Microaggression Experiences and Related Health Stress." *Global Qualitative Nursing Research* 2: 2333393615591569–2333393615591569.

Halpern, Jodi. 1993. "Empathy: Using Resonance Emotions in the Service of Curiosity." In *Empathy and the Practice of Medicine: Beyond Pills and the Scalpel*, edited by Mary G. McCreacurnen, Howard M. Spiro, Enid Peschel, and Deborah St. James, 160–173. New Haven, CT: Yale University Press.

Hammell, Karen W. 2004. "Dimensions of Meaning in the Occupations of Daily Life." *Canadian Journal of Occupational Therapy* 71 (5): 296–305.

Hanson, Erin, Daniel P. Gamez, and Alexa Manuel. 2020. "The Residential School System." *Indigenous Foundations*, September. https://indigenousfoundations.arts.ubc.ca/the_residential_school_system/.

Hansson, Sven O. 2020. "Neuroethics for Fantasyland or for the Clinic? The Limitations of Speculative Ethics." *Cambridge Quarterly of Healthcare Ethics* 29 (4): 630–641.

Harris, Janet. 2018. "Reviewing the Effectiveness of Participatory Health Research: Challenges and Possible Solutions." In *Participatory Health Research: Voices from Around the World*, edited by Michael T. Wright and Krystyna Kongats, 71–91. Cham, Switzerland: Springer International Publishing.

Hartman, Laura, Giulia Inguaggiato, Guy Widdershoven, Annelijn Wensing-Kruger, and Bert Molewijk. 2020. "Theory and Practice of Integrative Clinical Ethics Support: A Joint Experience within Gender Affirmative Care." *BMC Med Ethics* 21 (1): 79.

Hartman, Laura, Suzanne Metselaar, Guy Widdershoven, and Bert Molewijk. 2019. "Developing a 'Moral Compass Tool' Based on Moral Case Deliberations: A Pragmatic Hermeneutic Approach to Clinical Ethics." *Bioethics* 33 (9): 1012–1021.

Hedgecoe, Adam. 2010. "Bioethics and the Reinforcement of Socio-Technical Expectations." *Social Studies of Science* 40 (2): 163–186.

Hegel, Gerg W. F. 1993. *Phénoménologie de l'esprit*. Saint-Amand, France: Gallimard.

Heidegger, Martin. 1927/1996. *Being and Time*. Albany, NY: State University of New York Press.

Helliwell, John F., Richard Layard, Jeffrey Sachs, and Jan-Emmanuel De Neve. 2020. *World Happiness Report 2020*. New York, NY: Sustainable Development Solutions Network.

Herman, Amy E. 2021. *Fixed: How to Perfect the Fine Art of Problem Solving*. New York, NY: Harper Wave.

Higgins, Chris. 2010. "A Question of Experience: Dewey and Gadamer on Practical Wisdom." *Journal of Philosophy of Education* 44 (2–3): 301–333.

Hirsch, Jameson K., Paul R. Duberstein, Kenneth R. Conner, Marnin J. Heisel, Anthony Beckman, Nathan Franus, and Yeates Conwell. 2006. "Future Orientation and Suicide Ideation and Attempts in Depressed Adults Ages 50 and Over." *American Journal of Geriatric Psychiatry* 14 (9): 752–757.

Hirsch, Jameson K., Paul R. Duberstein, Kenneth R. Conner, Marnin J. Heisel, Anthony Beckman, Nathan Franus, and Yeates Conwell. 2007. "Future Orientation Moderates the Relationship between Functional Status and Suicide Ideation in Depressed Adults." *Depression and Anxiety* 24 (3): 196–201.

Hitler, Adolf. 1925. *Mein Kampf*. Germain Reich: Franz Eher Nachfolger GmbH.

Hitlin, Steven. 2003. "Values as the Core of Personal Identity: Drawing Links between Two Theories of Self." *Social Psychology Quarterly* 66 (2): 118–137.

Hoffman, Martin L. 1981. "Is Altruism Part of Human Nature?" *Personality Processes and Individual Differences* 40 (1): 121–137.

Hoffmaster, Barry. 1992. "Can Ethnography Save the Life of Medical Ethics?" *Social Science and Medicine* 35 (12): 1421–1431.

Hoffmaster, Barry. 2018. "From Applied Ethics to Empirical Ethics to Contextual Ethics." *Bioethics* 32 (2): 119–125.

Hoffmaster, Barry, and Cliff Hooker. 2009. "How Experience Confronts Ethics." *Bioethics* 23 (4): 214–225.

Hoffmeyer, Jesper. 2011. "The Natural History of Intentionality. A Biosemiotic Approach." In *The Symbolic Species Evolved*, edited by Theresa Schilhab, Frederik Stjernfelt, and Terrence Deacon, 97–116. Dordrecht, Netherlands: Springer.

Hojat, Mohammadreza, Michael J. Vergare, Kaye Maxwell, Geroge Brainard, Steven K. Herrine, Gerald A. Isenberg, Jon Veloski, and Joseph S. Gonnella. 2009. "The Devil Is in the Third Year: A Longitudinal Study of Erosion of Empathy in Medical School." *Academic Medicine* 84 (9): 1182–1191.

Hollinger, David A. 1980. "The Problem of Pragmatism in American History." *Journal of American History* 67 (1): 88–107.

Holt, Svetlana, and Joan Marques. 2012. "Empathy in Leadership: Appropriate or Misplaced? An Empirical Study on a Topic That is Asking for Attention." *Journal of Business Ethics* 105 (1): 95–105.

Honderich, Ted. 1995. *The Oxford Companion to Philosophy*. Oxford, UK: Oxford University Press.

Honneth, Axel. 1992. "Integrity and Disrespect: Principles of a Conception of Morality Based on the Theory of Recognition." *Political Theory* 20 (2): 187–201.

Huber, Beatrice, Kuehlmeyer, Katja. 2011. "Perspectives of Family Caregivers on the Vegetative State." In *Vegetative State: A Paradigmatic Problem of Modern Society*, edited by Ralf J. Jox, Katja Kuehlmeyer, Georg Marckmann, Eric Racine, 97–109. Muenster, Germany: LIT Verlag.

Hughes, Claire, and Judy Dunn. 2000. "Hedonism or Empathy?: Hard-to-Manage Children's Moral Awareness and Links with Cognitive and Maternal Characteristics." *British Journal of Developmental Biology* 18 (2): 227–245.

Hume, David. 1739/1975. *A Treatise of Human Nature*. Oxford, UK: Clarendon Press.

Hunt, Matthew R., and Franco A. Carnevale. 2011. "Moral Experience: A Framework for Bioethics Research." *Journal of Medical Ethics* 37 (11): 658–662.

Hunter, James D. 1992. *Culture Wars: The Struggle to Define America*. New York, NY: Basic Books.

Hurst, Samia A., and Alex Mauron. 2003. "Assisted Suicide and Euthanasia in Switzerland: Allowing a Role for Non-Physicians." *BMJ* 326 (7383): 271.

Ibbitson, John. 2017. "Your Rights, My Rights: Academic Freedom Faces Off with a Clarion Call for Safety on Campus." *Globe and Mail,* June 10. https://www.theglobeandmail.com/news/national/education/free-speech-university-campus/article35174756/.

Illes, Judy, and Eric Racine. 2005. "Neuroethics: A Dialogue on a Continuum from Tradition to Innovation." *American Journal of Bioethics* 5 (2): W3–W4.

Inguaggiato, Giulia, Guy A. Widdershoven, and Suzanne Metselaar. 2021. "Teaching Theories as Tools: A Pragmatist Approach to Ethics Education." *Bioethica Forum* 14 (1): 5–19.

Inguaggiato, Giulia, Suzanne Metselaar, Guy Widdershoven, and Bert Molewijk. 2019a. "Clinical Ethics Expertise as the Ability to Co-Create Normative Recommendations by Guiding a Dialogical Process of Moral Learning." *American Journal of Bioethics* 19 (11): 71–73.

Inguaggiato, Giulia, Suzanne Metselaar, Rouven Porz, and Guy Widdershoven. 2019b. "A Pragmatist Approach to Clinical Ethics Support: Overcoming the Perils of Ethical Pluralism." *Medicine, Health Care and Philosophy* 22 (3): 427–438.

Ingvar, David H. 1985. "'Memory of the Future': An Essay on the Temporal Organization of Conscious Awareness." *Human Neurobiology* 4 (3): 127–136.

Ives, Jonathan. 2014. "A Method of Reflexive Balancing in a Pragmatic, Interdisciplinary and Reflexive Bioethics." *Bioethics* 28 (6): 302–312.

Jacobs, David C. 2004. "A Pragmatist Approach to Integrity in Business Ethics." *Journal of Management Inquiry* 13 (3): 215–223.

James, William. 1890/1910. *The Principles of Psychology*. New York, NY: Henry Holt and Company.

Jasanoff, Sheila, and Sang-Hyun Kim. 2015. *Dreamscapes of Modernity: Sociotechnical Imaginaries and the Fabrication of Power*. Chicago, IL: University of Chicago Press.

John Paul II. 2004. *Address of John Paul II to the Participants in the International Congress on "Life-Sustaining Treatments and Vegetative State: Scientific Advances and Ethical Dilemmas."* International Congress on "Life-Sustaining Treatments and Vegetative State: Scientific Advances and Ethical Dilemmas." Rome: Vatican. https://www.vatican.va/content/john-paul-ii/en/speeches/2004/march/documents/hf_jp-ii_spe_20040320_congress-fiamc.html.

Johnson, Mark. 1994. *Moral Imagination: Implications of Cognitive Science for Ethics*. Chicago, IL: University of Chicago Press.

Johnson, Sarah Lindstrom, Robert W. Blum, and Tina L. Cheng. 2014. "Future Orientation: A Construct with Implications for Adolescent Health and Wellbeing." *International Journal of Adolescent Medicine and Health* 26 (4): 459–468.

Johnston, Bridget, Sally Lawton, Catriona McCaw., Emma Law, Joyce Murray, John Gibb, Jan Pringle, Gillian Munro, and Cesar Rodriguez. 2016. "Living Well with Dementia: Enhancing Dignity and Quality of Life, Using a Novel Intervention, Dignity Therapy." *International Journal of Older People Nursing* 11 (2): 107–120.

Jones, Thomas M. 1991. "Ethical Decision Making by Individuals in Organizations: An Issue-Contingent Model." *Academy of Management Review* 16 (2): 366–395.

Jonsen, Albert R. 1991. "Casuistry as Methodology in Clinical Ethics." *Theoretical Medicine and Bioethics* 12 (4): 295–307.

Jonsen, Albert R. 1995. "Casuistry: An Alternative or Complement to Principles?" *Kennedy Institute of Ethics Journal* 5 (3): 237–251.

Jonsen, Albert R., Mark Siegler, and William T. Winslade. 2010. *Clinical Ethics: A Practical Approach to Ethical Decisions in Clinical Medicine*. New York, NY: McGraw-Hill.

Jonsen, Albert R., and Stephen Toulmin. 1988. *The Abuse of Casuistry: A History of Moral Reasoning*. Berkeley, CA: University of California Press.

Kahneman, Daniel. 2011. *Thinking, Fast and Slow*. New York, NY: Farrar, Straus and Giroux.

Kant, Emmanuel. 1991. *Vers la paix perpétuelle. Que signifie s'orienter dans la pensée. Qu'est-ce que les lumières? Et autres textes*. Paris, France: GF-Flammarion.

Kant, Immanuel. 1781/1787/1996. *Critique of Pure Reason: Unified Edition with All Variants From the 1781 and 1787 Editions*. Indianapolis, IN: Hackett Publishing Company.

Kant, Immanuel. 1785/1998. *Groundwork of the Metaphysics of Morals*. Cambridge, UK: Cambridge University Press.

Kapitan, Lynn 2018. *Introduction to Art Therapy Research*. New York, NY: Routledge.

Kegley, Jacquelyn A. K. 2013. "Mary Mahowald: Removing Blinders and Crossing Boundaries." *Pluralist* 8 (3): 114–121.

Kekes, John. 1984. "Moral Sensitivity." *Philosophy* 59 (1): 3–19.

Kern, David E., Scott M. Wright, Joseph A. Carrese, Mack Lipkin Jr., Julia M. Simmons, Dennis H. Novack, Adina Kalet, and Richard Frankel. 2001. "Personal Growth in Medical Faculty: A Qualitative Study." *Western Journal of Medicine* 175 (2): 92–96.

Kestenbaum, Victor. 1992. "'Meaning and the Model of Truth': Dewey and Gadamer on Habit and Vorurteil." *Journal of Speculative Philosophy* 6 (2): 25–66.

Kestler-D'Amours, Jillian. 2019. "What's Behind Quebec's Ban on Religious Symbols." *Atlantic,* July 16. https://www.theatlantic.com/international/archive/2019/07/quebec-bans-religious-symbols/593998/.

Kihlstrom, John F. 2013. "The Person-Situation Interaction." In *The Oxford Handbook of Social Cognition,* edited by Donal E. Carlston, 786–805. New York, NY: Oxford University Press.

Killen, Melanie, and Judith G. Smetana. 2013. *Handbook of Moral Development*. London, UK: Psychology Press.

Kim, Eric S., Ashley V. Whillans, Matthew T. Lee, Ying Chen, and Tyler J. VanderWeele. 2020. "Volunteering and Subsequent Health and Well-Being in Older Adults: An Outcome-Wide Longitudinal Approach." *American Journal of Preventive Medicine* 59 (2): 176–186.

Kiosses, Vassilios N., Vassilios T. Karathanos, and Athina Tatsioni. 2016. "Empathy Promoting Interventions for Health Professionals: A Systematic Review of RCTs." *Journal of Compassionate Health Care* 3: Article 7. https://doi.org/10.1186/s40639-016-0024-9.

Klein, Nadav. 2017. "Prosocial Behavior Increases Perceptions of Meaning in Life." *Journal of Positive Psychology* 12 (4): 354–361.

Knoll, Michael, Robert Lord, Lars-Eric Petersen, and Oliver Weigelt. 2016. "Examining the Moral Grey Zone: The Role of Moral Disengagement, Authenticity, and Situational Strength in Predicting Unethical Managerial Behavior." *Journal of Applied Social Psychology* 46 (1): 65–78.

Kohlberg, Lawrence. 1968. "The Child as a Moral Philosoher." *Psychology Today* 2 (4): 25–30.

Kolb, David A. 1984. *Experiential Learning*. Englewood Cliffs, NJ: Prentice Hall.

Kopelman, Loretta M. 1994. "Case Method and Casuistry: The Problem of Bias." *Theoretical Medicine* 15 (1): 21–37.

Kouchaki, Maryam, and Isaac H. Smith. 2013. "The Morning Morality Effect: The Influence of Time of Day on Unethical Behavior." *Psychological Science* 25 (1): 95–102.

Kramer-Roy, Debbie. 2015. "Using Participatory and Creative Methods to Facilitate Emancipatory Research with People Facing Multiple Disadvantage: A Role for Health and Care Professionals." *Disability & Society* 30 (8): 1207–1224.

Krys, Kuba, C.-Melanie Vauclair, Colin A. Capaldi, Vivian Miu-Chi Lun, Michael Harris Bond, Alejandra Domínguez-Espinosa, Claudio Torres, Ottmar V. Lipp, L. Sam S. Manickam, Cai Xing, Radka Antalíková, Vassilis Pavlopoulos, Julien Teyssier, Taekyun Hur, Karolina Hansen, Piotr Szarota, Ramadan A. Ahmed, Eleonora Burtceva, Ana Chkhaidze, Enila Cenko, Patrick Denoux, Márta Fülöp, Arif Hassan, David O. Igbokwe, İdil Işık, Gwatirera Javangwe, María Malbran, Fridanna Maricchiolo, Hera Mikarsa, Lynden K. Miles, Martin Nader, Joonha Park, Muhammad Rizwan, Radwa Salem, Beate Schwarz, Irfana Shah, Chien-Ru Sun, Wijnand van Tilburg, Wolfgang Wagner, Ryan Wise, and Angela Arriola Yu. 2016. "Be Careful Where You Smile: Culture Shapes Judgments of Intelligence and Honesty of Smiling Individuals." *Journal of Nonverbal Behavior* 40 (1): 101–116.

Kuehlmeyer, K. Katja, Nicole Palmour, Richard J. Riopelle, James L. Bernat, Ralf J. Jox, and Eric Racine. 2014. "Physicians' Attitudes Toward Medical and Ethical Challenges for Patients in the Vegetative State: Comparing Canadian and German Perspectives in a Vignette Survey." *BMC Neurology* 14: 119.

Kuehlmeyer, Katja, Eric Racine, Nicole Palmour, Eva Hoster, Gian Domenico Borasio, and Ralf J. Jox. 2012. "Diagnostic and Ethical Challenges in Disorders of Consciousness and Locked-in Syndrome: A Survey of German Neurologists. *Journal of Neurology* 259 (10): 2076–2089.

Kurtz, Stanley. 2003. "Culture and Values in the 1960s." In *Never a Matter of Indifference: Sustaining Virtue in a Free Republic,* edited by Peter Berkowitz, 29–58. Stanford, CA: Hoover Press.

Lai, Calvin K., Maddalena Marini, Steven A. Lehr, Carlo Cerruti, Jiyun-Elizabeth L. Shin, Jennifer A. Joy-Gaba, Arnold K. Ho, Bethany A. Teachman, Sean P. Wojcik, and Spassena P. Koleva. 2014. "Reducing Implicit Racial Preferences: A Comparative Investigation of 17 Interventions." *Journal of Experimental Psychology: General* 143 (4): 1765–1785.

Lakoff, George, and Mark Johnson. 2003. *Metaphors We Live By*. Chicago, IL: University of Chicago Press.

Lambie, John. 2014. "Effects of Open-Mindedness on Decision Making, Morality, and Well-Being." In *How to be Critically Open-Minded: A Psychological and Historical Analysis*, edited by John Lambie, 77–91. London, UK: Palgrave Macmillan.

Lang, Jack. 2004. *Nelson Mandela: Leçons de vie pour l'avenir. Préface de Nadine Gordimer.* Paris, France: Perrin.

Larivière-Bastien, Danaë, Annette Majnemer, Michael Shevell, and Eric Racine. 2011. "Perspectives of Adolescents and Young Adolescents with Cerebral Palsy on the Ethical and Social Challenges Encountered in Healthcare Services." *Narrative Inquiry on Bioethics: A Journal of Qualitative Research* 1 (1): 43–54.

Lašas, Ainius, Rocio Garcia-Retamero, Vaida Jankauskaitė, and Vitalija Simonaitytė. 2017. "Living in the Past: The Impact of Victimization Memory on Threat Perceptions." *Memory Studies* 11 (4): 405–421.

Laureys, Steven, Gastone G. Celesia, François Cohadon, Jan Lavrijsen, José León-Carrión, Walter G Sannita, Leon Sazbon, Erich Schmutzhard, Klaus R. von Wild, Adam Zeman, Giuliano Dolce, and the European Task Force on Disorders of Consciousness. 2010. "Unresponsive Wakefulness Syndrome: A New Name for the Vegetative State or Apallic Syndrome." *BMC Medicine* 8 (1): 68.

Legg, Catherine, and Christopher Hookway. 2019. "Pragmatism." In *Stanford Encyclopedia of Philosophy*, first published August 16, 2008, substantive revision March 14, 2020. https://plato.stanford.edu/entries/pragmatism/.

Lekan, Todd. 2006. "Pragmatist Metaethics: Moral Theory as a Deliberative Practice." *Southern Journal of Philosophy* 44 (2): 253–271.

Leopold, Aldo. 1949. *A Sand County Almanac: And Sketches Here and There.* Oxford, UK: Oxford University Press.

Lethbridge, Emma M., Paul Richardson, Lisa Reidy, and Naira A. Taroyan. 2017. "Exploring the Relationship between Callous-Unemotional Traits, Empathy Processing and Affective Valence in a General Population." *Europe's Journal of Psychology* 13 (1): 162–172.

Levy, Neil. 2007. *Neuroethics: Challenges for the 21st Century.* Cambridge, UK: Cambridge University Press.

Levy, Neil. 2017. "Implicit Bias and Moral Responsibility: Probing the Data." *Philosophy and Phenomenological Research* 94 (1): 3–26.

Lewin, Kurt. 1939. "Field Theory and Experiment in Social Psychology: Concepts and Methods." *American Journal of Sociology* 44 (6): 868–896.

Lewin, Kurt. 1946. "Action Research and Minority Problems." *Journal of Social Issues* 2 (4): 34–46.

Limone, Pierpaolo, and Giusi A. Toto. 2022. "Origin and Development of Moral Sense: A Systematic Review." *Frontiers in Psychology* 13 (1): 887537.

Linkeviciute, Alma, Kris Dierickx, Virginia Sanchini, and Giovanni Boniolo. 2016. "Potential Pitfalls in the Evaluation of Ethics Consultation: The Case of Ethical Counseling." *American Journal of Bioethics* 16 (3): 56–57.

Lippman-Hand, Abby, and Frank C. Fraser. 1979a. "Genetic Counseling—The Post-counseling Period: Making Reproductive Choices." *American Journal of Medical Genetics* 4 (1): 73–87.

Lippman-Hand, Abby, and Frank C. Fraser. 1979b. "Genetic Counseling: Parents' Responses to Uncertainty." *Birth Defects Original Article Series* 9 (1): 325–339.

Lippman-Hand, Abby, and Frank C. Fraser. 1979c. "Genetic Counseling: Provision and Reception of Information." *American Journal of Medical Genetics* 3 (1): 13–27.

Little, Simon. 2018. "Canadians Love Firefighters and Nurses. Politicians and Car Salespeople. Not So Much: Survey." *Global News*, March 15 2018. https://globalnews.ca/news/4085069/most-respected-professions/#:~:text=How%20does%20your%20job%20stack,the%20annual%20Insights%20West%20survey.

Locock, Louise, and Sue Ziébland. 2015. "Mike Bury: Biographical Disruption and Long-Term and Other Health Conditions." In *The Palgrave Handbook of Social Theory in Health, Illness and Medicine*, edited by Fran Collyer, 582–598. London, UK: Palgrave Macmillan.

Luthar, Suniya S. 2006. "Resilience in Development: A Synthesis of Research across Five Decades." In *Developmental Psychopathology: Risk, Disorder, and Adaptation*, edited by Dante Cicchetti and Donald J. Cohen, 740–795. New York, NY: Wiley.

Lynam, M. Judith, and Sarah Cowley. 2007. "Understanding Marginalization as a Social Determinant of Health." *Critical Public Health* 17 (2): 137–149.

Macdonald, Mary E., Stephen Liben, Franco A. Carnevale, and S. Robin Cohen. 2008. "Signs of Life and Signs of Death: Brain Death and Other Mixed Messages at the End of Life." *Journal of Child Health Care* 12 (2): 92–105.

MacIntyre, Alasdair. 1984. *After Virtue: A Study in Moral Theory*. Notre Dame, IN: University of Notre Dame Press.

Mackenzie, Catriona, and Natalie Stoljar. 2000. *Relational Autonomy: Feminist Perspectives on Automony, Agency, and the Social Self.* Oxford, UK: Oxford University Press.

MacKenzie, Michael J., Kathleen D. Vohs, and Roy F. Baumeister. 2014. "You Didn't Have to Do That: Belief in Free Will Promotes Gratitude." *Personality and Social Psychology Bulletin* 40 (11): 1423–1434.

MacLean, Paul D. 1967. "The Brain in Relation to Empathy and Medical Education." *Journal of Nervous and Mental Disease* 144 (5): 374–382.

Mahowald, Mary B. 1994. "So Many Ways to Think. An Overview of Approaches to Ethical Issues in Geriatrics." *Clinics in Geriatric Medicine* 10 (3): 403–417.

Mahsud, Rubina, Gary Yukl, and Greg Prussia. 2010. "Leader Empathy, Ethical Leadership, and Relations-Oriented Behaviors as Antecedents of Leader-Member Exchange Quality." *Journal of Managerial Psychology* 25 (6): 561–577.

Manning, Rachel, Mark Levine, and Alan Collins. 2007. "The Kitty Genovese Murder and the Social Psychology of Helping: The Parable of the 38 Witnesses." *American Psychological Association* 62 (6): 555–562.

Marin, Stephanie. 2018. "MMA and Former Employees Plead Guilty, Fined $1.25 Million in Lac-Megantic Case." *CTV News,* February 5 2018. https://www.ctvnews.ca/canada/mma-and-former-employees-plead-guilty-fined-1-25-million-in-lac-megantic-case-1.3790377?cache=walqrkeg#:~:text=MONTREAL%20%2D%2D%20The%20bankrupt%20railway,given%20a%20conditional%20jail%20term.

Marshall, Albert, Murdena Marshall, Cheryl Bartlett. 2015. Two-Eyed Seeing in Medicine. In *Determinants of Indigenous Peoples' Health in Canada; Beyond the Social,* edited by Margo Greenwood, Sarah de Leeuw, and Nicole Marie Lindsay, 16–24. Toronto, Canada: Canadian Scholars' Press.

Marshall, David. 1986. *The Figure of Theater: Shaftesbury, Defoe, Adam Smith, and George Eliot.* New York, NY: Columbia University Press.

Martela, Frank. 2015. "Pragmatism as an Attitude." In *Action, Belief and Inquiry: Pragmatist Perspectives on Science, Society and Religion,* edited by Ulf Zackariasson, 187–207. Helsinki, Finland: Nordic Pragmatism Network.

Martela, Frank. 2017. "Moral Philosophers as Ethical Engineers: Limits of Moral Philosophy and a Pragmatist Alternative." *Metaphilosophy* 48 (1–2): 58–78.

Martineau, Joé T., Jean Decety, and Eric Racine. 2020. "The Social Neuroscience of Empathy Its Implication for Business Ethics." In *Organizational Neuroethics: Reflections on the Contributions of Neuroscience to Management Theories and Business Practices,* edited by Joé T. Martineau and Eric Racine, 167–189. Cham, Switzerland: Springer International Publishing.

Martineau, Joé T., Kevin Johnson, and Thierry C. Pauchant. 2017. "The Pluralist Theory of Ethics Programs Orientations and Ideologies: An Empirical Study Anchored in Requisite Variety." *Journal of Business Ethics* 142 (4): 791–815.

Marx, Karl, Friedrich Engels, and Gareth Stedman Jones. 2002. *The Communist Manifesto.* London, UK: Penguin Books.

Maslen, Hannah. 2015. "Shrewder Speculation: The Challenge of Doing Anticipatory Ethics Well." *Neuroethics* (blog), November 3 2015. http://www.theneuroethicsblog.com/2015/11/shrewder-speculation-challenge-of-doing.html 2021.

Maslow, Abraham. 1970. *Motivation and Personality*. New York, NY: Harper & Row.

Maxwell, Bruce, and Eric Racine. 2010. "Should Empathic Development Be a Priority in Biomedical Ethics Teaching? A Critical Perspective." *Cambridge Quarterly of Healthcare Ethics* 19 (4): 433–445.

Maxwell, Bruce, and Eric Racine. 2016. "Two Cases in Neuroeducation Knowledge Transfer: Behavioral Ethics and Responsive Parenting." In *Neuroscience and Education: A Philosophical Appraisal*, edited by Clarence Joldersma, 50–69. New York, NY: Routledge.

McClelland, James L., and Timothy T. Rogers. 2003. "The Parallel Distributed Processing Approach to Semantic Cognition." *Nature Reviews Neuroscience* 4 (4): 310–322.

McClung, John A., Russell S. Kamer, Margaret DeLuca, and Harlan J. Barber. 1996. "Evaluation of a Medical Ethics Consultation Service: Opinions of Patients and Health Care Providers." *American Journal of Medicine* 100 (4): 456–460.

McCumber, John. 2001. *Time in the Ditch: American Philosophy and the McCarthy Era*. Evanston, IL: Northwestern University Press.

McFarland, Craig P., and Jennifer J. Vasterling. 2017. "Prospective Memory in Depression: Review of an Emerging Field." *Archives of Clinical Neuropsychology* 33 (7): 912–930.

Mead, George H. 1934/2015. *Mind, Self, and Society*. Chicago, IL: University of Chicago Press.

Medical Assistance in Dying Research Group. 2017. *Exploring the Psychological Suffering of a Person Requesting Medical Assistance in Dying*. Montreal, Canada: Department of Psychiatry, CHUM, and CRCHUM.

Menand, Louis. 2001. *The Metaphysical Club: A Story of Ideas in America*. London, UK: MacMillan.

Mencl, Jennifer, and Douglas R. May. 2009. "The Effects of Proximity and Empathy on Ethical Decision-Making: An Exploratory Investigation." *Journal of Business Ethics* 85 (2): 201–226.

Messick, David M. 1999. "Alternative Logics for Decision Making in Social Settings." *Journal of Economic Behavior & Organization* 39 (1): 11–28.

Metselaar, Suzanne, Guy Widdershoven, Rouven Porz, and Bert Molewijk. 2017. "Evaluating Clinical Ethics Support: A Participatory Approach." *Bioethics* 31 (4): 258–266.

Mignacca Franca, Sarah Leavitt. 2021. Maimonides Residents Threaten Legal Action to Get 2nd Dose of Covid-19 Vaccine. CBC News, January 6 2021. https://www.cbc.ca/news/canada/montreal/maimonides-residents-covid-19-vaccine-legal-quebec-government-1.5862878.

Milgram, Stanley. 1963. "Behavioral Study of Obedience." *Journal of Abnormal Psychology* 67 (1): 371–378.

Mill, John Stuart. 1863/2001. *Utilitarianism*. Kitchener, Canada: Batoche Books.

Mill, John Stuart. 1869. *The Subjection of Women*. London, UK: Longmans, Green, Reader and Dyer.

Mill, John Stuart. 1859/1956. *On Liberty*. New York, NY: Bobbs-Merrill.

Miller, Franklin G., Joseph J. Fins, and Matthew D. Bacchetta. 1996. "Clinical Pragmatism: John Dewey and Clinical Ethics." *Journal of Contemporary Health Law and Policy* 13 (9): 27–51.

Miller, Franklin G., Robert D. Truog, and Dan W. Brock. 2010. "The Dead Donor Rule: Can It Withstand Critical Scrutiny?" *Journal of Medicine and Philosophy* 35 (3): 299–312.

Miller, Henry, and Kate Miller. 1996. "Language Policy and Identity: The Case of Catalonia." *International Studies in Sociology of Education* 6 (1): 113–128.

Miller, Jared A., Zachariah J. Rodgers, and John B. Bingham. 2014. "Moral Awareness." In *Research Companion to Ethical Behavior in Organizations: Constructs and Measures*, edited by Bradley R. Agle, David W. Hart, Jeffery A. Thompson, and Hilary M. Hendricks, 1–43. Northampton, MA: Edward Elgar Publishing.

Mills, Charles W. 1997. *The Racial Contract*. Ithaca, NY: Cornell University Press.

Misak, Cheryl. 2013. *The American Pragmatists*. Oxford, UK: Oxford University Press.

Mischel, Walter. 1977. "The Interaction of Person and Situation." In *Personality at the Crossroads: Current Issues in Interactional Psychology*, edited by David Magnusson and Norman S. Endler, 333–352. Hillsdale, NJ: Lawrence Erlbaum.

Molewijk, Bert, Dick Kleinlugtenbelt, and Guy Widdershoven. 2011. "The Role of Emotions in Moral Case Deliberation: Theory, Practice, and Methodology." *Bioethics* 25 (7): 383–393.

Mollaret, Pierre, and Mollaret Goulon. 1959. "Le coma dépassé (mémoire prélininaire)"*Revue Neurologique* 101 (9): 3–15.

Monti, Martin M., Audrey Vanhaudenhuyse, Martin R. Coleman, Melanie Boly, John D. Pickard, Luaba Tshibanda, Adrian M. Owen, and Steven Laureys. 2010. "Willful Modulation of Brain Activity in Disorders of Consciousness." *New England Journal of Medicine* 362 (7): 579–589.

Montreuil, Marjorie, and Franco A. Carnevale. 2016. "A Concept Analysis of Children's Agency within the Health Literature." *Journal of Child Health Care* 20 (4): 503–511.

Montreuil, Marjorie, and Franco A. Carnevale. 2018. "Participatory Hermeneutic Ethnography: A Methodological Framework for Health Ethics Research with Children." *Qualitative Health Research* 28 (7): 1135–1144.

Montreuil, Marjorie, Gail Teachman, and Franco Carnevale. 2019. "Recognizing the Voices of Children with 'Cognitive Impairments' in Research." In *Research Involving Participants with Impaired Cognition: Ethics, Autonomy, Inclusion, and Innovation*, edited by Ariel Cascio and Eric Racine, 135–148. Oxford, UK: Oxford University Press.

Mooney, Gavin. 2009. "Some Problems in Existing Health Economics." In *Challenging Health Economics*, 16–33. Oxford, UK: Oxford University Press.

Moradi, Saleh, Niels V. Quaquebeke, and John A. Hunter. 2018. "Flourishing and Prosocial Behaviors: A Multilevel Investigation of National Corruption level as a Moderator." *PLOS One* 13 (7): e0200062.

Moreno, Jonathan D. 2004. "Bioethics Imperialism." *ASBH Exchange* 7 (3): 2.

Moskowitz, Gordon B., and Peizhong Li. 2011. "Egalitarian Goals Trigger Stereotype Inhibition: A Proactive Form of Stereotype Control." *Journal of Experimental Social Psychology* 47 (1): 103–116.

Muaygil, Ruaim A. 2018. "Beyond Sacredness: Why Saudi Arabian Bioethics Must Be Feminist." *International Journal of Feminist Approaches to Bioethics* 11 (1): 125–143.

Muirhead, Alice, and Sarah de Leeuw. 2012. "Art and Wellness: The Importance of Art for Aboriginal Peoples' Health and Healing." *National Collaborating Centre for Aboriginal Health.* https://www.ccnsa-nccah.ca/docs/emerging/FS-ArtWellness-Muirhead-deLeeuw-EN.pdf.

Müller, Leonardo A. P. 2016. "The Philosophy of Adam Smith: Imagination and Speculation." PhD diss., Université Panthéon-Sorbonne, France. https://theses.hal.science/tel-01398360/document.

Multi-Society Task Force on PVS. 1994a. "Medical Aspects of the Persistent Vegetative State (1)." *New England Journal of Medicine* 330 (21): 1499–1508.

Multi-Society Task Force on PVS. 1994b. "Medical Aspects of the Persistent Vegetative State (2)." *New England Journal of Medicine* 330 (22): 1572–1579.

Mumford, Michael D., Ethan P. Waples, Alison L. Antes, Ryan P. Brown, Shane Connelly, Stephen T. Murphy, and Lynn D. Devenport. 2010. "Creativity and Ethics: The Relationship of Creative and Ethical Problem-Solving." *Creativity Research Journal* 22 (1): 74–89.

Murphy, Fionnuala C., Polly V. Peers, Simon E. Blackwell, Emily A. Holmes, and Tom Manly. 2019. "Anticipated and Imagined Futures: Prospective Cognition and Depressed Mood Following Brain Injury." *British Journal of Clinical Psychology* 58 (1): 91–109.

Nadeau, Jessica. 2021. "La préposée qui a tenu des propos dégradants envers Joyce Echaquan se voulait 'bienveillante.'" *Le Devoir*, May 20 2021. https://www.ledevoir.com/societe/602708/le-deces-de-joyce-echaquan-a-cree-un-plus-grand-fosse-entre-les-atikamekws-et-le-personnel-de-l-hopital-de-joliette.

Nagel, Saskia K., and Peter B. Reiner. 2013. "Autonomy Support to Foster Individuals' Flourishing." *American Journal of Bioethics* 13 (6): 36–37.

Nagel, Thomas. 1986. *The View from Nowhere*. Oxford, UK: Clarendon Press.

National Center for State Courts (Race & Ethnic Fairness in the Courts). 2012. *Strategies to Reduce the Influence of Implicit Bias*. https://horsley.yale.edu/sites/default/files/files/IB_Strategies_033012.pdf.

Naue, Ursula. 2008. "'Self-Care without a Self': Alzheimer's Disease and the Concept of Personal Responsibility for Health." *Medicine, Health Care and Philosophy* 11 (3): 315–324.

Nelkin, Dorothy. 2001. "Beyond Risk. Reporting about Genetics in the Post-Asilomar Press." *Perspectives in Biology and Medicine* 44 (2): 199–207.

Nelson, S. Katherine, Kristin Layous, Steven W. Cole, and Sonja Lyubomirsky. 2016. "Do Unto Others or Treat Yourself? The Effects of Prosocial and Self-Focused Behavior on Psychological Flourishing." *Emotion* 16 (6): 850–861.

Nietzsche, Friedrich. 1887/1971. *La généalogie de la morale*. Paris, France: Gallimard.

Nietzsche, Friedrich. 1889/1997. *Twilight of the Idols, or, How to Philosophize with a Hammer*. Translated by Richard Polt. Indianapolis, IN: Hackett Publishing Company.

Nijhof, André, Stephan Cludts, Olaf Fisscher, and Albertus Laan. 2003. "Measuring the Implementation of Codes of Conduct. An Assessment Method Based on a Process Approach of the Responsible Organisation." *Journal of Business Ethics* 45 (1): 65–78.

Noddings, Nel. 1982. *Caring: A Feminine Approach to Ethics and Moral Education*. Berkeley, CA: University of California Press.

Norton, Bryan G. 1996. "Moral Naturalism and Adaptive Management." *Hastings Center Report* 26 (6): 25.

Nussbaum, Martha C. 2011. *Creating Capabilities: The Human Development Approach*. Cambridge, MA: Harvard University Press.

Nussbaum, Martha C. 1997. "Capabilities and Human Rights." *Fordham Law Review* 66 (2): 273–300.

O'Donnell, Patrick, Diarmuid O'Donovan, and Khalifa Elmusharaf. 2018. "Measuring Social Exclusion in Healthcare Settings: A Scoping Review." *International Journal for Equity in Health* 17 (1): 15.

Orfali, Kristina. 2004. "Parental Role in Medical Decision-Making: Fact or Fiction? A Comparative Study of Ethical Dilemmas in French and American Neonatal Intensive Care Units." *Social Science and Medicine* 58 (10): 2009–2022.

Orfali, Kristina, and Elisa J. Gordon. 2004. "Autonomy Gone Awry: A Cross-Cultural Study of Parents' Experiences in Neonatal Intensive Care Units." *Theoretical Medicine and Bioethics* 25 (4): 329–365.

Owen, Adrian M., Martin R. Coleman, Melanie Boly, Matthew H. Davis, Steven Laureys, and John D. Pickard. 2006. "Detecting Awareness in the Vegetative State." *Science* 313 (5792): 1402.

Pais, Abraham. 1982. *"Subtle is the Lord": The Science and the Life of Albert Einstein.* Oxford, UK: Oxford University Press.

Pappas, Gregory F. 2008. *John Dewey's Ethics: Democracy as Experience*. Bloomington, IN: Indiana University Press.

Pâquet, Martin, and Stéphane Savard. 2021. *Brève histoire de la révolution tranquille.* Montréal, Canada: Éditions du Boréal.

Paquin, Stéphane, and Pier-Luc Lévesque. 2014. *Social-démocrate 2.0: Le Québec comparé aux pays scandinaves*. Montréal, Canada: Les Presses de l'Université de Montréal.

Parens, Erik. 2006. "Creativity, Gratitude, and the Enhancement Debate." In *Neuroethics: Defining the Issues in Theory, Practice, and Policy,* edited by Judy Illes, 75–86. Oxford, UK: Oxford University Press.

Parizeau, Marie-Hélène. 1995. "Compétence éthique, expertise éthique et modèles de comités d'éthique clinique." In *Hôpital & éthique: rôles et défis des comités d'éthique clinique,* edited by Marie-Hélène Parizeau, 139–166. Saint-Nicolas, Canada: Les Presses de l'Université Laval.

Parry, Richard and Harald Thorsrud. 2021. "Ancient Ethical Theory." In *Stanford Encyclopedia of Philosophy,* first published August 3, 2004, revision February 5, 2021. https://plato.stanford.edu/entries/ethics-ancient/.

Pasupathi, Monisha, Ursula M. Staudinger, and Paul B. Baltes. 2001. "Seeds of Wisdom: Adolescents' Knowledge and Judgment about Difficult Life Problems." *Developmental Psychology* 37 (3): 351.

Patterson, Orlando. 1985. *Slavery and Social Death: A Comparative Study*. Cambridge, MA: Harvard University Press.

Payne, B. Keith, Alan J. Lambert, and Larry L. Jacoby. 2002. "Best Laid Plans: Effects of Goals on Accessibility Bias and Cognitive Control in Race-Based Misperceptions of Weapons." *Journal of Experimental Social Psychology* 38 (4): 384–396.

Peguero, Eva, Anna Berenguera, Enriqueta Pujol-Ribera, Begoña Roman, Carmen M. Prieto, and Núria Terribas. 2015. "The Workers Opinions Have a Value in the Code of Ethics: Analysis of the Contributions of Workers in Virtual Forum Catalan Institute of Health." *BMC Medical Ethics* 16 (1): 90–90.

Peirce, Charles S. 1878. "How to Make Our Ideas Clear." *Popular Science Monthly* 12 (1): 286–302.

Peirce, Charles Sanders. 1877. "The Fixation of Belief." *Popular Science Monthly* 12 (1): 1–15.

Pekarsky, Daniel. 1990. "Dewey's Conception of Growth Reconsidered." *Educational Theory* 40 (9): 283–294.

Pellegrino, Edmund D. 2001. "The Internal Morality of Clinical Medicine: A Paradigm for the Ethics of the Helping and Healing Professions." *Journal of Medicine and Philosophy* 26 (6): 559–579.

Pereira, Ken, and Luc Bouchard. 2015. *Bras de fer.* Montréal, Canada: VLB Éditeur.

Perkins, Douglas D., and Marc A. Zimmerman. 1995. "Empowerment Theory, Research, and Application." *American Journal of Community Psychology* 23 (5): 569–579.

Perry, Cheryl L., Kelli A. Komro, Resa M. Jones, Karen Munson, Carolyn L. Williams, and Leonard Jason. 2002. "The Measurement of Wisdom and Its Relationship to Adolescent Substance Use and Problem Behaviors." *Journal of Child & Adolescent Substance Abuse* 12 (1): 45–63.

Pescosolido, Bernice A., Tait R. Medina, Jack K. Martin, and J. Scott Long. 2013. "The 'Backbone' of Stigma: Identifying the Global Core of Public Prejudice Associated with Mental Illness." *American Journal of Public Health* 103 (5): 853–860.

Peterson, Christopher, Martin Seligman, and Steven Maier. 1995. *Learned Helplessness: A Theory for the Age of Personal Control.* New York, NY: Oxford University Press.

Philipse, Herman. 2021. *Heidegger's Philosophy of Being: A Critical Interpretation.* Princeton, NJ: Princeton University Press.

Piaget, Jean. 1932. *The Moral Judgment of the Child.* London, UK: Kegan, Paul, Trench, Trubner & Company.

Pilapil, Renante D. 2012. "From Psychologism to Personhood: Honneth, Recognition, and the Making of Persons." *Res Publica* 18 (1): 39–51.

Pitts-Taylor, Victoria. 2010. "The Plastic Brain: Neoliberalism and the Neuronal Self." *Health* 14 (6): 635–652.

Plato. 1981. "Meno." In *Five Dialogues: Euthyphro, Apology, Crito, Meno, Phaedo,* edited by John Cooper, 59–88. Indianapolis, IN: Hackett.

Pluhar, Evelyn. 1988. "Moral Agents and Moral Patients." *Between the Species* 4 (1): 32–45.

Polkinghorne, Donald. 1988. *Narrative Knowing and the Human Sciences.* Albany, NY: SUNY Press.

Poulin, Richard. 1997. *Les fondements du marxisme.* Gatineau, Canada: Vents d'Ouest.

Prairat, Eirick. 2014. "Valuation et évaluation dans la pensée de Dewey." *Le Télémaque* 46 (2): 167–176.

Pyeritz, Reed E. 1990. "Must Ethics Consultants See Patients?" *Journal of Clinical Ethics.* 1990 1 (2): 168–169.

Quality Standards Subcommittee of the American Academy of Neurology. 1995. "Practice Parameters: Assessment and Management of Patients in the Persistent Vegetative State." *Neurology* 45 (5): 1015–1018.

Quintal, Ariane. 2023. "Un projet de recherche-action participative abordant les situations moralement problématiques vécues par les adultes atteints de maladies rares." PhD diss., Université de Montréal, Canada. https://espum.umontreal.ca/lespum/evenements/evenement/news/eventDetail/Event/soutenance-de-these-de-ariane-quintal/.

Quintal, Ariane, Élissa Hotte, Annie-Danielle Grenier, Caroline Hébert, Isabelle Carreau, Yves Berthiaume, and Eric Racine. 2023. "Morally Problematic Situations Encountered by Adults Living with Rare Diseases." *AJOB Empirical Bioethics* 15 (3): 192–205.

Racine, Eric. 2004. "L'éthique clinique, les émotions et le processus d'analyse de cas: Une étude qualitative et multi-site de comités d'éthique clinique québécois. " PhD diss., Université de Montréal, Canada. https://papyrus.bib.umontreal.ca/xmlui/handle/1866/14382.

Racine, Eric. 2005. "Pourquoi et comment doit-on tenir compte des neurosciences en éthique ? Esquisse d'une approche neurophilosophique émergentiste et interdisciplinaire." *Laval théologique et philosophique* 61 (1): 77–105.

Racine, Eric. 2007. "HEC Member Perspectives on the Case Analysis Process: A Qualitative Multi-Site Study." *HEC Forum* 19 (3): 185–206.

Racine, Eric. 2008a. "Enriching Our Views on Clinical Ethics: Results of a Qualitative Study of the Moral Psychology of Healthcare Ethics Committee Members." *Journal of Bioethical Inquiry* 5 (1): 57–67.

Racine, Eric. 2008b. "Interdisciplinary Approaches for a Pragmatic Neuroethics." *American Journal of Bioethics* 8 (1): 52–53.

Racine, Eric. 2008c. "Which Naturalism for Bioethics? A Defense of Moderate (pragmatic) Naturalism." *Bioethics* 22 (2): 92–100.

Racine, Eric. 2010. *Pragmatic Neuroethics: Improving Treatment and Understanding of the Mind-Brain.* Cambridge, MA: MIT Press.

Racine, Eric. 2013. "Pragmatism and the Contribution of Neuroscience to Ethics." *Essays in the Philosophy of Humanism* 21 (1): 13–20.

Racine, Eric. 2016. "Can Moral Problems of Everyday Clinical Practice Ever Be Resolved? A Proposal for Integrative Pragmatist Approaches." In *Ethics in Child Health: Principles and Cases in Neurodisability*, edited by Peter L. Rosenbaum, Gabriel M. Ronen, Eric Racine, Jennifer Johannesen, and Bernard Dan, 33–48. London, UK: Mac Keith Press.

Racine, Eric. 2017. "A Proposal for a Scientifically-Informed and Instrumentalist Account of Free Will and Voluntary Action." *Frontiers in Psychology* 8 (1): 754.

Racine, Eric. 2022. "How Ethics Liberates Experience: Insights from Pragmatist Theory and Contemporary Research." *Journal of Speculative Philosophy* 36 (4): 517–536.

Racine, Eric, Rakesh Amaram, Matthew Seidler, Marta Karczewska, and Judy Illes. 2008. "Media Coverage of the Persistent Vegetative State and End-of-Life Decision-Making." *Neurology* 71 (13): 1027–1032.

Racine, Eric, John Aspler, Cynthia Forlini, and Jennifer Chandler. 2017b. "Contextualized Autonomy and Liberalism: Broadening the Lenses on Complementary and Alternative Medicines in Preclinical Alzheimer's Disease." *Kennedy Institute of Ethics Journal* 27 (1): 1–41.

Racine, Eric, Emily Bell, Barbara Farlow, Steven Miller, Antoine Payot, Lisa Anne Rasmussen, Michael I. Shevell, Donna Thomson, and Pia Wintermark. 2017a. "The 'ouR-HOPE' Approach for Ethics and Communication about Neonatal Neurological Injury." *Developmental Medicine & Child Neurology* 59 (2): 125–135.

Racine, Eric, Emily Bell, Allison Yan, Gail Andrew, Lorraine E Bell, Margaret Clarke, Veljko Dubljević, Dan Goldowitz, Annie Janvier, Kaitlyn McLachlan, Nazeem Muhajarine, David Nicholas, Maryam Oskoui, Carmen Rasmussen, Lisa Anne Rasmussen, Wendy Roberts, Michael Shevell, Lucie Wade, and Jerome Y Yager. 2014a. "Ethics Challenges of Transition from Paediatric to Adult Health Care Services for Young Adults with Neurodevelopmental Disabilities." *Paediatrics & Child Health* 19 (2): 65–68.

Racine, Eric, and Ariel Cascio. 2020. "The False Dichotomy between Empirical and Normative Bioethics." *AJOB Empirical Bioethics* 11 (1): 5–7.

Racine, Eric, Ariel Cascio, Marjorie Montreuil, and Aline Bogossian. 2019. "Instrumentalist Analyses of the Functions of Ethics Concepts: A Proposal to Bridge Conceptual and Empirical Ethics Methodology." *Theoretical Medicine and Bioethics* 40 (4): 253–278.

Racine, Eric, Ilinca M. Cioaba, and Wren Boehlen. 2023. "Ethical Aspects of the Work Conditions of Public Safety Personnel: A Need for Attention and Solidarity." *Canadian Journal of Public Health* 114 (3): 502–506.

Racine, Eric, and Veljko Dubljević. 2016. "Porous or Contextualized Autonomy? Knowledge Can Empower Autonomous Moral Agents." *American Journal of Bioethics* 16 (2): 48–50.

Racine, Eric, and Veljko Dubljević. 2017. "Behavioral and Brain-Based Research on Free Moral Agency: Threat or Empowerment?" In *Neuroethics: Anticipating the Future*, edited by Judy Illes and Sharmin Hossain, 388–410. Oxford, UK: Oxford University Press.

Racine, Eric, Sophie Ji, Valérie Badro, Aline Bogossian, Claude Julie Bourque, Marie-Ève Bouthillier, Vanessa Chenel, Clara Dallaire, Hubert Doucet, Caroline Favron-Godbout, Marie-Chantal Fortin, Isabelle Ganache, Anne-Sophie Guernon, Marjorie Montreuil, Catherine Olivier, Ariane Quintal, Abdou Simon Senghor, Michèle Stanton-Jean, Joé T Martineau, Andréanne Talbot, and Nathalie Tremblay. 2024. "Living Ethics: A Stance and Its Implications in Health Ethics." *Medicine, Health and Philosophy* 27 (2): 137–154.

Racine, Eric, Sarah Kusch, Ariel Cascio, and Aline Bogossian. 2021. "Making Autonomy an Instrument: A Pragmatist Account of Contextualized Autonomy." *Humanities and Social Sciences Communications* 8 (139).

Racine, Eric, Danaë Larivière-Bastien, Emily Bell, Annette Majnemer, and Michael Shevell. 2013. "Respect for Autonomy in the Healthcare Context: Observations from a Qualitative Study of Young Adults with Cerebral Palsy." *Child: Care, Health and Development* 39 (6): 873–879.

Racine, Eric, and Nathalie Prud'homme. 2019. "La conciliation entre connaissances et préférences est-elle possible en matière d'alimentation?" *Nutrition—Science en évolution* 17 (2): 14–18.

Racine, Eric, Catherine Rodrigue, James L. Bernat, Richard Riopelle, and Sam D. Shemie. 2010. "Observations on the Ethical and Social Aspects of Disorders of Consciousness." *Canadian Journal of Neurological Science* 37 (6): 758–768.

Racine, Eric, Tristana M. Rubio, Jennifer Chandler, Cynthia Forlini, and Jayne Lucke. 2014b. "The Value and Pitfalls of Speculation about Science and Technology in Bioethics: The Case of Cognitive Enhancement." *Medicine, Health Care and Philosophy* 17 (3): 325–337.

Racine, Eric, and Matthew Sample. 2018. "Two Problematic Foundations of Neuroethics and Pragmatist Reconstructions." *Cambridge Quarterly of Healthcare Ethics* 27 (4): 566–577.

Racine, Eric, Sebastian Sattler, and Wren Boehlen. 2021. "Cognitive Enhancement: Unanswered Questions about Human Psychology and Social Behavior." *Science and Engineering Ethics* 27 (2): 19.

Racine, Eric, and Abdou Simon Senghor. 2023. "Diversity in Neuroethics: Which Diversity and Why It Matters?" In *Neuroethics and Cultural Diversity*, edited by Michele Farisco, 55–75. London, UK: ISTE/Wiley.

Racine, Eric, and Emma Zimmerman. 2012. "Pragmatic Neuroethics and Neuroscience's Potential to Radically Change Ethics." In *Neuroscientific Turn: Transdisciplinarity in the Age of the Brain*, edited by Melissa M. Littlefield and Jenell Johnson, 135–151. Ann Arbor, MI: University of Michigan Press.

Radio-Canada. 2020. Une Femme Autochtone Meurt à l'Hôpital de Joliette dans des Circonstances Troubles. *Radio-Canada*, September 28 2020. https://ici.radio-canada.ca/espaces-autochtones/1737180/femme-atikamekw-hopital-joliette-video-facebook.

Ralston, Shane J. 2010. "Dewey's Theory of Moral (and Political) Deliberation Unfiltered." *Education and Culture* 26 (1): 23–43.

Rathert, Cheryl, Douglas R. May, and Hye S. Chung. 2016. "Nurse Moral Distress: A Survey Identifying Predictors and Potential Interventions." *International Journal of Nursing Studies* 53 (1): 39–49.

Rauthmann, John F., and Ryne A. Sherman. 2019. "Toward a Research Agenda for the Study of Situation Perceptions: A Variance Componential Framework." *Personality and Social Psychology Review* 23 (3): 238–266.

Rauthmann, John F., Ryne A. Sherman, and David C. Funder. 2015. "Principles of Situation Research: Towards a Better Understanding of Psychological Situations." *European Journal of Personality* 29 (3): 363–381.

Rauthmann, John F., and Ryne A. Sherman. 2016. "Situation Change: Stability and Change of Situation Variables between and Within Persons." *Frontiers in Psychology* 6 (1): 1938.

Rawls, John. 1971. *A Theory of Justice*. Cambridge, MA: Harvard University Press.

Raynault, Marie-France, Dominique Côté, and Sébastien Chartrand. 2013. *Le bon sens à la scandinave: Politiques et inégalités sociales de santé*. Montréal, Canada: Presses de l'Université de Montréal.

Read, Carveth. 1898. *Logic: Deductive and Inductive*. London, UK: Grant Richards.

Reason, Peter, and Hilary Bradbury. 2008. *The Sage Handbook of Action Research: Participative Inquiry and Practice*. Thousand Oaks, CA: Sage Publishing.

Reese, James J., Jr., and Phillip L. Pearl. 2016. "Discussing Sudden Unexpected Death in Newly Diagnosed Epilepsy." In *Ethics in Child Health: Principles and Cases in Neurodisability*, edited by Peter L. Rosenbaum, Gabriel M. Ronen, Eric Racine, Jennifer Johannesen, and Bernard Dan, 303–310. London, UK: Mac Keith Press.

Rest, James R. 1982. "A Psychologist Looks at the Teaching of Ethics." *Hastings Center Report* 12 (1): 29–36.

Rest, James. 1986. *Moral Development: Advances in Research and Theory*. New York, NY: Praeger.

Reynolds, Scott. J. 2006. "Moral Awareness and Ethical Predispositions: Investigating the Role of Individual Differences in the Recognition of Moral Issues." *Journal of Applied Psychology* 91 (1): 233–243.

Reynolds, Scott J. 2008. "Moral Attentiveness: Who Pays Attention to the Moral Aspects of Life?" *Journal of Applied Psychology* 93 (5): 1027–1041.

Reynolds, Scott J., and Jared A. Miller. 2015. "The Recognition of Moral Issues: Moral Awareness, Moral Sensitivity and Moral Attentiveness." *Current Opinion in Psychology* 6 (1): 114–117.

Reynolds, Scott J., Keith Leavitt, and Katherine A. DeCelles. 2010. "Automatic Ethics: The Effects of Implicit Assumptions and Contextual Cues on Moral Behavior." *Journal of Applied Psychology* 95 (4): 752–760.

Reynolds, Scott J., and Tara L. Ceranic 2007. "The Effects of Moral Judgment and Moral Identity on Moral Behavior: An Empirical Examination of the Moral Individual." *American Psychological Association* 92(6):1610–24.

Ricoeur, Paul. 1992. *Oneself as Another*. Chicago, IL: The University of Chicago Press.

Riedweg, Christoph. 2002. *Pythagoras: His Life, Teachings, and Influence*. Ithaca, NY: Cornell University Press.

Riess, Helen, John M. Kelley, Robert W. Bailey, Emily J. Dunn, and Margot Phillips. 2012. "Empathy Training for Resident Physicians: A Randomized Controlled Trial of a Neuroscience-Informed Curriculum." *Journal of General Internal Medicine* 27 (10): 1280–1286.

Rigoni, Davide, Simone Kuhn, Gennaro Gaudino, Giuseppe Sartori, and Marcel Brass. 2012. "Reducing Self-Control by Weakening Belief in Free Will." *Consciousness and Cognition* 21 (3): 1482–1490.

Robertson, David W. 1996. "Ethical Theory, Ethnography, and Differences between Doctors and Nurses in Approaches to Patient Care." *Journal of Medical Ethics* 22 (1): 292–299.

Robles, Pablo H. 1999. "Evaluation of Healthcare Ethics Committees: The Experience of an HEC in Spain." *HEC Forum* 11 (3): 263–276.

Rodogno, Raffaele, Katrine Krause-Jensen, and Richard E. Ashcroft. 2016. "'Autism and the Good Life': A New Approach to the Study of Well-Being." *Journal of Medical Ethics* 42 (6): 401–408.

Rogers, Carl. 1961. *On Becoming a Person: A Therapist's View of Psychotherapy*. Boston, MA: Houghton Miflfin.

Rosch, Eleanor, Carolyn B. Mervis, Wayne D. Gray, David M. Johnson, and Penny Boyes-Braem. 1976. "Basic Objects in Natural Categories." *Cognitive Psychology* 8 (3): 382–439.

Rose, Flemming. 2017. "Safe Spaces on College Campuses Are Creating Intolerant Students." *Huffington Post,* March 30 2017. https://www.huffpost.com/entry/safe-spaces-college-intolerant_b_58d957a6e4b02a2eaab66ccf.

Rota, Gian-Carlo. 1991. "The Pernicious Influence of Mathematics Upon Philosophy." *Synthese* 88 (2): 165–178.

Rousse, B. Scot. 2011. "Thrown Projection: An Interpretation and Defense of the Hermeneutic Conception of the Self in Heidegger's 'Being and Time.'" PhD diss., Northwestern University, United States. https://www.proquest.com/docview/915716258?pq-origsite=gscholar&fromopenview=true.

Rousseau, Jean-Jacques. 1819. *Oeuvres complètes de J.J. Rousseau.* Paris, France: Dupont.

Rowland, Jesseca E., Meelah K. Hamilton, Nicholas Vella, Bianca J. Lino, Philip B. Mitchell, and Melissa J. Green. 2012. "Adaptive Associations between Social Cognition and Emotion Regulation Are Absent in Schizophrenia and Bipolar Disorder." *Frontiers in Psychology* 3 (1): 607.

Roy, David J., John R. Williams, and Bernard M. Dickens. 1994. *Bioethics in Canada.* Scarborough, Canada: Prentice Hall.

Roy, Eleanor A. 2019. "New Zealand 'Wellbeing' Budget Promises Billions to Care for Most Vulnerable." *Guardian,* May 30 2019. https://www.theguardian.com/world/2019/may/30/new-zealand-wellbeing-budget-jacinda-ardern-unveils-billions-to-care-for-most-vulnerable.

Ruger, Jennifer P. 2005. "Democracy and Health." *QJM: An International Journal of Medicine* 98 (4): 299–304.

Rumelhart, David E., James L. McClelland, and PDP Research Group. 1986. *Parallel Distributed Processing, Volume 1: Explorations in the Microstructure of Cognition: Foundations.* London, UK: MIT Press.

Russell, Bertrand. 2011. *Political Ideals.* New York, NY: Prometheus Books.

Ryan, Richard M., and Edward L. Deci. 2000. "Self-Determination Theory and the Facilitation of Intrinsic Motivation, Social Development, and Well-Being." *American Psychologist Journal* 55 (1): 68–78.

Ryan, Richard M., and Edward L. Deci. 2017. *Self-Determination Theory: Basic Psychological Needs in Motivation, Development, and Wellness.* New York, NY: Guilford Publishing.

Ryan, Sharon. 2014. "Wisdom." *The Stanford Encyclopedia of Philosophy* (Fall 2023 Edition), edited by Edward N. Zalta & Uri Nodelman. https://plato.stanford.edu/archives/fall2023/entries/wisdom/.

Ryff, Carol D. 1989. "Happiness is Everything, or Is It? Explorations on the Meaning of Psychological Well-Being." *Journal of Personality and Social Psychology* 57 (6): 1069–1081.

Ryff, Carol D. 2014a. "Psychological Well-Being Revisited: Advances in the Science and Practice of Eudaimonia." *Psychotherapy and Psychosomatics* 83 (1): 10–28.

Ryff, Carol D. 2014b. "Self Realization and Meaning Making in the Face of Adversity: A Eudaimonic Approach to Human Resilience." *Journal of Psychology in Africa* 24 (1): 1–12.

Ryff, Carol D., and Burton H. Singer. 1998. "The Contours of Positive Human Health." *Psychological Inquiry* 9 (1): 1–28.

Ryff, Carol D., and Burton H. Singer. 2008. "Know Thyself and Become What You Are: A Eudaimonic Approach to Psychological Well-Being." *Journal of Happiness Studies* 9 (1): 13–39.

Saigle, Victoria, Veljko Dubljević, and Eric Racine. 2018. "The Impact of a Landmark Neuroscience Study on Free Will: A Qualitative Analysis of Articles Using Libet and Colleagues' Method." *American Journal of Bioethics—Neuroscience* 9 (1): 29–41.

Salvador, Rommel, and Robert G. Folger. 2009. "Business Ethics and the Brain." *Business Ethics Quarterly* 19 (1): 1.

Sample, Matthew, Marjorie Aunos, Stefanie Blain-Moraes, Christoph Bublitz, Jennifer A Chandler, Tiago H Falk, Orsolya Friedrich, Deanna Groetzinger, Ralf J Jox, Johannes Koegel, Dennis McFarland, Valerie Neufield, David Rodriguez-Arias, Sebastian Sattler, Fernando Vidal, Gregor Wolbring, Andreas Wolkenstein, and Eric Racine. 2019. "Brain–Computer Interfaces and Personhood: Interdisciplinary Deliberations on Neural Technology." *Journal of Neural Engineering* 16 (6): 063001.

Sandel, Michael. 1998. *Liberalism and the Limits of Justice*. Cambridge, UK: Cambridge University Press.

Santi, Jenny. 2017. "The Secret to Happiness is Helping Others." *TIME*, August 4 2017. https://time.com/4070299/secret-to-happiness/.

Sarbin, Theodore R. 1986. *Narrative Psychology: The Storied Nature of Human Conduct*. Westport, CT: Praeger.

Sartre, Jean-Paul. 1946/1996. *L'existentialisme est un humanisme*. Saint-Amand, France: Gallimard.

Saul, John R. 2014. *The Comeback: How Aboriginals are Reclaiming Power and Influence*. Toronto, Canada: Penguin Random House Canada.

Schacter, Daniel L., Donna R. Addis, and Randy L. Buckner. 2007. "Remembering the Past to Imagine the Future: The Prospective Brain." *Nature Reviews Neuroscience* 8 (9): 657–661.

Schacter, Daniel L., Joan Y. Chiao, and Jason P. Mitchell. 2003. "The Seven Sins of Memory: Implications for Self." *Annals of the New York Academy of Sciences* 1001 (1): 226–239.

Schechtman, Marya. 1996. *The Constitution of Selves*. Ithaca, NY: Cornell University Press.

Schechtman, Marya. 2009. "Getting Out Stories Straight: Self-Narrative and Personal Identity." In *Personal Identity and Fractured Selves*, edited by Debra J. H. Matthews, Hilary Bok, and Peter V. Rabins, 65–92. Baltimore, MD: John Hopkins University Press.

Schechtman, Marya. 2014. *Staying Alive: Personal Identity, Practical Concerns, and the Unity of a Life*. Oxford, UK: Oxford University Press.

Schmidt, Richard A. 1975. "A Schema Theory of Discrete Motor Skill Learning." *Psychological Review* 82 (4): 225–260.

Schurz, Matthias, Martin Kronbichler, Sebastian Weissengruber, Andrew Surtees, Dana Samson, and Joseph Perner. 2015. "Clarifying the Role of Theory of Mind Areas during Visual Perspective Taking: Issues of Spontaneity and Domain-Specificity." *Neuroimage* 117 (1): 386–396.

Seginer, Rachel. 2009. *Future Orientation: Developmental and Ecological Perspective*. New York, NY: Springer.

Seligman, Martin E. 1975. *Helplessness: On Depression, Development, and Death*. San Francisco, CA: W. H. Freeman.

Seligman, Martin E. 2012. *Flourish: A Visionary New Understanding of Happiness and Well-Being*. Toronto, Canada: Free Press.

Seligman, Martin E., and Mihaly Csikszentmihalyi. 2000. "Positive Psychology: An Introduction." *American Psychologist* 55 (1): 5–14.

Sellars, Wilfred. 1963. *Science, Perception, and Reality*. New York, NY: Humanities Press.

Sen, Amartya. 1985. "Well-Being, Agency and Freedom: The Dewey Lectures 1984." *Journal of Philosophy* 82 (4): 169–221.

Sen, Amartya. 1989. "Development as Capability Expansion." *Journal of Development Planning* 19 (1): 41–58.

Sen, Amartya. 1999. *Development as Freedom*. Oxford, UK: Oxford University Press.

Senghor, Abdou Simon, and Eric Racine. 2022. "How to Evaluate the Quality of an Ethical Deliberation? A Pragmatist Proposal for Evaluation Criteria and Collaborative Research." *Medicine, Health Care and Philosophy* 25 (3): 309–326.

Serra, Juan P. 2010. "What Is and What Should Pragmatic Ethics Be? Some Remarks on Recent Scholarship." *European Journal of Pragmatism and American Philosophy* 2 (2): 1–15.

Shahaeian, Ameneh, Candida C. Peterson, Virginia Slaughter, and Henry M. Wellman. 2011. "Culture and the Sequence of Steps in Theory of Mind Development." *Developmental Psychology* 47 (5): 1239–1247.

Shalin, Dmitri N. 1991. "The Pragmatic Origins of Symbolic Interactionism and the Crisis of Classical Science." *Studies in Symbolic Interaction* 12 (1): 223–251.

Shaw, William H. 2005. *Business Ethics*. Belmont, CA: Thomson Wadsworth.

Sheng, Feng, and Shihui Han. 2012. "Manipulations of Cognitive Strategies and Intergroup Relationships Reduce the Racial Bias in Empathic Neural Responses." *NeuroImage* 61 (4): 786–797.

Sherwin, Susan. 1999. "Foundations, Frameworks, Lenses: The Role of Theories in Bioethics." *Bioethics* 13 (3–4): 198–205.

Shevell, Michael I. 1999. "Neurosciences in the Third Reich: From Ivory Tower to Death Camps." *Canadian Journal of Neurological Sciences* 26 (1): 132–138.

Shusterman, Richard. 2008. "Dewey's Somatic Philosophy." *Revue internationale de philosophie* 3 (245): 293–311.

Sikandar, Aliya. 2016. "John Dewey and His Philosophy of Education." *Journal of Education and Educational Development* 2 (1): 191.

Simcock, Gabrielle, Larisa T. McLoughlin, Tamara De Regt, Kathryn M. Broadhouse, Denise Beaudequin, Jim Lagopoulos, and Daniel F. Hermens. 2020. "Associations between Facial Emotion Recognition and Mental Health in Early Adolescence." *International Journal of Environmental Research and Public Health* 17 (1): 330.

Simm, Louise A., Robert D. Jamieson, Ben Ong, Mark W. J. Garner, and Glynda J. Kinsella. 2017. "Making Sense of Self in Alzheimer's Disease: Reflective Function and Memory." *Aging and Mental Health* 21 (5): 501–508.

Simmons, Aaron. 2014. "In Defense of the Moral Significance of Empathy." *Ethical Theory and Moral Practice* 17 (1): 97–111.

Simon, Herbert. 1955. "A Behavioral Model of Rational Choice." *Quarterly Journal of Economics* 69 (1): 99–118.

Sims, Ronald R., and Johannes Brinkmann. 2003. "Enron Ethics (or: Culture Matters More Than Codes)." *Journal of Business Ethics* 45 (3): 243–256.

Singer, Peter. 2011. *Practical Ethics*. Cambridge, UK: Cambridge University Press.

Siouï, Marjolaine, and Nancy Gros-Louis McHugh. 2021. *Notre vision du mieux-être et de la santé*. Montreal Health Ethic Seminar Series, April 22, 2021. Montréal, Canada, Pragmatic Health Ethics Research Unit, IRCM.

Skinner, Burrhus F. 1938. *The Behavior of Organisms: An Experimental Analysis*. Cambridge, MA: B. F. Skinner Foundation.

Sloan, Danetta H., Karlynn BrintzenhofeSzoc, Erin Mistretta, Jennifer Cheng, and Ann Berger. 2017. "The Influence of Relationships on the Meaning Making Process: Patients' Perspectives." *Annals of Palliative Medicine* 6 (3): 220–226.

Slowther, Anne. 2000. "Clinical Ethics Committees: They Can Change Clinical Practice but Need Evaluation." *British Medical Journal* 321 (9): 649–650.

Slowther, Anne, Chris Bunch, Brian Woolnough, and Tony Hope. 2001. "Clinical Ethics Support Services in the UK: An Investigation of the Current Provision of Ethics Support to Health Professionals in the UK." *Journal of Medical Ethics* 27 (9): 2–8.

Smilansky, Saul. 1996. "The Ethical Dangers of Ethical Sensitivity." *Journal of Applied Philosophy* 13 (1): 13–20.

Smith, Adam. 1759. *The Theory of Moral Sentiments*. Edinburgh, Scotland: Printed for Andrew Millar, Alexander Kincaid, and J. Bell.

Smith, Jacqui, and Paul B. Baltes. 1990. "Wisdom-Related Knowledge: Age/Cohort Differences in Response to Life-Planning Problems." *Developmental Psychology* 26 (3): 494.

Sparks John R., and Shelby D. Hunt. 1998. "Marketing Researcher Ethical Sensitivity: Conceptualization, Measurement, and Exploratory Investigation." *Journal of Marketing* 62 (2): 92–109.

Spencer, Herbert. 1892. *The Principles of Ethics*. London, UK: Williams and Northgate.

Stanton-Jean, Michèle. 2017. Les pièges des mots rendent-ils impossible une vision globale de la bioéthique ? *Ethics, Medicine, and Public Health* 3 (1): 195–202.

Staudinger, Ursula M., David F. Lopez, and Paul B. Baltes. 1997. "The Psychometric Location of Wisdom-Related Performance: Intelligence, Personality, and More?" *Personality and Social Psychology Bulletin* 23 (11): 1200–1214.

Stebbins, Robert A. 1981. "The Social Psychology of Selfishness." *Canadian Revue of Sociology and Anthropology* 18 (1): 82–92.

Stengel, Barbara S. 2009. "More than 'Mere Ideas': Deweyan Tools for the Contemporary Philosopher." *Education and Culture* 25 (2): 89–100.

Stolper, Margreet, Bert Molewijk, and Guy Widdershoven. 2016. "Bioethics Education in Clinical Settings: Theory and Practice of the Dilemma Method of Moral Case Deliberation." *BMC Medical Ethics* 17 (45).

Strauss, Leo. 1952. *Persecution and the Art of Writing*. Chicago, IL: University of Chicago Press.

Stroessner, Steven J., and Charles W. Green. 2001. "Belief in Free Will or Determinism on Attitudes toward Punishment and Locus of Control." *Journal of Social Psychology* 130 (6): 789–799.

Strohminger, Nina, Brendan Caldwell, Daryl Cameron, Jana S. Borg, and Walter Sinnott-Armstrong. 2014. "Implicit Morality: A Methodological Survey." In *Experimental Ethics,* edited by Christoph Luetge, Hannes Rusch, and Matthias Uhl, 133–156. London, UK: Palgrave Macmillan.

Strohminger, Nina, and Shaun Nichols. 2014. "The Essential Moral Self." *Cognition* 131 (1): 159–171.

Sugarman, Jeremy, Philip W. Lavori, Maryann Boeger, Carole Cain, Robert Edsond, Vicki Morrison, and Shing S. Yeh. 2005. "Evaluating the Quality of Informed Consent." *Clinical Trials* 2 (1): 34–41.

Sullivan, Shannon. 2001. *Living Across and Through Skins: Transactional Bodies, Pragmatism, and Feminism*. Bloomington, IN: Indiana University Press.

Sullivan, Shannon. 2017. "On the Harms of Epistemic Injustice: Pragmatism and Transactional Epistemology." In *The Routledge Handbook of Epistemic Injustice*, edited by Ian James Kidd, José Medina, and Gaile Pohlhaus Jr., 205–212. New York, NY: Routledge.

Svantesson, Mia, Jan Karlsson, Pierre Boitte, Jan Schildman, Linda Dauwerse, Guy Widdershoven, Reidar Pedersen, Martjin Huisman, and Bert Molewijk. 2014. "Outcomes of Moral Case Deliberation—The Development of an Evaluation Instrument for Clinical Ethics Support (the Euro-MCD)." *BMC Medical Ethics* 15 (1): 30.

Szpunar, Karl K., Donna R. Addis, Victoria C. McLelland, and Daniel L. Schacter. 2013. "Memories of the Future: New Insights into the Adaptive Value of Episodic Memory." *Frontiers in Behavioral Neuroscience* 7 (1): 47.

Tait, Caroline, William Mussell, and Robert Henry. 2019. "Micro-Reconciliation as a Pathway for Transformative Change: Applying a Reconciliation Strategy to the Everyday Relationships Indigenous Peoples have with the Human Service Sector." *International Journal of Indigenous Health* 14 (1): 19–38.

Tanner, Daniel. 2016. "Jean Piaget's debt to John Dewey." *AASA Journal of Scholarship and Practice* 13 (1): 6–25.

Tapper, Elliot B. 2013. "Consults for Conflict: The History of Ethics Consultation." *Proceedings Baylor University Medical Center* 26 (4): 417–422.

Taylor, Charles. 1971. "Interpretation and the Sciences of Man." *Review of Metaphysics* 25 (1): 3–51.

Taylor, Charles. 1989. *Sources of the Self: The Making of Modern Identity*. Cambridge, MA: Harvard University Press.

Taylor, Charles. 1992. "The Politics of Recognition." In *Multiculturalism and "the Politics of Recognition": An Essay with Commentary*, edited by Charles Taylor. Princeton, NJ: Princeton University Press.

Temple, Jeromey B., Margaret Kelaher, and Ruth Williams. 2018. "Discrimination and Avoidance Due to Disability in Australia: Evidence from a National Cross Sectional Survey." *BMC Public Health* 18 (1): 1347.

Tenbrunsel, Ann E., and David M. Messick. 2004. "Ethical Fading: The Role of Self-Deception in Unethical Behavior." *Social Justice Research* 17 (2): 223–236.

Tenbrunsel, Ann E., and Kristin Smith-Crowe. 2008. "Ethical Decision Making: Where We've Been and Where We're Going." *Academy of Management Annals* 2 (1): 545–607.

Thaler, Richard H., and Cass R. Sunstein. 2008. *Nudge: Improving Decisions about Health, Wealth, and Happiness*. New Haven, CT: Yale University Press.

Tiberius, Valerie. 2008. *The Reflective Life: Living Wisely with Our Limits*. Oxford, UK: Oxford University Press.

Tigard, Daniel W. 2019. "The Positive Value of Moral Distress." *Bioethics* 33 (5): 601–608.

Tollefsen, Christopher. 2008. *Artificial Nutrition and Hydration: The New Catholic Debate*. Philosophy and Medicine. Dordrecht, Netherlands: Springer.

Toulmin, Stephen. 1982. "How Medicine Saved the Life of Ethics." *Perspectives in Biology and Medicine* 25 (4): 736–750.

Trowbridge, Richard H., and Michel Ferrari. 2011. "Sophia and Phronesis in Psychology, Philosophy, and Traditional Wisdom." *Research in Human Development* 8 (2): 89–94.

Truth and Reconciliation Commission of Canada. 2015. *Honouring the Truth, Reconciling for the Future: Summary of the Final Report of the Truth and Reconciliation of Canada*. Ottawa, Canada: Government of Canada.

Tulsky, James A., and Ellen Fox. 1996. "Evaluating Ethics Consultation: Framing the Questions." *Journal of Clinical Ethics* 7 (2): 109–115.

Turner, Leigh. 2004. "Bioethics Needs to Rethink Its Agenda." *BMJ* 328 (7432): 175.

Tversky, Amos, and Daniel Kahneman. 1974. "Judgment under Uncertainty: Heuristics and Biases." *Science* 185 (4157): 1124–1131.

Tversky, Amos, and Daniel Kahneman. 1981. "The Framing of Decisions and the Psychology of Choice." *Science* 211 (4481): 453–458.

UNESCO. 2005. *Universal Declaration on Bioethics and Human Rights,* UNESCO, Paris, France.

United Nations. 1948. *Universal Declaration of Human Rights*, United Nations, New York, NY.

United Nations. 2016. Identifying Social Inclusion and Exclusion. In *Leaving No One Behind: The Imperative of Inclusive Development, Report on the Social Situation,* edited by United Nations, 17–28. New York, NY: Department of Economic and Social Affairs.

United Nations. 2021. "Ageism is a Global Challenge: UN." *World Health Organization*, March 18 2021. https://www.who.int/news/item/18-03-2021-ageism-is-a-global-challenge-un.

Upton, Candace L. 2009. "Virtue Ethics and Moral Psychology: The Situationism Debate." *Journal of Ethics* 13 (1): 103–115.

Urry, Heather L., Jack B. Nitschke, Isa Dolski, Daren C. Jackson, Kim M. Dalton, Corrina J. Mueller, Melissa A. Rosenkranz, Carol D. Ryff, Burton H. Singer, and

Richard J. Davidson. 2004. "Making a Life Worth Living: Neural Correlates of Well-Being." *Psychological Science* 15 (6): 367–372.

van der Scheer, Lieke, and Guy Widdershoven. 2004. "Integrated Empirical Ethics: Loss of Normativity?" *Medicine, Health Care and Philosophy* 7 (1): 71–79.

VanderWeele, Tyler J. 2017. On the Promotion of Human Flourishing. *Proceedings of the National Academy of Sciences of the United States of America* 114 (31): 8148–8156.

VanderWeele, Tyler J., Brendan W. Case, Ying Chen, Richard G. Cowden, Byron Johnson, Matthew T. Lee, Tim Lomas, and Katelyn G. Long. 2022. "Flourishing in Critical Dialogue." *Social Science & Medicine—Mental Health* 3 (1): 100172.

Veatch, Robert M. 2019. "Controversies in Defining Death: A Case for Choice." *Theoretical Medicine and Bioethics* 40 (5): 381–401.

Veyne, Paul. 2016. *Sexe et pouvoir à Rome.* Paris, France: Éditions Tallandier.

Voarino, Nathalie, Veljko Dubljević, and Eric Racine. 2017. "tDCS for Memory Enhancement: Analysis of the Speculative Aspects of Ethical Issues." *Frontiers in Human Neuroscience* 10 (1): 678.

Vohs, Kathleen D., and Jonathan W. Schooler. 2008. "The Value of Believing in Free Will: Encouraging a Belief in Determinism Increases Cheating." *Psychological Science* 19 (1): 49–54.

Waal, Frans de. 1996. *Good Natured: The Origins of Right and Wrong in Humans and Other Animals.* Cambridge, MA: Harvard University Press.

Wagner, Nils-Frederic, and Georg Northoff. 2014. "Habits: Bridging the Gap between Personhood and Personal Identity." *Frontiers in Human Neuroscience* 8 (1): 330.

Walker, Margaret U. 1993. "Keeping Moral Space Open New Images of Ethics Consulting." *Hastings Center Report* 23 (2): 33–40.

Walker, Robert M., Steven H. Miles, Carol B. Stocking, and Mark Siegler. 1991. "Physicians' and Nurses' Perceptions of Ethics Problems on General Medical Services." *Journal of General Internal Medicine* 6 (5): 424–429.

Walzer, Michael. 1994/2019. *Thick and Thin: Moral Argument at Home and Abroad.* Notre Dame, IN: University of Notre Dame Press.

Wang, Xiangling, Jiahui Han, Fuhong Li, and Bihua Cao. 2018. "Both Rewards and Moral Praise Can Increase the Prosocial Decisions: Revealed in a Modified Ultimatum Game Task." *Frontiers in Psychology* 9 (1): 1865.

Wangmo, Tenzin, Sirin Hauri, Eloise Gennet, Evelyn Anane-Sarpong, Veerle Provoost, and Bernice S. Elger. 2018. "An Update on the "Empirical Turn" in Bioethics: Analysis of Empirical Research in Nine Bioethics Journals." *BMC Medical Ethics* 19 (1): 6.

Ward, Julia, Julianne Cody, Mary Schaal, and Mohammadreza Hojat. 2012. "The Empathy Enigma: An Empirical Study of Decline in Empathy among Undergraduate Nursing Students." *Journal of Professional Nursing* 28 (1): 34–40.

Washington, Booker T. 1900 [1997]. *Up from Slavery: An Autobiography*. Electronic Edition. Text scanned by Don Sechler and text encoded by Natalia Smith, University Library, UNC-Chapel Hill, University of North Carolina at Chapel Hill, https://docsouth.unc.edu/fpn/washington/washing.html

Watson, John B. 1913. "Psychology as the Behaviorist Views It." *Psychological Review* 20 (1): 158–177.

Weathers, Elizabeth, Rónán O'Caoimh, Nicola Cornally, Carol Fitzgerald, Tara Kearns, Alice Coffey, Edel Daly, Ronan O'Sullivan, Ciara McGlade, and D. William Molloy. 2016. "Advance Care Planning: A Systematic Review of Randomised Controlled Trials Conducted with Older Adults." *Maturitas* 91 (1): 101–109.

Weaver, Kathryn. 2007. "Ethical Sensitivity: State of Knowledge and Needs for Further Research." *Nursing Ethics* 14 (2): 141–155.

Weaver, Kathryn, and Janice M. Morse. 2006. "Pragmatic Utility: Using Analytical Questions to Explore the Concept of Ethical Sensitivity." *Research and Theory for Nursing Practice* 20 (3): 191–214.

Webster, George C., and Françoise E. Baylis. 2000. "Moral Residue." In *Margin of Error: The Ethics of Mistakes in the Practice of Medicine*, edited by Susan B. Rubin and Laurie Zoloth, 217–230. Hagerstown, MD: University Publishing Group.

Webster, Jeffrey D. 2003. "An Exploratory Analysis of a Self-Assessed Wisdom Scale." *Journal of Adult Development* 10 (1): 13–22.

Webster, Jeffrey D. 2007. "Measuring the Character Strength of Wisdom." *International Journal of Aging and Human Development* 65 (2): 163–183.

Webster, Jeffrey D. 2010. "Wisdom and Positive Psychosocial Values in Young Adulthood." *Journal of Adult Development* 17 (2): 70–80.

Weeks, Carly. 2022. "What Access to Abortion Looks Like Across Canada." *Globe and Mail*, May 3 2022. https://www.theglobeandmail.com/canada/article-abortion-access-laws-canada/#:~:text=Abortion%20clinics%20are%20mainly%20located,expenses%20to%20access%20the%20service.

Wegener, Joel R., Clark E. Ludlow, Ammon J. Olsen, Marta Tortosa, and Philip H. Wintch. 2007. "Ego Depletion: A Contributing Factor of Hopelessness Depression." *Intuition* 3 (1): 12–17.

Weidema, Froukje C., Bert Molewijk, Frans Kamsteeg, and Guy Widdershoven. 2013. "Aims and Harvest of Moral Case Deliberation." *Nursing Ethics* 20 (6): 617–631.

West, Cornel. 1989. "The Decline and Resurgence of American Pragmatism: W. V. Quine and Richard Rorty." In *The American Evasion of Philosophy: A Genealogy of Pragmatism*, edited by Cornel West, 182–210. London, UK: Palgrave Macmillan.

West, Thomas G., and Plato. 1979. *Plato's Apology of Socrates: An Interpretation, With a New Translation*. Ithaca, NY: Cornell University Press.

Westwood, Sue. 2019. *Ageing, Diversity and Equality: Social Justice Perspectives*. London, UK: Routledge.

Wexler, Anna, and Laura Specker Sullivan. 2021. "Translational Neuroethics: A Vision for a More Integrated, Inclusive, and Impactful Field." *AJOB Neuroscience* 14 (4): 388–399.

White, Jocelyn C., Patrick M. Dunn, and Lou Homer. 1997. "A Practical Instrument to Evaluate Ethics Consultations." *HEC Forum* 9 (3): 228–246.

Whitehead, Jack, and Jean McNiff. 2006. *Action Research: Living Theory*. Thousand Oaks, CA: Sage Publications.

Whyte, William F. 1991. *Participatory Action Research*. Thousand Oaks, CA: Sage Publications.

Widdershoven, Guy, Tineke Abma, and Bert Molewijk. 2009. "Empirical Ethics as Dialogical Practice." *Bioethics* 23 (4): 236–248.

Willen, Sarah S. 2022. "Flourishing and Health in Critical Perspective: An Invitation to Interdisciplinary Dialogue." *SSM—Mental Health* 2 (1): 100045.

Willen, Sarah S., Abigail Fisher Williamson, Colleen C. Walsh, Mikayla Hyman, and William Tootle. 2022. "Rethinking Flourishing: Critical Insights and Qualitative Perspectives from the US Midwest." *SSM—Mental Health* 2 (1): 100057.

Willette, Christine. 1998. "Practical Discourse as Policy making: An application of Habermas's Discourse Ethics within a Community Mental Health Setting." *Canadian Journal of Community Mental Health* 17 (2): 27–38.

Williams, Bernard. 1985. *Ethics and the Limits of Philosophy*. London, UK: Fontana Press/Collins.

Williams, Nicole. 2021 "Committee Backs Right to Freedom of Expression After Ottawa Prof Uses N-Word." *CBC News,* November 4 2021. https://www.cbc.ca/news/canada/ottawa/ottawa-professor-derogatory-word-freedom-of-academic-1.6237256.

Wilson, Edward O. 1975. *Sociobiology: The New Synthesis*. Cambridge, MA: Belknap Press.

Winkler, Earl R. 1998. "Applied Ethics Overview." In *Encyclopedia of Applied Ethics (Second Edition)*, edited by Ruth Chadwick, 174–78. Cambridge, MA: Academic Press.

Winner, Langdon. 1980. "Do Artifacts Have Politics?" *Daedalus* 109 (1): 121–136.

Witten, Heidi, Shazly Savahi, and Sabirah Adams. 2019. "Adolescent Flourishing: A Systematic Review." *Cogent Psychology* 6 (1): 1.

Wong, Paul T. 1998. "Implicit Theories of Meaningful Life and the Development of the Personal Meaning Profile." In *The Human Quest for Meaning: A Handbook of Psychological Research and Clinical Applications,* edited by Paul T. Wong and Prem S. Fry, 111–140. Mahwah, NJ: Erlbaum.

Wong, Paul T. 2014. "Meaning in Life." In *Encyclopedia of Quality of Life and Well-Being Research,* edited by Alex C. Michalos, 3894–3898. New York, NY: Springer.

Wooldridge, Mike. 2013. "Mandela Death: How he Survived 27 Years in Prison." *BBC News,* December 11. https://www.bbc.com/news/world-africa-23618727.

Wren, Thomas E. 1990. *The Moral Domain: Essays in the Ongoing Discussion Between Philosophy and the Social Sciences.* Cambridge, MA: MIT Press.

Yakin, Halina S. M., and Andreas Totu. 2014. "The Semiotic Perspectives of Peirce and Saussure: A Brief Comparative Study." *Procedia—Social and Behavioral Sciences* 155 (6): 4–8.

Yarrow, Chantalle L. n.d. "Heideggerian Hermeneutics: Understanding as Thrown Projection." https://chantallelouise.com/philosophywritings/heideggerian-hermeneutics-understanding-as-thrown-projection.

You, Di, Yukiko Maeda, and Muriel J. Bebeau. 2011. "Gender Differences in Moral Sensitivity: A Meta-Analysis." *Ethics & Behavior* 21 (4): 263–282.

Zika, Sheryl, and Kerry Chamberlain. 1992. "On the Relation between Meaning in Life and Psychological Well-Being." *British Journal of Psychology* 83 (1): 133–145.

Zimbardo, Philip. 2007. *The Lucifer Effect: Understanding How Good People Turn Evil.* New York, NY: Random House Trade Paperbacks.

Zizzo, Natalie, Emily Bell, Anne-Louise Lafontaine, and Eric Racine. 2016. "Examining Chronic Care Patient Preferences for Involvement in Healthcare Decision Making: The Case of Early Stage Parkinson's Disease Patients in a Patient-Centred Clinic." *Health Expectations* 20 (4): 655–664.

Zizzo, Natalie, Emily Bell, and Eric Racine. 2016. "What is Everyday Ethics? A Review and a Proposal for an Integrative Concept." *Journal of Clinical Ethics* 27 (2): 117–128.

Index

Basic Bioethics

Arthur Caplan, editor

Books Acquired under the Editorship of Glenn McGee and Arthur Caplan

Peter A. Ubel, *Pricing Life: Why It's Time for Health Care Rationing*

Mark G. Kuczewski and Ronald Polansky, eds., *Bioethics: Ancient Themes in Contemporary Issues*

Suzanne Holland, Karen Lebacqz, and Laurie Zoloth, eds., *The Human Embryonic Stem Cell Debate: Science, Ethics, and Public Policy*

Gita Sen, Asha George, and Piroska Östlin, eds., *Engendering International Health: The Challenge of Equity*

Carolyn McLeod, *Self-Trust and Reproductive Autonomy*

Lenny Moss, *What Genes Can't Do*

Jonathan D. Moreno, ed., *In the Wake of Terror: Medicine and Morality in a Time of Crisis*

Glenn McGee, ed., *Pragmatic Bioethics, 2d edition*

Timothy F. Murphy, *Case Studies in Biomedical Research Ethics*

Mark A. Rothstein, ed., *Genetics and Life Insurance: Medical Underwriting and Social Policy*

Kenneth A. Richman, *Ethics and the Metaphysics of Medicine: Reflections on Health and Beneficence*

David Lazer, ed., *DNA and the Criminal Justice System: The Technology of Justice*

Harold W. Baillie and Timothy K. Casey, eds., *Is Human Nature Obsolete? Genetics, Bioengineering, and the Future of the Human Condition*

Robert H. Blank and Janna C. Merrick, eds., *End-of-Life Decision Making: A Cross-National Study*

Norman L. Cantor, *Making Medical Decisions for the Profoundly Mentally Disabled*

Margrit Shildrick and Roxanne Mykitiuk, eds., *Ethics of the Body: Post-Conventional Challenges*

Alfred I. Tauber, *Patient Autonomy and the Ethics of Responsibility*

David H. Brendel, *Healing Psychiatry: Bridging the Science/Humanism Divide*

Jonathan Baron, *Against Bioethics*

Michael L. Gross, *Bioethics and Armed Conflict: Moral Dilemmas of Medicine and War*

Karen F. Greif and Jon F. Merz, *Current Controversies in the Biological Sciences: Case Studies of Policy Challenges from New Technologies*

Deborah Blizzard, *Looking Within: A Sociocultural Examination of Fetoscopy*

Ronald Cole-Turner, ed., *Design and Destiny: Jewish and Christian Perspectives on Human Germline Modification*

Holly Fernandez Lynch, *Conflicts of Conscience in Health Care: An Institutional Compromise*

Mark A. Bedau and Emily C. Parke, eds., *The Ethics of Protocells: Moral and Social Implications of Creating Life in the Laboratory*

Jonathan D. Moreno and Sam Berger, eds., *Progress in Bioethics: Science, Policy, and Politics*

Eric Racine, *Pragmatic Neuroethics: Improving Understanding and Treatment of the Mind-Brain*

Martha J. Farah, ed., *Neuroethics: An Introduction with Readings*

Jeremy R. Garrett, ed., *The Ethics of Animal Research: Exploring the Controversy*

Books Acquired under the Editorship of Arthur Caplan

Sheila Jasanoff, ed., *Reframing Rights: Bioconstitutionalism in the Genetic Age*

Christine Overall, *Why Have Children? The Ethical Debate*

Yechiel Michael Barilan, *Human Dignity, Human Rights, and Responsibility: The New Language of Global Bioethics and Bio-Law*

Tom Koch, *Thieves of Virtue: When Bioethics Stole Medicine*

Timothy F. Murphy, *Ethics, Sexual Orientation, and Choices about Children*

Daniel Callahan, *In Search of the Good: A Life in Bioethics*

Robert Blank, *Intervention in the Brain: Politics, Policy, and Ethics*

Gregory E. Kaebnick and Thomas H. Murray, eds., *Synthetic Biology and Morality: Artificial Life and the Bounds of Nature*

Dominic A. Sisti, Arthur L. Caplan, and Hila Rimon-Greenspan, eds., *Applied Ethics in Mental Healthcare: An Interdisciplinary Reader*

Barbara K. Redman, *Research Misconduct Policy in Biomedicine: Beyond the Bad-Apple Approach*

Russell Blackford, *Humanity Enhanced: Genetic Choice and the Challenge for Liberal Democracies*

Nicholas Agar, *Truly Human Enhancement: A Philosophical Defense of Limits*

Bruno Perreau, *The Politics of Adoption: Gender and the Making of French Citizenship*

Carl Schneider, *The Censor's Hand: The Misregulation of Human-Subject Research*

Lydia S. Dugdale, ed., *Dying in the Twenty-First Century: Towards a New Ethical Framework for the Art of Dying Well*

John D. Lantos and Diane S. Lauderdale, *Preterm Babies, Fetal Patients, and Childbearing Choices*

Harris Wiseman, *The Myth of the Moral Brain*

Arthur L. Caplan and Jason Schwartz, eds., *Vaccine Ethics and Policy: An Introduction with Readings*

Tom Koch, *Ethics in Everyday Places: Mapping Moral Stress, Distress, and Injury*

Nicole Piemonte, *Afflicted: How Vulnerability Can Heal Medical Education and Practice*

Abigail Gosselin, *Mental Patient: Ethics from a Patient's Perspective*

Laurie Zoloth, *May We Make the World?*

Robert Baker, *Making Modern Medical Ethics*

Margaret Battin, *Sex and the Planet: What "Opt In" Reproduction Could Do for the Globe*

John R. Shook and James Giordano, *Bioethics and Brains: A Disciplined and Principled Neuroethics*

Eric Racine, *The Theory of Deliberative Wisdom*